THE PHYSICAL CHEMISTRY OF AQUEOUS SYSTEMS

A Symposium in Honor of
Henry S. Frank
on His Seventieth Birthday

THE PHYSICAL CHEMISTRY
OF AQUEOUS SYSTEMS

A Symposium in Honor of Henry S. Frank

on His Seventieth Birthday

Edited by

ROBERT L. KAY

Department of Chemistry
Carnegie-Mellon University
Pittsburgh, Pennsylvania

PLENUM PRESS • NEW YORK AND LONDON

Library of Congress Cataloging in Publication Data

Symposium on the Physical Chemistry of Aqueous Systems, University of Pittsburgh, 1972.
The physical chemistry of aqueous systems.

"Originally published in the Journal of solution chemistry, vol. 2, no. 2/3."
1. Solution (Chemistry)—Congresses. 2. Water—Congresses. 3. Frank, Henry S., 1902-   I. Frank, Henry S., 1902-   II. Kay, Robert L., 1924- ed. III. Title.

| QD540.S95 | 1972 | 541'.34 | 74-1384 |

ISBN-13:978-1-4613-4513-8   e-ISBN-13:978-1-4613-4511-4
DOI: 10.1007/978-1-4613-4511-4

Symposium on The Physical Chemistry of Aqueous Systems
held in Pittsburgh, Pennsylvania, June 12-14, 1972
Originally published in the *Journal of Solution Chemistry*, Vol. 2, Nos. 2/3, 1973

© 1973 Plenum Press, New York
Softcover reprint of the hardcover 1st edition 1973

A Division of Plenum Publishing Corporation
227 West 17th Street, New York, N.Y. 10011

United Kingdom edition published by Plenum Press, London
A Division of Plenum Publishing Company, Ltd.
4a Lower John Street, London W1R 3PD, England

Contents

Dedication to Henry S. Frank: *"Esse Quam Videri"*

"To be rather than to seem"; one would be hard pressed to find a motto more representative of the essence of Henry S. Frank's character. As a matter of fact, this motto, extolling substance in contrast to appearance, is one he has admired since his early days in China although he likes to point out, with a chuckle, that he first encountered it 50 years ago as a part of the symbol with which a British shipping company decorated the plates in the dining saloons of their steamers plying the China coast trade. Regardless of its origin, the motto is certainly a fitting opening to this special double issue as a dedication to the man who has lead the scientific community in its efforts to elucidate the structure of water and aqueous solutions over the last twenty-five years.

Henry S. Frank was born in Pittsburgh and attended the University of Pittsburgh, receiving the B. Chem. degree in 1922. Upon graduation he made

Henry S. Frank

his way as a missionary teacher to Lingnan University, an American-founded institution in southern China near Canton, where he devoted most of his next thirty years, holding the titles of Professor, Chairman of the Chemistry Department, Dean of the College of Sciences, and finally Provost.

During this period, which most men would consider a complete career in itself, Dr. Frank experienced some of the turbulence that characterized that era of Chinese history. In 1938 Canton was captured by the Japanese and the University transferred its operations to Hong Kong. After the capture of Hong Kong by the Japanese on Christmas Day, 1941, Dr. Frank was interned but had the good fortune to be repatriated in June of 1942. He returned with his wife and three children to Lingnan University following the war, but in 1951, after the Communist take-over, he was forced to return to the United States, where he became Chairman of the Chemistry Department of the University of Pittsburgh.

Professor Frank's professional career as a research chemist began with an undergraduate summer research assistantship at the Westinghouse Research Laboratory. He worked there on the penetration of moisture through bakelite micarta and this suggested a problem on water adsorption onto glass which later became his Ph.D. thesis topic while working under Professor G. N. Lewis at Berkeley. It is interesting now to list some of his fellow students in graduate school—Henry Eyring, F. H. Spedding, J. E. Mayer, and F. D. Rossini. It was clear that Dr. Frank's field of endeavor would be chemical thermodynamics, although it was his interest in the physical interpretation of what is now called statistical thermodynamics that lead to his most important work. In 1939, while on furlough from Lingnan University, he collaborated with A. L. Robinson of the University of Pittsburgh in his first publication dealing with structural effects, a paper entitled, "The Entropy of Dilution of Strong Electrolytes in Aqueous Solution." On his return to China, he commenced work on the classical study he finally published with Marjorie W. Evans in 1945 concerning the effect of water structure on both thermodynamic and transport properties of electrolytes. This work went comparatively unnoticed until Robinson and Stokes published their important book on electrolytes in 1955 in which, in the first chapter, they outlined Professor Frank's model of water structure and ionic hydration. Its influence was reinforced by Ronald W. Gurney's book, *Ionic Processes in Solution*, which also employed the concept of solvent structural effects and which received considerable attention at that time. In 1957, at a Faraday Society Discussion, Frank and Wen outlined the qualitative features of their theory and showed that the large tetraalkylammonium ions possessed the peculiar properties which have made them model ions for the study of hydrophobic effects in aqueous solutions. As is known, during the last decade the framework of Henry S. Frank's original ideas has been verified by a wide variety of measurements and has lead to deeper probings into the problems of aqueous systems than had heretofore been thought of.

This symposium, "The Physical Chemistry of Aqueous Systems," was organized by a committee consisting of Professor E. M. Arnett of the University of Pittsburgh, chairman, with co-chairmen J. Coetzee of the University of Pittsburgh and myself as the representative from Carnegie-Mellon University. Through the efforts of Dean J. L. Rosenberg, a generous supporting grant was obtained from the Army Office of Ordinance Research, Durham, and financial assistance was also given by National Science Foundation University Science Development Grant No. GU-3184. About 100 invited scientists from many parts of the world attended the scientific sessions, which covered three days commencing June 12, 1972. Included here with the original papers are the discussions which were recorded at the meeting and edited by the speakers. One interesting feature of the symposium was the afternoon devoted to papers by three of Professor Frank's former students. Having reached retirement age, Professor Frank has been made Distinguished Service Professor Emeritus of Chemistry, but he is continuing to direct his research group in his laboratory at the University of Pittsburgh.

With this special double issue we all wish Professor Frank many years of continued membership in the scientific community which holds him in such high esteem.

Robert L. Kay, Editor

Conformation and Hydration of Sugars and Related Compounds in Dilute Aqueous Solution[1]

F. Franks,[2] D. S. Reid,[2] and A. Suggett[2]

Received November 27, 1972

The effect of hydroxyl substitution on the nature of the hydration of an alkane chain has been studied by calorimetric techniques. Static permittivities (ε_0) of a range of monosaccharides and related compounds in aqueous solution have also been determined. The ε_0 data, suitably processed, have provided information about the solute dipole moments. In conjunction with earlier results from volumetric, compressibility, and relaxation studies, the specific hydration model is further developed and the relationships between solute molecular conformations and solute–water interactions are discussed.

KEY WORDS: Heats of solution; heat capacity; dipole moments; static permittivity; carbohydrates; dielectric relaxation; sugar conformation; alcohols.

1. INTRODUCTION

In terms of solution thermodynamics, nonionic solutes in aqueous solutions fall roughly into two well characterized classes[1] according to whether deviations from ideal (Raoult's law) behavior are determined by the solution excess enthalpy or entropy. Thus, for monofunctional solutes (alcohols, ethers, ketones, amines) $|\Delta H^E| < T|\Delta S^E|$, giving rise to entropy-dominated solution properties, whereas polyfunctional solutes exhibit the type of behavior, $|\Delta H^E| > T|\Delta S^E|$, more commonly encountered in nonaqueous systems.

The phenomenon of entropy-dominated solution behavior, sometimes referred to as hydrophobic hydration,[2] has been extensively studied in terms of thermodynamic, structural, and dynamic properties, and its manifestations,

[1] This paper was presented at the symposium, "The Physical Chemistry of Aqueous Systems," held at the University of Pittsburgh, Pittsburgh, Pennsylvania, June 12–14, 1972, in honor of the 70th birthday of Professor H. S. Frank.
[2] Biophysics Division, Unilever Research Laboratory Colworth/Welwyn, Colworth House, Sharnbrook, Bedford, England.

1

if not its origin, are now reasonably well understood (for a review, see, for instance, ref. 3). This is not the case with solutions of more complex polyfunctional molecules where the observed properties most likely reflect in part the direct hydrogen-bonding interaction with the solvent (hydration), but some contribution from longer-range water–water interactions cannot be ruled out *a priori*.

The work described in this paper forms part of a general program of studies aimed at elucidating how the nature of the solute molecule affects its interactions with water, whether water in turn affects the solute conformation, and whether long-range hydration effects, i.e., beyond the first or second molecular hydration shell, are indicated. In particular, we have tried to chart the transition form "typical aqueous" (entropy directed) to "typical non-aqueous" (enthalpy directed) behavior, as an apolar solute molecule, such as propane or a cyclic ether, is gradually modified by the addition of polar groups (usually OH groups).

Our studies are based mainly on thermodynamic, static dielectric, and various relaxation methods, and some of the results have recently been published in this journal.[4, 5]

2. EXPERIMENTAL

2.1. Thermodynamic Studies

Calorimetric measurements of the limiting heat of solution ΔH_s° of 1-propanol, 1,2- and 1,3-propanediol, and glycerol have been performed at 5, 25, and 40°C. The differential calorimeter used was a modified version of that described by Franks and Watson.[6] The original glass cells have been replaced by stainless steel ones, the calorimeter is now self-balancing, and a phase-sensitive detector has been incorporated. The control operates by producing a frequency which is proportional to the off-balance output of the Wheatstone bridge. Standard heat pulses are applied at this frequency to the dummy calorimeter cell, and the number of pulses is counted.

2.2. Dielectric Measurements

The static and time domain (TDS) methods employed have been previously described.[5, 7]

2.3. Sample Preparation

1-Propanol was refluxed with iodine and magnesium and distilled. Glycerol and the propanediols were dried over molecular sieves prior to distillation, and their purity was checked by gas chromatography. All solutes had a purity >99.5% and a water content of <0.3%. The purification of the cyclic ether derivatives and monosaccharides has been reported elsewhere.[4, 8]

3. RESULTS

3.1. Calorimetry

In Table I are listed the experimental ΔH_s^0 and ΔC_p^0 results for the propane derivatives; for comparison, previously published data[9] are shown in parentheses. For 1-propanol and glycerol the corresponding free-energy data are available, and therefore ΔS_s^0 may be evaluated as -13.4 and -3 eu, respectively. The trend in the limiting thermodynamic functions clearly shows the changing nature of the solute molecule, i.e., a common carbon backbone with increasing OH substitution. This transition from an apolar to a typical hydrophilic behavior is best seen in the effect of OH substitution on ΔC_p^0 (Fig. 1). The choice of ΔC_p^0 rather than $\bar{C}_{p_2}^0$ for this comparison is, we believe, justified for the following reasons: the hydroxylic solutes are associated in the pure liquid state, and replacement of a solute–solute hydrogen bond by a solute–solvent hydrogen bond in the solution process should not have a significant effect on the total heat capacity of the system, i.e., this process should not contribute to ΔC_p. Since consideration of ΔC_p^0 also eliminates to a large extent the intrinsic contribution of the carbon backbone, the magnitude of ΔC_p^0 should reflect the "structural" influence of the solute on the solvent. It appears that the substitution of an OH group almost completely removes the ΔC_p^0 contribution of the carbon center to which the OH has been attached. In this connection it is relevant that Konicek and Wadsö[10] have observed that, within experimental error, $\bar{C}_{p_2}^0$ of members of various homologous series (alcohols, amines, carboxylic acids, amides) depends *only* on the number of carbon atoms in the alkyl chain but *not* on the nature of the polar head group.

Table I. Limiting Heats of Solution ΔH_s^0 as Functions of Temperature and Heat Capacities ΔC_p^0 at 25°C of the Hydroxy Derivatives of n-Propane

	$-\Delta H_s^0$, cal-mole^{-1}			
Compound	5°	25°	40°	ΔC_p^0, cal-(deg-mole)$^{-1}$
Propane				(70)[c]
1-Propanol	3447 (3489)[a]	2413 (2422)[a]	1740 (1729)[a]	48.3
2-Propanol				(50.0)[a]
1,2-Propanediol	3042	2505	2210	23.3
1,3-Propanediol	2453	2008	1730	20.4
Glycerol	1333	1281 (1253)[b]	1295	1.0

[a] Reference 9.
[b] R. C. Wilhoit, personal communication.
[c] D. M. Alexander, D. J. T. Hill, and L. R. White, *Australian J. Chem.* **24**, 1143 (1971).

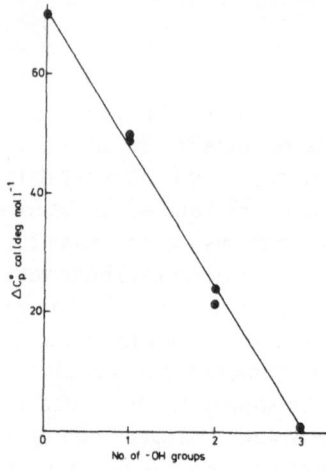

Fig. 1. ΔC_p^0 (25°C) of hydroxy-substituted propanes as a function of the degree of substitution. See Table I.

3.2. Dielectric Studies

Table II lists the limiting dielectric decrements δ defined by

$$\delta = [\varepsilon_0(\text{water}) - \varepsilon_0(\text{soln})]/c$$

Table II. Dielectric Decrements at 5, 25, and 40°C and Limiting Partial Molal Volumes at 5°C of Polyhydroxy Compounds and Their Analogs[a]

Compound	δ			ϕ_v^0, cm³-mole⁻¹
	5°	25°	40°	at 5°
Tetrahydrofuran (THF)	4.94	4.85	4.83	75.7
1,4-Dioxane	7.36	7.26	6.80	78.9
Acetone	3.05	3.25	2.78	—
Tetrahydrofurfuryl alcohol (THFA)	5.08	4.74	4.46	92.4
Glucose	4.51	4.27	3.69	109.5
Mannose	4.43	4.25		109.2
Galactose	3.31	3.28		107.7
Ribose	3.07	2.72		93
Myo-inositol[b]	2.65	2.17		97
Mannitol	2.44	2.48		117.1
Sorbitol	2.75	2.75		117.0
α-Methyl glucoside	6.79	6.35		130.9

[a] See text for the method of calculating μ_{app} from ε_0.
[b] This molecule has five equatorial and one axial OH groups and had to be used because of the very low solubility of scyllo-inositol which has six equatorial OH groups. Limited solubility also prevented the inclusion of tetrahydropyran and tetrahydropyran-2-carbinol in this study.[4, 8]

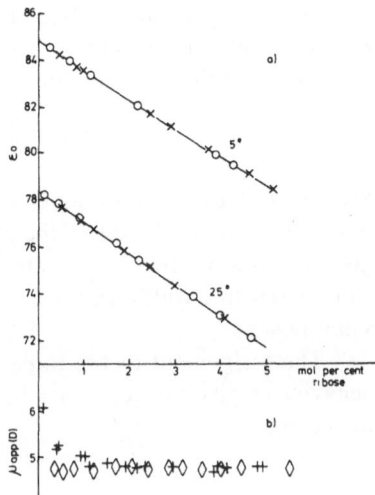

Fig. 2. (a) Static permittivities ε_0 of aqueous solutions of D-ribose as function of concentration. Circles and crosses refer to independent experimental runs; (b) μ_{app} of ribose, calculated from the above results by Eq. (3). + refers to 5° and ◇ to 25°C data.

where ε_0 is the static permittivity and c the molar concentration. Solution concentrations ranged up to 6 mole % and only in one case—dioxane—was a marked, nonlinear trend with concentration observed. Figure 2 provides a graphical illustration of selected experimental results, indicating the degree of scatter and error limits.

Dielectric relaxation studies for glucose and ribose (at much higher concentrations) have already been reported.[5] The relaxation of a 2.8 m glucose solution at 5°C is shown on a complex plane diagram (Cole–Cole plot) in Fig. 3. It is clear that the asymmetric nature of this plot is far from the semicircular arc expected for a single Debye-type relaxation. In fact, the general features of the experimental results can be summarized as follows.

The data can be accounted for in terms of 3 overlapping Debye-type relaxations. Confidence in such an interpretation of $\varepsilon(t)$ is provided by the fact

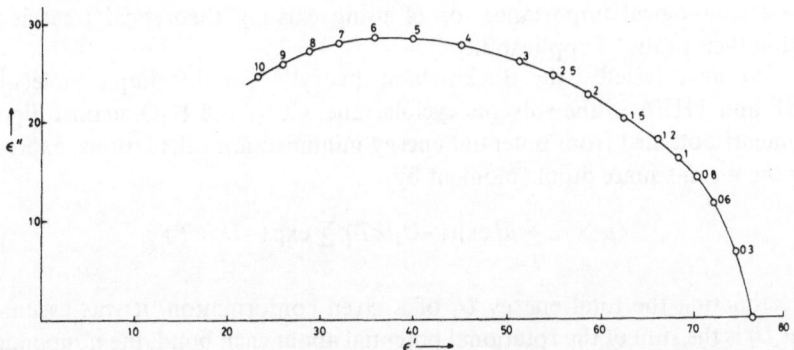

Fig. 3. Complex plane representation of the dielectric relaxation of 2.8 m glucose at 5°C; indicated frequencies are in GHz.

that two independent analyses of the data, i.e., Fourier transformation to $\varepsilon(\omega)$ with subsequent minimization analysis in terms of multiple relaxations and the direct fitting of the time domain data in the form

$$\varepsilon(t) = \varepsilon_0 - \sum_i A_i \exp\left[-t/\tau_i\right] \tag{1}$$

are in very satisfactory agreement. Furthermore, the concentration and temperature dependences of the amplitudes A_i and relaxation times τ_i do not give rise to any irregularities and are also consistent with inferences derived from NMR line-width measurements of ^{17}O-enriched H_2O in solutions of the same sugars.

The assignment of the three relaxation processes in terms of hydration behavior as given in ref. 5 has been somewhat revised and will be discussed in Sec. 6.1.

4. COMPUTATION OF APPARENT DIPOLE MOMENTS FROM STATIC PERMITTIVITY MEASUREMENTS ON AQUEOUS SOLUTIONS

In order to develop a model for the interaction of carbohydrates with water and, in particular, to account for the differences in the solution properties of closely similar solute molecules (e.g., glucose and galactose), it is highly desirable to have information regarding the conformation of the solute molecule. Of special interest is the question whether particular solvents can influence this conformation and to what extent. Knowledge of the solute dipole moment would presumably provide such information but, for obvious reasons, few attempts have been made to determine dipole moments of simple carbohydrates.[11, 12, 13] Unfortunately these molecules only dissolve in strongly interacting solvents, in particular, water, and current dielectric theories are not well suited to this type of situation.[14] One therefore has the alternative of ignoring such solutions, in spite of their undoubted biological and technological importance, or of using existing theoretical treatments within their limits of applicability.

We have tested[7] the Buckingham theory[15] for the simple molecules THF and THFA in the solvents cyclohexane, CCl_4, and H_2O against dipole moments obtained from potential-energy-minimization calculations, expressing the mean-square dipole moment by

$$\langle \mu^2 \rangle = \sum_i \mu_i^2 \exp(-U_i/kT) / \sum_i \exp(-U_i/kT) \tag{2}$$

In estimating the total energy U_i of a given conformation, it was assumed that U_i is the sum of the rotational potential about each bond, the nonbonded interactions being represented by a Lennard-Jones function—the dipole–dipole and, where applicable, the hydrogen bond energies. The dipole moment for any given conformation was obtained by vector addition of effective bond moments.

According to the Buckingham theory, the orientation polarization of the solute species $_\mu P_2$ is expressed in terms of static permittivities by

$$
_\mu P_2 = x_2^{-1} \left[\frac{\varepsilon_{12} + (n_2^2 - \varepsilon_{12}) B_2}{1 + (n_2^2 - 1) B_2} \right]^2 \left\{ \frac{\varepsilon_{12} - n_{12}^2}{\varepsilon_{12}(2\varepsilon_{12} + n_{12}^2)} \frac{x_1 M_1 + x_2 M_2}{d_{12}} \right.
$$

$$
\left. - x_1 {}_\mu P_1 \left[\frac{1 + (n_1^2 - 1) B_1^2}{\varepsilon_{12} + (n_1^2 - \varepsilon_{12}) B_2} \right]^2 \right\}
\tag{3}
$$

Subscripts 1, 2, and 12 refer to solvent, solute, and solution, respectively; x_i is the mole-fraction concentration, n_i is the optical refractive index, M_i is the molecular weight, and B_i are factors relating to the molecular symmetry. For the sake of simplicity, we put $B_1 = B_2 = 0.333$, thus assuming spherical symmetry.[3]

There has been extensive discussion whether the substitution of n^2 for ε_∞ in Eq. (3) is justified. This substitution is based on the assumption that the atomic polarization contribution can be neglected. In the case of solid solutes, there is the further problem of choosing the correct refractive index; the n value of the crystalline sugar may not be appropriate, and the glassy state might resemble more closely the situation in a concentrated solution. With the solutes THF and THFA, this particular difficulty did not arise, and the comparison of μ_{app}, as given by experiment and Eq. (3) with $\langle \mu^2 \rangle^{1/2}$, as calculated by Eq. (2), shows excellent agreement in the case of nonaqueous solutions, lending credibility to the set of potential functions used in the calculation. For aqueous solutions it was found that $\mu_{app} > \langle \mu^2 \rangle^{1/2}$, and calculations showed that for THFA this discrepancy could not originate wholly from changes in the conformation.

Buckingham comments on the poor accuracy of his theory when applied to solutes with low dipole moments in high permittivity, associated solvents. We believe that the Buckingham theory is far superior to any other available theory and that it is not a case of poor accuracy but of a systematic deviation, since μ_{app} is always in excess of μ_{gas} and $(\mu_{app} - \mu_{gas}) \sim 1$ D. Table III shows μ calculated for various solutes by different theories.

Some of the discrepancies are undoubtedly of an experimental origin, e.g., neither Mizutani nor Krishna et al. deionized the solutions prior to permittivity determination. We have found that undeionized solutions of monosaccharides have high conductivities, and this affects the accuracy of ε_0. The high μ_{app} values for ribose are difficult to account for and are probably

---

[3] It is of interest to note that for nonaqueous solutions μ_{app}, as calculated by the Buckingham equation, is not particularly sensitive to changes in B_i, but the opposite is the case for aqueous solutions.

Table III. Solute Dipole Moments Calculated from the Onsager (O) and Buckingham (B) Theories[a]

Compound	Pyridine			Water			Gas or calcd.
	$O^{(13)}$	$O^{(12)}$	$B^{(12)b}$	$O^{(13)}$	$B^{(11)}$	B (this work)	
THF				2.5		2.5	1.6
THFA				3.6[c]		3.8	2.2
Acetone						3.8	2.9
Glucose	8.0	14.1	4.9[b]	12.1		4.5	
Galactose	11.4	11.3	4.7[b]	11.6		5.3	
Mannose		11.3	5.4[b]	11.9		4.8	
Ribose	12.7	40.0	40.0[b]	39.2		4.8	
Arabinose	9.2	9.6		11.5	4.3[c]	—	
Dioxane						?	0.3
Myo-inositol						5.0	
Mannitol						5.8	
Sorbitol						5.8	
α-Me-glucoside						4.5	

[a] The data by Krishna et al.[13] are presented in such a manner that a recalculation in terms of the Buckingham theory is impossible.

[b] Calculated by us from Mizutani's original data.

[c] The values given by Oehme[11] are much lower because he used an extrapolation technique which gives most weight to low-concentration data which have large experimental uncertainties. The THFA value $\mu = 3.6$ D is derived by linear extrapolation from higher-concentration results.

also due to experimental effects. Approximate calculations on the six known conformers [α- and β-furanose and pyranose (1C and C1)] indicate the improbability of such high dipole moments.

The case of 1,4-dioxane requires some comment. Table II shows that this essentially apolar substance gives rise to a large δ, but substitution of all the relevant data into Eq. (3) leads to a negative $_\mu P_2$. One way of rationalizing these observations is to assume that dioxane in solution exists as hydrate complex dioxane $\cdot r$ H$_2$O. This has of course been suggested by many authors, and different types of experiments have produced different estimates of r.[1, 14] There is now some more direct evidence for a 1:1 complex of 1,3- and 1,4-dioxane with water, as well as with H$_2$O$_2$, as shown in the solid–liquid phase diagrams (Fig. 4). In the aqueous systems the hydrates appear to be metastable (peritectic points at $x_2 \simeq 0.5$),[16] but with H$_2$O$_2$ a stable 1:1 compound is formed.[17] If one assumes the existence of a hydrated dioxane species, then Eq. (3) can be applied in the normal manner, and μ_{app} calculated.

The postulate of a dioxane hydrate in solution poses the question whether compounds such as THF and THFA might not also be hydrated in solution. In the case of THF, diffusion (both translational and rotational) measurements

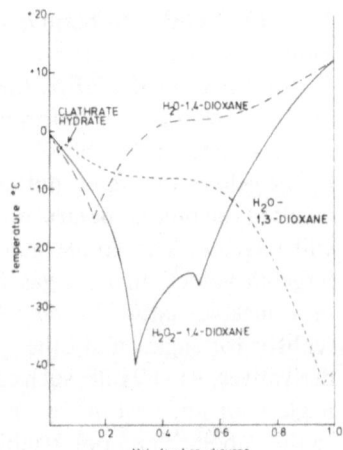

Fig. 4. Solid–liquid phase diagrams of the systems H_2O–1,3-dioxane, H_2O–1,4-dioxane, and H_2O_2–1,4-dioxane.

have shown quite convincingly that no long-lived hydration complexes exist,[18, 19] i.e., there is no correlated rotation or diffusion, nor is there any indication that THF forms a hydrate other than the well-known clathrate hydrate THF·17H_2O in which the oxygen atom is not hydrogen-bonded to the clathrate cage.

To return to the main argument, we believe that the high μ_{app} values obtained by the use of Eq. (3) are largely due to polarizability effects which are not properly accounted for in Buckingham's treatment of aqueous solutions. Consider a THF molecule in aqueous solution: There will be a field at this molecule due to the dipole, quadrupole, and higher electric moments in the surrounding water molecules. Since it appears that the THF molecule is free (more so than the H_2O molecules) to rotate in aqueous solution,[18] it will orient itself with its dipole moment in the direction of this field. Because the THF molecule is polarizable, its dipole moment will increase due to this field. The water molecules will now be polarized both by their neighbors and by the THF molecule, and this mutual polarization process will continue. Coulson and Eisenberg[20] have calculated the effective field at a water molecule due to the surrounding H_2O molecules in an ice lattice and find that the effective dipole moment is considerably enhanced. The environment of a THF molecule will certainly be different from that of an ice lattice, but if one applies Coulson's results in an attempt to ascertain the order of magnitude of the mutual polarization effect, then one can write

$$m = \mu_{gas} + \alpha E$$

where E is the effective field due to the water neighbors. For E we take Coulson's value of 0.5×10^6 esu-cm^{-2}. Assuming the polarizability of THF to be about

1.6×10^{-24} cm^3, we obtain $m = 2.5$ D, which is very close to the experimental value of 2.4 D.

In the case of THFA, the first effect of the field of the water molecules may be to change the conformation of the solute molecule to that correspond-ing to the largest possible dipole moment. This molecule will then be polarized by the field of the water molecules, and the dipole moment will increase. As a result of the mutual polarization, the effective field due to the water molecules will increase. The dipole moment of THFA will therefore be considerably larger in water than in the gas. For reasons already discussed,[5] the assumption of a quasi-ice lattice in the calculation of polarizability effects may be more realistic for sugars and other polyhydroxy compounds than for THF and its derivatives. At any rate, such calculations indicate that the Buckingham theory neglects or underestimates the large effective field at the solute molecule due to the "apolar" and polarizable water molecules.

5. MOLECULAR CONFORMATION AND SPECIFIC HYDRATION

The above discussion suggests that μ_{app}, as calculated from Eq. (3), provides a means of comparing different monosaccharides and related model compounds, and some information can also be derived by a comparison of δ values and the effect of temperature on δ. A comparative analysis of δ values must take cognizance of three effects:

(1) The contribution to ε_0 of the dipole moment of the solute
(2) The size of the solute molecule, i.e., the volume of water displaced, related to the limiting partial molar volume, ϕ_v^0,
(3) Specific hydration effects arising from solute–water hydrogen bonding and subsequent volume changes.

The trend in δ of the substances listed in Table I shows an approximate correlation with μ_{app} and ϕ_v^0. In fact, it is possible to calculate δ from an estimate of the individual solute and solvent contributions to the measured ε_0. Here again, the lack of knowledge of the magnitude of the atomic polarization introduces uncertainties. In view of the fact that, as a first approximation, δ can be calculated on the basis of points (1) and (2) above, any effect on δ arising from specific, conformation-dependent hydration interactions must be of a minor nature. Nevertheless, increasing evidence supports the existence of such effects.[4, 5, 8, 21, 22] It is found, for instance, that δ calculated for mannitol and sorbitol is not in good agreement with the experimental values, and this is consistent with the view that these compounds find it easier to adapt them-selves to the existing solvent environment than do the rigid sugar molecules.

Substitution of the CHOH group by CHOMe produces an increase in δ (although μ_{app} is not affected!), reflecting the decreased affinity of the sugar ester for the existing aqueous environment. The emphasis on conformational compatibility between the solute molecule and water is particularly well illustrated by myo-inositol on the one hand and THF on the other. Thus, it is generally accepted that THF, like other essentially hydrophobic solutes, considerably perturbs the local water environment[8, 18] both structurally and dynamically, and this is associated with a large δ, whereas myo-inositol (very small δ) fits almost ideally (apart from the one axial OH group) to the normal aqueous environment in which second-nearest neighbor oxygen atoms are located at a mean distance of 4.85 Å from one another. This also happens to be the distance between equatorial OH groups in cyclic polyhydroxy compounds. We have drawn attention to the abnormally large negative partial compressibility of myo-inositol[4]—almost as negative as that of aqueous NaCl—and its excellent cryoprotective properties as arising from the same conformational compatibility with water.

It is found experimentally (Table II) that $(d\delta/dT) \leqslant 0$, but $(d\mu_{app}/dT) = 0$. This observation supports the claim to internal consistency of Eq. (3) as a reliable measure of $\langle \mu^2 \rangle^{1/2}$.[14] It also suggests that mean solute conformations are independent of temperature but that the compatibility with the solvent increases at higher temperatures (see also ref. 5). This may well be due to the fact that with increasing temperature O–O correlations in water decrease, so that the 4.85 Å distance which is so critical for hydrophilic hydration at low temperatures becomes less important.

In this connection, it is worth noting that for THF $(d\delta/dT) = 0$, supporting the suggestion that the interactions of this molecule with water do not rely on the 4.85 Å spacing. Indeed, if the hydration of THF and other hydrophobic molecules is of a quasi-clathrate type, then the second nearest neighbor O–O distances in water would be expected to be considerably perturbed, and other correlations would be favored. The nature of the calorimetric results (Sec. 3.1) for propane derivatives supports this line of reasoning. Thus the large ΔC_p^0 effects observed for propane and its monofunctional derivatives suggest the existence of a long-range, thermally labile enhancement of water correlations. However, since the effect of –OH substitution on ΔC_p^0 appears to be confined to the associated carbon center (its contribution to ΔC_p^0 being lost), this indicates that the polar group interactions are short-range in nature; and indeed glycerol, with a near-zero ΔC_p^0, appears to have only interactions of a short-range nature, the enhancement of water correlations being completely absent. Thus ΔH_s^0 for both types of compounds are of similar magnitudes in the particular temperature range studied; the fundamental differences only become apparent in $\Delta H_s^0(T)$. It may be relevant here to note that the glycerol molecule can adopt a conformation in which the oxygen atoms attached to the first and third carbon atoms are spaced at a distance of 4.85 Å.

6. HYDRATION NUMBERS AND DYNAMICS FROM RELAXATION

6.1. Relaxation Experiments

The three dielectric relaxation processes observed in solutions of glucose and ribose were previously[5] assigned to the following processes (characteristic relaxation times τ_i at 5°C for 2.8 m solution):

I $\tau_1 = 18.5$ psec   bulk water
II $\tau_2 = 69$ psec   solute molecule
III $\tau_3 = 250$ psec   hydration shell ("bound" water)

It was thus assumed that the total solvent contribution to the permittivity was included in processes I and III, and the solute contribution in process II. We now believe this assumption to be an oversimplification for the following reasons.

1. It is possible to estimate the permittivity contribution of the water in an aqueous solution of given solute volume fraction f by a suitable permittivity mixture model. For solutions of small molecules, for example, the static contribution of the water can be calculated from a formula given by Brown[23]:

$$\varepsilon_{12} - \varepsilon_1 [2\varepsilon_1 + n_2^2 - 2(\varepsilon_1 - n_2^2) f]/[2\varepsilon_1 + n_2^2 + (\varepsilon_1 - n_2^2) f] \qquad (4)$$

where the symbols have the same significance as in Eq. (3). Pottel and Kaatze[24] have used Eq. (4) to calculate ε_{12} for a range of tetraalkylammonium bromides and compared the results with the measured values. In all cases,

$$(\varepsilon_{meas} - \varepsilon_{calc}/\varepsilon_{meas})$$

lies within about 1–2%, independent of solute concentration and size of the alkyl chains. We have therefore used Eq. (4) to calculate the expected contribution of the water to the permittivity of monosaccharide solutions and have found in every case that the sum of the amplitudes $A_1 + A_3$ [see Eq. (1)] does not account, even approximately, for the total concentration of water in the system.

2. There is also an anomaly regarding process II. Estimates of the dipole moments of monosaccharides as discussed in Sec. 4 fall within the range 3–5 D (Table III). On the other hand, A_2 in Eq. (1) for glucose solutions is consistent with a μ_{app} value significantly in excess of this range.

The conclusion to be drawn from both the above factors is that there is a contribution from the solvent to process II, i.e., the relaxation observed is due to the reorientation of the solute together with its associated water of hydration. The extent of this contribution can be estimated if some assumptions are made regarding the solute conformation (as indicated previously[5]). However, in estimating the extent of monosaccharide hydration we prefer at this time to minimize the number of assumptions and use the following alternative method: The total permittivity contribution of the water in the solutions can be calculated using Eq. (4) and compared with the relaxation amplitude A_1 for bulk water. The difference between these values is expressed in terms of the molar concentration of water, and hence the number of water molecules associated on a time average with one molecule of monosaccharide is calculated. The procedure makes no assumptions about the nature of other relaxation processes. Hydration numbers (r_h) of 6 ± 1 and 2.5 ± 1 are obtained in the above manner for D-glucose and D-ribose, respectively, at 5°C, thus demonstrating the significant difference between the hydration properties of the hexose and pentose as suggested previously.[5]

Hydration estimates using ^{17}O NMR relaxation times depend upon the assumption of a correlation time τ_c for the hydration water which can be taken, as a first approximation, to be equal to $\frac{1}{3}$ of the equivalent dielectric relaxation time. Unfortunately, this latter quantity cannot be obtained in a straightforward manner, as there appear to be two mechanisms by which the hydration water relaxes—corresponding to process II (by reorientation with the solute) and process III (by chemical exchange with the bulk solvent, followed by reorientation?). We therefore define a mean dielectric relaxation time $\bar{\tau}_h$ for the hydration water

$$\bar{\tau}_h = \{[(\Delta - A_3)/\Delta]\tau_2 + (A_3/\Delta)\tau_3\}$$

where Δ is the difference between the predicted amplitude of the bulk water relaxation for zero hydration [from Eq. (4)] and the observed amplitude A_3. Thus $(\Delta - A_3)$ and A_3 correspond to the permittivity contributions of the hydration water to the two alternative relaxation mechanisms. The combination of ^{17}O NMR relaxation times and $\bar{\tau}_h$ leads to $r_h = 5 \pm 1$ and 2.5 ± 1 for D-glucose and D-ribose at 5°C, i.e., in good agreement with the values obtained from the dielectric relaxation amplitudes.

Although these numbers constitute an increase over our previous estimates,[5] the above analysis does not require any modification of the specific hydration model described in ref. 5.

6.2. Hydration Numbers from Solution Compressibility Measurements

By assuming that, to a first approximation, the water directly affected by sugar molecules ("bound" water) has the compressibility of ice, Shiio[25] has calculated r_h for several mono- and disaccharides from adiabatic compressibility measurements; for monosaccharides, $2.5 < r_h < 4.0$ at 25°C. Thus, for glucose at 25°C, $r_h = 3.5$ mole/mole, in good agreement with our estimates derived from relaxation experiments. It is found that r_h is very sensitive to temperature changes, and the temperature dependence of r_h (in the case of glucose) is also in accord with our findings.

We have subjected some of our own, rather more precise and numerous compressibility data[4] to a similar analysis and find that, although r_h for glucose [obtained by an extrapolation procedure involving compressibility (β) and density (d) data] agrees with Shiio's value, the concentration dependences of both β and d are very different, so that the agreement in the extrapolated value is probably fortuitous.

A similar analysis on our ribose data[4] yields $r_h = 6$ at 5°C and 1.5 at 25°C, and for myo-inositol $r_h \simeq 16$—a rather surprising result even allowing for the abnormally large negative partial compressibility of this substance.

7. SOLVENT-INDUCED CONFORMATIONS

It is interesting to speculate which of the two related postulates is more realistic:

(1) that the conformation of the sugars is largely determined by energetic considerations independent of the solvent, and that the nature and extent of hydration is a function of the relative compatibility of that conformation with the aqueous environment, or

(2) that the conformation of the sugars is controlled to a significant degree by the intermolecular order in the aqueous component in a way which maximizes the extent of interaction (H-bonding) between the components.

An increasing amount of experimental information points to the second alternative as being the more reasonable. On the basis of the specific hydration model for monosaccharides,[5] we can successfully account for shifts in sugar conformational equilibria which occur with changes in temperature, and it has been demonstrated by NMR measurements[5] that the sugar conformation which is most compatible (in terms of H-bond formation) with a tetrahedral arrangement of water molecules becomes increasingly the most favored form as the temperature is lowered. Further, examinations of the sugar conformations by optical-rotation measurements[21, 26] have indicated that the conformation adopted by many sugars in aqueous solution is significantly different from that in nonaqueous solvents. It is concluded that the aqueous solvent exerts a controlling influence on the observed solution conformation.

Finally, recent NMR pseudocontact shift measurements on transition-metal complexes of adenosine monophosphate (AMP) indicate clearly that in solution the ribose ring takes up a well-defined orientation with respect to the adenine residue and that this orientation differs in aqueous and dimethyl sulfoxide solutions.[22]

We are presently investigating the possibility of solvent-induced conformational equilibria in solutions of disaccharides, where the mutual orientations of the sugar residues can be studied as functions of type of linkage and the nature of the solvent. At the present time it does appear that in polar molecules, where low-energy barrier conformational transitions are possible, the nature of the solvent and, in particular, water at low temperatures markedly affects the conformational states.

ACKNOWLEDGMENTS

We are indebted to Drs. D. A. Weyl and B. de Nooijer, both formerly of this Division, for initiating the static permittivity studies and developing methods for calculating μ_{app}. We acknowledge our gratitude to Mrs. P. Quickenden who was responsible for all the experimental dielectric studies reported in this communication.

REFERENCES

1. F. Franks, *Hydrogen-Bonded Solvent Systems*, A. K. Covington and P. Jones, eds. (Taylor and Francis, London, 1968).
2. F. Franks and D. G. J. Ives, *Quart. Rev.* 20, 1 (1966).
3. W.-Y. Wen, *J. Solution Chem.* 2, 253 (1973).
4. F. Franks, J. R. Ravenhill, and D. S. Reid, *J. Solution Chem.* 1, 3 (1972).
5. M. J. Tait, A. Suggett, F. Franks, S. Áblett, and P. A. Quickenden, *J. Solution Chem.* 1, 131 (1972).
6. F. Franks and B. Watson, *J. Phys. E.* 1, 940 (1968).
7. B. de Nooijer, D. Spencer, S. G. Whittington, and F. Franks, *Trans. Faraday Soc.* 67, 1315 (1971).
8. F. Franks, M. A. J. Quickenden, D. S. Reid, and B. Watson, *Trans. Faraday Soc.* 66, 582 (1970).
9. D. M. Alexander and D. J. T. Hill, *Australian J. Chem.* 22, 347 (1969).
10. J. Konicek and I. Wadsö, *Acta Chem. Scand.* 25, 1571 (1971).
11. F. Oehme and M. Feinauer, *Chemiker Zg.* 86, 71 (1962).
12. M. Mizutani, *Osaku Baigaku Igaku Zasshi* 8, 1334 (1956).
13. B. Krishna, M. L. Bala, and S. C. Srivastava, *J. Sci. Ind. Res. (India)* 24, 626 (1965).
14. J. B. Hasted, in *Dielectric and Related Molecular Processes* (Chemical Society, London, 1972), p. 121.
15. A. D. Buckingham, *Australian J. Chem.* 6, 93, 323 (1953).
16. K. W. Morcom and R. W. Smith, *J. Chem. Thermodyn.* 3, 507 (1971).
17. P. A. Giguère, personal communication.
18. E. von Goldammer and M. D. Zeidler, *Ber. Bunsenges.* 73, 4 (1969).
19. E. von Goldammer and H. G. Hertz, *J. Phys. Chem.* 74, 3734 (1970).
20. C. A. Coulson and D. Eisenberg, *Proc. Roy. Soc.* A291, 445 (1966).
21. D. A. Thom and D. A. Rees, personal communication.
22. R. J. P. Williams, personal communication.
23. W. F. Brown, *Handbuch der Physik* (Springer, Berlin, 1956), Vol. 73, p. 437.
24. R. Pottel and U. Kaatze, *Ber. Bunsenges. Phys. Chem.* 73, 437 (1969).
25. H. Shiio, *J. Am. Chem. Soc.* 80, 70 (1958).
26. A. J. Hannaford, *Carbohydrate Res.* 3, 295 (1967).

DISCUSSION

Professor H. S. Frank (*University of Pittsburgh*). To tie this in with what Friedman was telling us earlier this morning, I wonder what Professor Robinson has to say about the fact that there is a real difference in the osmotic coefficient for solutions of mannitol and sorbitol. He found some years ago that mannitol and sorbitol showed real differences in the osmotic coefficients in measurements up to 1 M. From the present point of view, the real significance of this is that it ties in the question of whatever causes osmotic coefficients to be different with the structural compatibility with the water.

Professor R. H. Stokes (*University of New England, Armidale, Australia*). It can indicate stronger interactions of sorbitol with the water molecule.

Dr. F. Franks (*Unilever Research Laboratories, England*). We have no evidence that the short-range sorbitol–water interactions are stronger than are the corresponding mannitol ones. In connection with Robinson's activity-coefficient studies, it has recently been shown that the standard enthalpy of transfer of NaCl from water to aqueous mannitol solutions exhibits an endothermic peak near 0.1 m mannitol, whereas the transfer to sorbitol solutions is exothermic at all concentrations. Thus mannitol behaves as do the alkanols, whereas sorbitol behaves similarly to simple carbohydrates [J. H. Stern and M. E. O'Connor, *J. Phys. Chem.* **76**, 3077 (1972)].

Professor E. M. Arnett (*University of Pittsburgh*). With regard to that, the conformations of the hexoses have been worked out very nicely in the last two years by Jeffrey (for the x-ray methods) and by Horton of Ohio State using NMR. If I am not mistaken, there is a difference between sorbitol and mannitol. One of them has a completely zigzag structure and the other is a sickle-shaped molecule, and this is true even in solution. If the spacing of the hydroxyl groups is the same, then the actual conformation for the two may be different in solution.

Professor Frank. In which case, mere structure would influence the short-range interaction of these sugars.

Dr. Franks. Yes. What is more, and as I have already indicated in my paper, there are now good indications that the particular conformation taken up by complex organic compounds is dictated by the nature of the solvent.

Professor G. A. Jeffrey (*University of Pittsburgh*). Actually, work related to the solution conformation was done on the acetylated alditols, whereas x-ray studies were done on the hydroxyl compounds.

Professor Arnett. Is it true that they were different?

Professor Jeffrey. Yes.

Professor Arnett. I have a question about your ^{17}O work. Since you are dealing with the hexoses and not their reduced form, it would seem to me that ^{17}O in the solvent is going to exchange with that in the aldehyde group since there is a dynamic equilibrium and some ^{17}O is going to end up in the hemiacetal part of the molecule. Does that explain anything or introduce any problems?

Dr. Franks. Oxygen exchange is very slow. We thought it was preferable to using deuterium where the exchange is very rapid.

Professor P. A. Giguère (*Université Laval*). In connection with the phase diagram of the binary system *p*-dioxane–water that you just showed, we have investigated recently the closely related system p-dioxane–hydrogen peroxide. A stable 1:1 compound is formed with congruent melting point at −25°C. From the vibrational spectra the structure must consist of chains of the two alternated molecules linked by hydrogen bonds of some 3 to 4 kcal. Both molecules retain their normal configuration. Does dioxane also form a hydrate?

Dr. Franks. Probably not a stable one, but both dioxane phase diagrams indicate peritectic behavior in the region of 0.5 mole fraction. Possibly unstable hydrates do exist.

Professor Giguère. Any attempt to explain the maximum density of water at 4°C must also account for the equally important fact that in heavy water the temperature range of increasing density is nearly double: viz. 3.8° to 11.5°C. One last remark again regarding hydrogen peroxide. The melting point of that crystal is very close (−0.4°C) to that of ice in spite of a nearly double molecular mass. Now, the packing forces are quite the same in both crystals—namely, four hydrogen bonds per molecule. However, in ice these factors are much better balanced (tetrahedral about each O atom) than in hydrogen peroxide.

Professor Stokes. Of the many fascinating things revealed in your work, I was most struck, I think, by the observation that the apparent molal volume of the hydroxyl group appears to be zero. There is surely an interesting comparison here, too, with the fact that the hydroxide *ion* has a negative apparent molal volume. The hydroxyl ion is remarkable, of course, being an ion with a dipole. One suspects that the disappearing trick of the hydroxyl group may also be connected with its local dipole.

Professor J. B. Hyne (*University of Calgary*). Might I ask Dr. Franks to develop a little more the interesting observation of similar compressibilities of myo-inositol and sodium chloride solutions?

Dr. Franks. If we take the negative partial compressibility as a measure of the protection which the solute affords the solution against the effect of pressure over and above what normally happens in pure water, then the inositols are of course the molecules that are the most symmetrical with respect to equatorial OH groups. Therefore, if one looks at the planar projection of the puckered cyclohexane ring, in the scyllo-inositol molecule there are three OH groups directed at right angles above, and three below the plane of the ring, and the O–O distances will be just right for hydrogen bonding with water, because the next nearest neighbor O–O distances in water correspond to the polar group separations in inositol.

Professor Frank. One trouble there is that as long as part of the compressibility of water is relaxational compressibility, if the structure equilibrium is shifted, the contribution of relaxational compressibility is shifted. That is something that is hard to disentangle.

Dr. Franks. That is, of course, true, but whatever takes place in sodium chloride solutions seems also to occur in solutions of inositol.

Professor Frank. However, sodium chloride is a structure breaker.

Professor Hyne. Aren't you really saying here that inositol is buttressing the water structure? True, in this case it is binding with the water through hydrogen-bond interactions, but if a hydrocarbon is put in there and the surrounding water molecules are forced to interact with one another, the water structure is being buttressed as well, to a very small extent.

Dr. Franks. Yes, to a very small extent. The point is that at low temperatures all organic compounds have negative compressibilities, but never of this magnitude.

Professor Hyne. Does that suggest then that there are two kinds of buttressing effects on the aqueous environment, one generated by hydrocarbon nonpolar-hydrophobic interaction and one by hydrogen-bonding interaction?

Dr. Franks. Yes indeed. The difference is that effects due to solute-water hydrogen bonding are as temperature-sensitive as those due to apolar groups. Of course, one can always say that placing a hard sphere into a liquid will reduce the compressibility in any case, but this is obviously a simple volume effect. I would not expect this to give rise to the observed differences which arise from subtle changes in the steric configuration of the solutes employed.

Dr. F. H. Stillinger (*Bell Laboratories, New Jersey*). Do these rather small hydration numbers also apply to inositol?

Dr. Franks. Inositol cannot be studied well by our relaxation methods because of its limited solubility. Compressibility results suggest a hydration number of 16, which may be stretching the imagination somewhat.

Dr. Stillinger. Can I ask a question generated from the first part of your talk? It has to do with the mean dipole moment of tetrahydrofurfurol. It is now generally known, quantum-mechanically, that the transformation of a hydrogen bond into an alcohol OH always tends to incorporate a certain degree of charge transfer whose net result is to increase very substantially the dipole moment of that group. The magnitude of the dipole moment quoted was about 3.4 to 3.8 D. Certainly this is well within the ball park from what one would estimate from *ab initio* quantum-mechanical calculations.

Dr. Franks. No, maybe I didn't make myself clear. What I said was that the estimated dipole moment of this molecule, using the potential functions I described, was 2.2 D. That agrees well with the measured value in carbon tetrachloride and cyclohexane solutions. In water, the apparent dipole moment is 3.8 D. We can achieve 3.4 D by adjusting the two angles ϕ_1 and ϕ_2 to give us the maximum possible dipole moment. However, this would lead to a very high potential energy and might make nonsense of the calculated $\langle \mu^2 \rangle^{1/2}$.

Dr. Stillinger. Yes, but my point is that surely pendant OH group is solvated through a hydrogen bond and that bonding does in fact incorporate this very considerable charge transfer which results in a significant enhancement of the dipole moment of that OH group. Whether or not the solute molecule is twisted into any of these postulated structures, it can still pick up a good deal more dipole moment by the charge-transfer effect, which I presume was not incorporated in your calculations.

Dr. Franks. In order to test such a possibility, we studied the tetrahydrofuran molecule where such rotational freedom does not exist. Here again the calculated dipole moment agreed well with that obtained from solutions in apolar solvents, but the apparent dipole moment in aqueous solution was again too high. Assuming a clathrate-like environment for the THF molecule, we calculated the dipole-moment enhancement to be expected at the center of the clathrate cavity due to 28 nearest-neighbor point dipole water molecules; however, the change in dipole moment so obtained was not large enough to make up the discrepancy.

Dr. Stillinger. There is another possibility here which I don't have quantitatively at my fingertips, and it has to do with the fact that a local orientational order can give rise to an apparent enhancement of the dipole moment, even if the individual molecules themselves do not suffer a change in dipole moment. A local cooperative orientational parallelism in dipole moments could conceivably produce an enhancement also.

Dr. Franks. This is encouraging.

Dr. Stillinger. It is not certain whether it happens in this compound or not. It is a very complicated situation. The other quantum-mechanical effect is certain.

Professor Arnett. I have a comment concerning the relation of electrolyte chemistry to carbohydrate chemistry. It has been known for 50 to 60 years that some of the carbohydrates form stable complexes with alkali and alkaline-earth metals, and this is quite dependent upon the conformation of the carbohydrates. A particular maverick among the ethers described was 1,3-dioxolane. We are studying right now the complexation of alkali metals with different ethers. This one is an extremely poor solvator for sodium ions. It is the only one of the group in which the p-orbitals are forced to point away from each other. That would also be true for its interaction with water.

Professor E. U. Franck (*University of Karlsruhe, West Germany*). I don't understand how your second relaxation can be caused by the rotation of the glucose molecule. Can one exclude internal relaxation problems?

Dr. Franks. In the glucose molecule in particular, the only other possibility would be $-CH_2OH$ group rotation which would be outside the measured frequency range. We have confirmed our assignment by studying cellobiose instead of glucose; this almost doubles the second relaxation time. Of course the water of hydration contributes to τ_2, but it is mainly due to solute reorientation.

Dr. J. C. Hindman (*Argonne National Laboratory*). An alternative interpretation for the relaxation process in pure water can be given. In this model it is assumed that there are two dynamic processes. The process involving the large activation energy is visualized as a rotational motion of a water molecule in a hydrogen-bonded lattice, i.e., it is both a hydrogen-bond breaking and rotational process. The experimental relaxation data can be explained by assuming that this is a very rapid process which is followed by a cooperative relaxation of the associated lattice molecules. The second, low-activation-energy, motion is interpreted as diffusional reorientation of a less strongly bound water molecule. If this model is correct, it will probably require an alternate explanation for Franks' observations.

Unidentified speaker. In the several years that I have been interested in thermodynamic data of biologically interesting material, yours is the first real strong plea for thermodynamic data based on well-defined standard states. I wonder if you would give some indication of what experiments you would be interested in and what you would do with the information.

Dr. Franks. I think to make any meaningful thermodynamic (calorimetric) measurement on a series of related compounds, one must have a common standard state, or else these comparisons become meaningless, especially in the case of solids where the lattice energies are not known. We have given this problem of standard states considerable thought in connection with interpretations of our data on liquid solutes, e.g., we have considered the pure liquids at the same reduced temperature, or dissolved in a model apolar solvent. With the sugars, of course, we are particularly lucky because they are isomers, and if one knew the heats of formation, I think a common standard state could be computed, and a meaningful comparison of the thermodynamic data could be made.

Professor Frank. But aren't a large number of significant figures needed in heats of formation for sugars?

Dr. Franks. Yes, good combustion calorimetry is needed. There are three or four centers around the world where such data could be generated.

Studies of Hydrophobic Bonding in Aqueous Alcohols: Enthalpy Measurements and Model Calculations[1,2]

Harold L. Friedman[3] and C. V. Krishnan[3]

Received December 4, 1972

The coefficients which measure the contribution of a pair of solute molecules to the excess enthalpy have been measured in water at 25°C for all pairs of alcohols which can be formed from the series methyl to n-butyl plus t-butyl as well as for ethanol with some of the higher alcohols and with the n-alkyl sulfonates through octyl. The methylene-group contribution to these coefficients is readily identifiable in suitable cases. These data and the corresponding free-energy and volume coefficients, where they are known, are analyzed in terms of a model which specifies the core repulsion and solvation-layer overlap terms in the potential for the interaction of two solute molecules. The latter term has an adjustable parameter, the so-called Gurney free-energy parameter which is adjusted for each solute pair to fit the free-energy data. Its temperature and pressure derivatives are adjusted to fit the enthalpy and volume data, respectively. These parameters are compared with the corresponding thermodynamic coefficients of solvation as far as possible.

KEY WORDS: Hydrophobic bonding; excess enthalpies; model calculations; aqueous alcohols; alkyl sulfonates.

1. INTRODUCTION

Hydrophobic solutes, or solutes with hydrophobic groups, in aqueous solutions are important systems for both theoretical and practical reasons.

Their solvation thermodynamics has some striking features compared with other systems. As Frank and Evans showed,[1] these features can be understood in terms of a shift in "a mobile equilibrium between different

---

[1] This paper was presented at the symposium, "The Physical Chemistry of Aqueous Systems," held at the University of Pittsburgh, Pittsburgh, Pennsylvania, June 12–14, 1972, in honor of the 70th birthday of Professor H. S. Frank.

[2] A portion of the talk presented at the H. S. Frank Symposium has been published elsewhere.[11] The discussion section at the end of this paper pertains mainly to that portion.

[3] Department of Chemistry, State University of New York, Stony Brook, New York 11790.

structures" in the water near the solute; the equilibrium referred to is the one postulated in the Bernal and Fowler description of liquid water.[2] Frank and Evans showed how such a perturbation of the water structure in the neighborhood of a hydrophobic solute provided a basis for interpreting the characteristic effects in the solvation entropies and heat capacities of hydrophobic groups. Their analysis may be described as the postulation and treatment of a *chemical model*; one postulates an equilibrium among solvent species, whether two species or a continuum, or something in between, which is perturbed in the neighborhood of the solute particle. Then they apply the methods of chemical thermodynamics to deduce the solvation properties.

Now, 27 years later, it is clear that the model has been spectacularly successful. In this period the experimental information about hydrophobic hydration has been enriched by the studies of many more systems and now includes relaxation and spectroscopic coefficients of solvation as well as many more thermodynamic and transport coefficients than were known in 1945. For each kind of measurement the experimental coefficient may be compared with the expectations from the chemical model by the use of the methods for estimating the same coefficient for a chemical mixture. Most often, as for the viscosity, one can only estimate the qualitative trends for systematic changes in the solute. There are few cases, notably the proton chemical shift of the solvent in solutions of hydrophobic solutes, in which the data do not seem to fit the expectations of the Frank–Evans model. In such cases it has been difficult to establish that the discrepancy is due to inadequacy of the model rather than to the difficulty of deducing the behavior of the particular coefficient from the model.

It is a challenge, of course, to find a model which specifies the intermolecular potential functions, a *Hamiltonian model*, which will cover the same ground as the chemical model. The study by Eley and Evans[3] of the role of the librational motion was an early step in this direction, but further advances apparently had to wait for the development of sufficiently powerful statistical-mechanical methods such as those recently applied by Rahman and Stillinger to the study of a Hamiltonian model for water itself.[4]

The practical importance of hydrophobic hydration is notable in its appearance as a basis for understanding several phenomena in biological systems. For these applications, however, the solvation aspect of hydrophobic hydration needs to be supplemented by knowledge of the interactions of hydrophobic solute species with each other in aqueous media as pointed out by Kauzmann, who termed this interaction hydrophobic bonding and showed that it has an attractive component.[5]

A chemical model for solvation is not necessarily a sufficient basis for prediction of the effect of solvation upon the solute–solute interaction. The problem is readily described in terms of the cosphere-solvation concept of R. W. Gurney,[6] as we have discussed in some detail elsewhere.[7] Figure 1(a) represents a solvent medium in which there are two solute particles, each

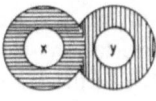

Fig. 1. Cospheres on solute particles x
and y.   a   b

surrounded by a *cosphere*, the region in which the solvent properties are changed significantly by the presence of the solute. In Fig. 1(b) the solute particles are close enough together so that the cospheres overlap. A certain amount of work needs to be done to move particles x and y from the configuration in Fig. 1a to that in 1b; this is the free-energy change in displacing some of the cosphere material and, in general, changing the properties of the solvent which remains in the cospheres. Knowledge of the local chemical properties of the material in the separated cospheres is not adequate to predict even the sign of this free-energy change because the solvent in between particles x and y in Fig. 1(b) may have local properties which are not simply related to those of the solvent in the separated cospheres.

However, there does seem to be a general rule which can be used to supplement a chemical model for solvation so that one can predict the effects of solvation on activity coefficients and other thermodynamic excess functions, at least in dilute solutions.[7] The rule is that the overlap of cospheres mainly results in the elimination of some of the material from one of the cospheres or the other. Pictorially, this means that Fig. 1(b) is realistic and the material within the cospheres remains unchanged. It seems that this rule was first noticed by Frank and Robinson in correlating excess entropies[8, 9]; however, they described it in somewhat different terms. Now it seems clear that this rule has the consequence that there be a quantitative relation between solvation thermodynamics and excess-function thermodynamics; the overlap in Fig. 1(b) merely removes part of a cosphere, while taking the solute particles from the solution annihilates both cospheres. From a model which incorporates the cosphere effects, one can calculate the overlap volume at any concentration,[10] and if the concentration is sufficiently low, we have to deal only with pairwise overlap, as illustrated in Fig. 1(b).

The thermodynamic excess functions of solutions up to about 1 M concentration can now be calculated quite accurately from Hamiltonian models[10] in which the solute–solute pair potentials are specified for each pair of solute species in the solution. Extensive studies have been carried out for a general model in which the pair potential is the sum of several terms. One of these, the so-called Gurney term, represents the cosphere overlap effect discussed above, while the others represent other physical interactions which are understood well enough so that they can be formulated without introducing parameters which can only be adjusted by fitting the model to experimental data for solutions. Thus, for the pair potential u_{xy} between solute x and solute y, there is one parameter A_{xy} which is in the Gurney term and which, most simply interpreted, is the free-energy change per mole of solvent displaced as cospheres of particles x and y are brought together to overlap, as in the passage from the configuration in Fig. 1(a) to that in Fig. 1(b). Further details of this model and the results of its application to a variety of aqueous ionic solutions and salting-out systems have recently been summarized.[11,12] The aspect which is of principal interest here is that, in spite of the expectation developed above, there is little correlation of A_{xy} and its derivatives with respect to T or P with coefficients of solvation, except possibly in aqueous solutions of tetraalkylammonium salts.[12]

Here we report the application of the same model to the study of the thermodynamic excess functions of aqueous alcohols. These systems have an advantage over the electrolyte solutions studied earlier in that each of the solute particle concentrations is independently variable, and so there should no longer be more parameters in the model than are required to fit the data.[11,12] However, the alcohol solutions have the disadvantage that the solute particles are not spherical, and so the solute–solute interaction cannot be represented realistically in terms of central forces, an aspect which is discussed again in Secs. 3 and 6. However, we were encouraged to carry through the present studies, in which orientation-dependent interactions are not explicitly accounted for, because of the degree of success in a study at the same level by Kozak, Knight, and Kauzmann.[13] There is a lot of overlap between their work and that reported here, but our model is somewhat more detailed, our method of calculation is considerably different, and we include comparison of the model with some new experimental data, also reported here, although not with all the data discussed by them. Our results are compared with theirs in Sec. 8.

Two other features of the background for this work should be mentioned here. The first has to do with enthalpy–entropy compensation. The generalized Barclay–Butler rule[14] states that for a series of processes $1, 2, \ldots, n, \ldots$ which are sufficiently similar, such as the solvation of a series of solutes in a given solvent, one finds

$$\Delta H(n) = H_0 + T^* \, \Delta S(n), \tag{1}$$

where H_0 and T^* are parameters which do not depend upon n. Most often the characteristic temperature T^* is found to be of the order of $1000°K$, but for a variety of processes involving aqueous solutions near $25°C$ one finds that Eq. (1) holds with $T^* = 280°K$, a regularity which we shall call Lumry's law.[15]4 In an earlier study of aqueous tetraalkylammonium salts, it was found possible to bring the Gurney parameters for the interaction of two

---

4 The qualitative content of this rule has been known for a long time, as pointed out by Lumry and Rajender.[15] Another example is developed in footnote 36 of ref. 13, based on the comment of a referee. The correlations giving rise to this rule have often been attributed to a kind of experimental error in the enthalpy coefficients derived from the Gibbs–Helmholz equation, but this possible error is avoided when calorimetric measurements are made, as is increasingly the case.[15]

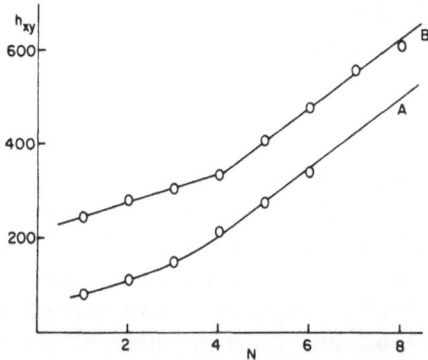

Fig. 2. Enthalpy interaction parameters h_{xy} (cal-mole^{-1}-molal^{-1}) as a function of the length of the alkyl chain R $= n$-$C_N H_{2N+1}$- for EtOH–ROH (A) and EtOH–RSO$_3$Na (B).

R$_4$N$^+$ into conformity with Lumry's law[12]; it is of interest to find the behavior of the Gurney parameters in the simpler situation encountered in this study where all the solute particle concentrations are independently variable.

The other feature to be mentioned is a regularity which is often found in data for coefficients of hydration of solutes with alkyl chains, namely that for long-enough alkyl chains one finds additive methylene-group contributions to the coefficients.[16] This regularity is expected if one assumes that the cosphere on each substituent group has properties which are determined mainly by that group and only to a minor extent by neighboring groups. Recent studies of the solvation enthalpies of a variety of solutes have led to fairly precise notions of both the validity[17] and the limitations[17-19] of the underlying concept. It is entirely remarkable that other coefficients of solvation do not closely follow the regularity shown by the enthalpies. Thus partial molal heat capacities[20] show much less, and molal solvent proton NMR shifts[21] show much more interaction among the cospheres of neighboring groups than do the solvation enthalpies.

Some new data (Sec. 5) in Fig. 2 show the result of looking for the same kind of regularity in a solute–solute interaction coefficient rather than in a coefficient of solvation. The linear and parallel portions of the graphs where $N > 3$ signify additive methylene-group contributions to the excess enthalpy coefficient h_{xy} (defined in Sec. 4), which is just what one expects if the effect is due to cosphere overlap and if the cosphere's properties depend mainly on the group it surrounds. These considerations encourage the effort to find a quantitative basis for comparing solvation coefficients with excess-function coefficients.

2. MODEL AND CALCULATIONS

The model used in the calculations reported here as well as the method of calculation are the same as in several studies[11, 12, 22] of models for electrolyte solutions. As specialized to solutions of nonelectrolytes, the model assumption is that the potential of the force between solute particles x and y in the pure solvent is

$$u_{xy}(r) = COR_{xy}(r) + GUR_{xy}(r) \tag{2}$$

where the COR term is a repulsive potential of the r^{-9} form.[22] The choice of radius parameters r_x^* and r_y^* in this term when x and y are nonspherical particles such as alcohol molecules is a matter of some difficulty, as discussed below. The GUR term represents the cosphere overlap effect (Fig. 1). It is[22] proportional to the overlap volume of the cospheres when r is the center-to-center distance between x and y and the thickness of an intact cosphere is 2.76 Å. It has a parameter A_{xy} which is adjusted to fit the data and which has the significance discussed in Sec. 1.

In this model only central forces between x and y are explicit, the dependence of the more detailed potential $u_{xy}(r, \theta)$ upon orientational variables θ having been suppressed by averaging as shown here.

$$\exp(-u_{xy}(r)/kT) = \int d\theta \exp(-u_{xy}(r, \theta)/kT) / \int d\theta \tag{3}$$

In $u_{xy}(r, \theta)$ the COR term is dependent upon the orientational variables, but the consequences of not representing this dependence explicitly are not very serious, as shown by Kozak et al.[13] The orientation dependence of the GUR term in $u_{xy}(r, \theta)$ may be pictured in terms of the cosphere structures shown in Fig. 3.

When two particles like those in Fig. 3 are close enough together for their cospheres to overlap, the contribution to the potential may depend upon the orientations which will determine the kind of cospheres which overlap each other, i.e., cospheres on the groups –OH, $-CH_2-$, or $-CH_3$. That even the difference between cospheres on the last two groups might be significant is implied by the large difference in hydration enthalpy of hexane compared to six methylene groups.[17]

Therefore, in the calculations using the model specified in Eq. (2) which are reported here, the Gurney A_{xy} parameters derived by fitting the models to the data are averages of overlap contributions of different kinds of cosphere pairs. Two different ways in which the consequences of this complication might be studied may be remarked.

1. The third and higher virial coefficients of the osmotic pressure [e.g., coefficient b in Eq. (4) below] depend upon different orientation averages from that specified in Eq. (3). Therefore, if the orientation effects are important, then a model with A_{xy} adjusted to fit the second virial coefficient of a solution will not fit the data over a modest concentration range, say, up to 1 M, in which the higher virial coefficients become experimentally significant. Unfortunately, this test cannot be made in the present study because of the lack of sufficiently accurate experimental data.

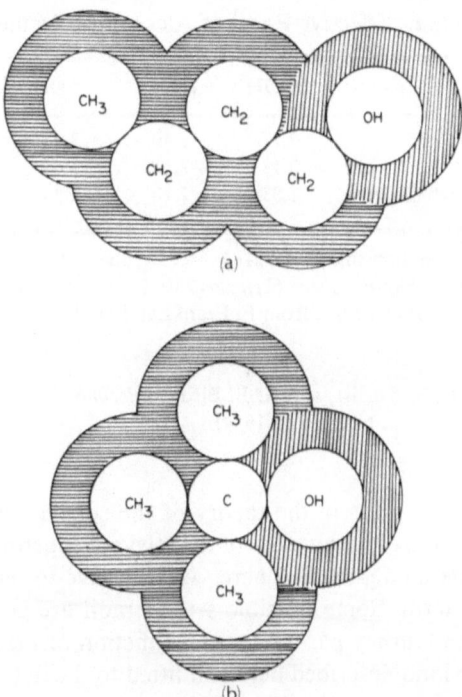

Fig. 3. Cospheres for n-BuOH and t-BuOH. Schematic two-dimensional representation to illustrate the source of orientation dependence in the Gurney term.

2. Assuming that the cosphere on an –OH group is quite different from that on –CH$_2$– or –CH$_3$ and that these are more like each other, one can see from Fig. 3 that the orientation effects are likely to be quite different for two n-BuOH than for two t-BuOH. If it transpires that A_{xx} is nearly the same in the two cases, then it is likely that the orientation effects are not important for either, except that some fraction of A_{xx} is due to overlap of cospheres on two –OH groups, a larger fraction to overlap of a cosphere on –OH with one on –CH$_2$– or –CH$_3$, and the main contribution to the overlap of cospheres on the aliphatic groups. This remark suggests a kind of analysis which can be best applied if one has determined A_{xy} coefficients for a collection of xy pairs with systematically varied structure; Fig. 2 shows the same principle applied to certain enthalpy coefficients. It is applied to the Gurney coefficients as far as possible in later sections.

Of course it would be of great interest to carry out calculations for the aqueous alcohols using more elaborate models in which orientation dependent potentials were specified. Unfortunately, this would require more elaborate computational procedures than the present study uses and also some basis, whether theoretical or empirical, for determining the much larger number of Gurney coefficients which would be involved. This in turn would seem to

Table I. Effective Radii of Alcohol Molecules, Å

Source	MeOH	EtOH	PrOH	n-BuOH	t-BuOH
From Conway's equation[a]	2.17	2.49	2.73	2.95	2.98
From molal volumes[b]	2.13	2.41	2.61	2.80	
From second virial coefficients[c]	1.83	2.18	2.35		

[a] Calculated using the relationship $\phi_v^0 = 2.51r^3 + 3.15r^2$ due to B. E. Conway, R. E. Verral, and J. E. Desnoyers, *Z. Physik. Chem.* (*Leipzig*) **230**, 157 (1965). The volume data are from ref. 26 except for t-BuOH which is from F. Franks and H. T. Smith, *Trans. Faraday Soc.* **64**, 2962 (1968).
[b] $r = (\phi_v^0 3/4\pi)^{1/3}$.
[c] J. O. Hirschfelder, C. F. Curtiss, and R. B. Bird, *Molecular Theory of Gases and Liquids* (John Wiley and Sons, Inc., New York, 1954).

depend on further progress in the theory of the solvent-averaged potentials or in the determination of many more of the relevant experimental coefficients.

For the simple model studied here, we still have to select radii to parametrize the COR term. Some possible sets of radii are given in Table I. In Fig. 4 we show the Gurney parameter as a function of the radius parameter for a model of the kind described here and fitted to the data for aqueous urea solutions.[11] It is apparent that there is a range of reasonable values of r_u^* between 2 and 3 Å in which there is only a small fractional change in the Gurney parameter. This result is typical.[12] In the rest of the calculations reported here the "Conway radii" in Table I have been used.

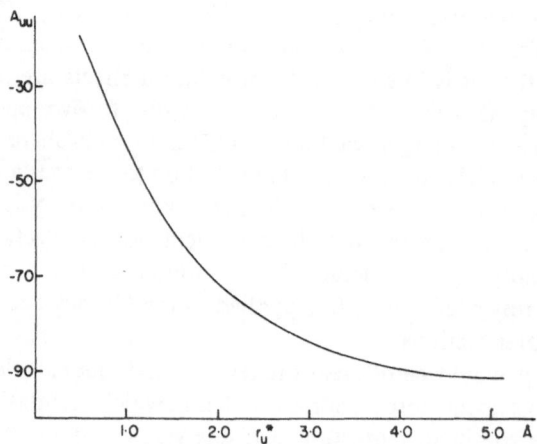

Fig. 4. The Gurney parameter A_{uu} for the interaction of two urea molecules obtained by fitting the model, with various choices of the radius parameter r_u^*, to the data for aqueous solutions of urea.

In most of the computations reported here the model's pair correlation function(s) were calculated by the HNC approximation from its pair potential(s) $u_{xy}(r)$, and then the thermodynamic properties were calculated from the pair correlation functions, all as in the earlier work.[10-12, 22, 23] This procedure was used because the computing programs were all available, but, in view of the limitations of the experimental data, it would have been entirely adequate to use a much simpler procedure in which one only evaluated the cluster integrals

$$\int [\exp(-u_{xy}(r)/kT) - 1] d^3 r$$

and their T and P derivatives.

3. EXCESS FREE ENERGIES AND GURNEY PARAMETERS

The experimental data used here are taken from Kozak, Knight, and Kauzmann[13] and from Knight's dissertation.[24] As these authors show, the excess free-energy coefficients they use can be expressed in terms of the coefficients of the virial expansion of the osmotic coefficient

$$\phi = 1 + am + bm^2 + \cdots \tag{4}$$

where m is the molality. Here ϕ is the Lewis–Randall osmotic coefficient ϕ_{LR}, while the model calculations give the McMillan–Mayer osmotic coefficient ϕ_{MM} which has the virial expansion

$$\phi = 1 + a'c + b'c^2 + \cdots \tag{5}$$

where c is the molarity of the solution in the McMillan–Mayer state.[25] The available data are for solutions of a single alcohol in water.

"Best values" of coefficient a at 25°C were estimated from the data[13, 24] at the freezing point and the corresponding enthalpy coefficients h_{xx} in the present notation (cf. Sec. 4). Values of h_{xx} have been determined at 25° (Table III, Sec. 5) and 0°C.[13] The a coefficients estimated in this way are given in Table II. From these the a' coefficients were calculated by the equation[25]

$$a' = a + V_x^0/V_w^0 \tag{6}$$

where V_x^0 is the limiting partial molar volume of the solute[26] and V_w^0 is the volume of a kilogram of solvent.

The Gurney parameter A_{xx} for the model for each of the aqueous alcohols was adjusted to fit the a' coefficient. The results are shown in Table II.

The results for A_{xx} may be compared with what one would expect if there were data for the a coefficients for solutions of the hydrocarbons methane through pentane in water and if the model were fitted to these. Then, assuming that the cosphere material was the same for the hydrocarbons from one end of the series to the other, one would expect A_{xx} to be constant with a negative value which reflects hydrophobic bonding. To get this simple result, one has

Table II. Excess-Free-Energy Coefficients at 25°C in Water

Coefficient	Solute				
	MeOH[a]	EtOH	PrOH	n-BuOH	t-BuOH
a	−0.005	−0.04	−0.08	−0.14	−0.02
a'	0.033	0.02	−0.01	−0.05	0.07
A_{xx}, cal-mole^{-1}	−85	−102	−110	−119	−104

[a] In this case it is particularly hard to pick the "best value." An equally good choice would be $a = 0.11$ which leads to $a' = 0.15$ and $A_{xx} = -38$ cal-mole^{-1}.

also to assume that orientation effects due to the nonspherical solute particles are negligible. Finally, one might expect the A_{xx} coefficients for the alcohols to reach a constant negative value as the alkyl chain gets longer so that the effect of the cosphere on –OH is dominated by the effect of the cosphere on the alkyl group. In fact, A_{xx} does seem to level off. Moreover, a reasonable extrapolation gives $A_{xx} = -140$ for hexanol, which is the same as A_{xy} found for the interaction of an R_4N^+ ion with a hydrocarbon R'H by fitting the same kind of model to the data of Wen and Hung[27, 28] and close to $A_{xx} = -190$ cal-mole^{-1} found for the interaction of two R_4N^+ ions.[12]

Taken together, these results indicate that the model studied here does serve to quantify the hydrophobic interaction, although, of course, it cannot lead directly to a deeper insight into the underlying molecular phenomena. But further insight might be obtained by extending the present studies in various ways. For example, for solutions in N-methyl acetamide one finds[28] $A_{xx} \simeq -60$ cal-mole^{-1} for the interaction of two sufficiently large R_4N^+ ions by fitting the model to the data of Kreis and Wood.[29] Clearly, there is a phenomenon like hydrophobic bonding in these solutions in the absence of water, as one can already see by examining the osmotic-coefficient data.[29] Because of the Lumry's law phenomena mentioned in the Introduction, one might expect that excess enthalpies and entropies would be more characteristic of solvation in water. This consideration led to the studies reported in the following sections.

It is remarked that the coefficient A_{xx} for t-BuOH is not very different from that for n-BuOH, which is what one must expect if the prior average over orientations does not make a distracting contribution to A_{xx} as discussed in Sec. 2.

Finally, the simplest interpretation of A_{xx} is that it measures the free-energy effect of displacing some of the solvent in the cosphere. At least for nonionic solutes the thermodynamics of solvation is determined by the interaction of the solute with solvent molecules next to it, i.e., in its cosphere, so one may expect A_{xx} to be related to the solvation free energy of a methylene group. This is not very well known; from several sources for different series of compounds, summarized by Cabani, Conti, and Lepori,[30] we find

$$(-CH_2-)_{aq} \rightarrow (-CH_2-)_g \qquad \Delta G^0 = -200 \pm 150 \text{ cal-mole}^{-1}$$

If this is attributed only to the relaxation of the cosphere water to its normal state when the CH_2 is removed and if, as estimated at the end of Sec. 6, there are 2.3 water molecules in the cosphere of a methylene group, then from the solvation data we estimate $A_{xx} = -87 \pm 65$ cal-mole^{-1} for the overlap of cospheres on methylene. The rather favorable comparison of this figure with the coefficients derived from the excess-function data is encouraging, so we turn to the enthalpy data.

4. THERMODYNAMICS

Some of the experiments and calculations reported in the following sections concern solutions of two nonelectrolytes. The thermodynamic coefficients for such systems need to be defined and related to experiment. The following discussion is specialized to the enthalpy but is readily extended to other thermodynamic properties except that the characteristic measurements are different for each.

We consider a solution of solutes of species x and y in a solvent of species w. For variation of the molalities m_x and m_y at fixed temperature and total pressure the excess enthalpy H^{ex} is given by

$$H^{ex}(m_x, m_y) = H(m_x, m_y) - H_w^0 - H_x^0 m_x - H_y^0 m_y \qquad (7)$$

$$= h_{xx} m_x^2 + 2h_{xy} m_x m_y + h_{yy} m_y^2 + h_{xxx} m_x^3 + 3h_{xxy} m_x^2 m_y + \cdots \qquad (8)$$

where $H(m_x, m_y)$ is the enthalpy of the quantity of solution containing a kilogram of solvent and H_x^0 and H_y^0 are the partial molal enthalpies at infinite dilution. The coefficients h_{xx}, etc., can be calculated from models for the interaction of the set of solute particles in the subscript when they are in the pure solvent. The integer coefficients 2, 3,... are required so that, in case species x is the same as species y except for some label which does not affect the intermolecular forces, we would have

$$h_{xx} = h_{xy} = h_{yy}$$

etc.

We need to know how to extract these coefficients from experimental data. Three cases are considered.

A. One measures the heat of dilution of a single nonelectrolyte.

$$\text{solution} \quad + \quad (m_x/m_x' - 1) \text{ kg of } w \quad \rightarrow \quad \text{mixture}$$

$$\begin{bmatrix} m_x \text{ moles of } x \\ 1 \text{ kg of } w \end{bmatrix} \qquad\qquad\qquad \begin{bmatrix} \text{molality} \\ = m_x' \end{bmatrix}$$

For this process ΔH may be expressed in terms of the coefficients in Eq. (7) as

$$\Delta H/m_x(m_x' - m_x) = h_{xx} + h_{xxx}(m_x + m_x') + \cdots \tag{9}$$

Lange and coworkers[31] made many measurements of this kind and reported the quantity on the left as "neigung." By plotting neigung as a function of $m_x + m_x'$ and extrapolating to zero abscissa, one obtains h_{xx} from their data.[5]

B. One adds pure solute x to a solution of x.

$$\text{solute } x \quad + \quad \text{solution} \quad \rightarrow \quad \text{mixture}$$

$$[m_x \text{ moles}] \quad \begin{bmatrix} m_x' - m_x \text{ moles } x \\ 1 \text{ kg } w \end{bmatrix} \quad \begin{bmatrix} \text{molality} \\ = m_x' \end{bmatrix}$$

For this process we have

$$\Delta H/m_x = H_x^0 - H_x^p + h_{xx}(2m_x' - m_x) + 3h_{xxx}(m_x' - m_x)m_x' + \cdots \tag{10}$$

where H_x^p is the molar enthalpy of pure solute x.

C. One adds pure solute x to a solution of solute y.

$$\text{solute } x \quad + \quad \text{solution} \quad \rightarrow \quad \text{mixture}$$

$$[m_x \text{ moles}] \quad \begin{bmatrix} m_y \text{ moles of } y \\ 1 \text{ kg of } w \end{bmatrix} \quad \begin{bmatrix} \text{molalities} \\ m_x, m_y \end{bmatrix}$$

For this process we have

$$\Delta H/m_x = H_x^0 - H_x^p + 2h_{xy}m_y + 3h_{xyy}m_y^2 + \cdots \tag{11}$$

where, on the right, terms proportional to m_x have been omitted, as is allowed if $m_x \ll m_y$.

In the present study, processes B and C have been studied to determine h_{xx} and h_{xy} as described in the following section.

5. CALORIMETRIC EXPERIMENTS AND RESULTS

The processes in Secs. 4B and 4C were studied at 25°C for several solutes by instrumental techniques which have been described elsewhere,[32] using materials which were prepared in earlier studies.[17, 33] The molality m_x which appears in Eqs. (10) and (11) was less than 10^{-3} m in each experiment. The coefficient $H_x^0 - H_x^p$, which is the standard enthalpy of solution, was taken from earlier studies.[17, 33] The coefficients h_{xx} and h_{xy} given in Table III were obtained from experiments in which $m_{x'}$ or m_y was 0.4 m and in each case is

---

[5] In these papers[31] Lange's results are reported in molar concentration scales and his "neigung" corresponds to Eq. (9) with molar rather than molal concentration units. The difference is about 4% for 0.5 m aqueous n-BuOH and is less for more dilute solutions and lower alcohols. The resulting difference in h_{xx} is negligible.

Table III. Enthalpy Interaction Coefficients h_{xy} for Aqueous Solutes at 25°C (Units: cal-mole^{-1}-molal^{-1})

Solute x	Solute y				
	MeOH	EtOH	n-PrOH	n-BuOH	t-BuOH
MeOH	48	80	97	117	125
	57 ± 1^a				
EtOH	75	110	137	178	165
		55 ± 3^a			
n-PrOH	102	150	225	304	300
			114 ± 5^a		
n-BuOH	140	215	340	500	445
				280^a	
t-BuOH		180		360	
				145^a	
				410 ± 50^b	
n-PeOH		275		575	

Solute y = EtOH

Solute x =	n-HexOH	MeSO$_3$Na	EtSO$_3$Na	n-PrSO$_3$Na	n-BuSO$_3$Na
h_{xy}	340	245	280	305	335

Solute x =	n-PeSO$_3$Na	n-HexSO$_3$Na	n-HeptSO$_3$Na	n-OctSO$_3$Na
h_{xy}	410	480	560	610

[a] Derived from measurements of Lange and coworkers.[31] The ± values are our estimated maximum errors, judging from the scatter of the data.
[b] At 26°C. Calculated from the data of Kenttamaa.[34]

the average of two or more experiments in which $\Delta H/m_x$ was the same within 50 cal-mole^{-1}. The contributions due to higher terms with coefficients such as h_{xxx} or h_{xyy} were neglected. The uncertainties in h_{xx} and h_{xy} due to the calorimetric uncertainty is less than $50/(2 \times 0.4) = 65$ cal-mole^{-1}-molal^{-1}.

In many cases the coefficients given in Table III were also obtained from experiments in which $m_{x'}$ or m_y was 0.8 m. The results agreed with those in Table III within 30 cal-mole^{-1}-molal^{-1} and are not reported separately.

In the absence of experimental errors or significant contributions due to the neglect of the higher terms in Eqs. (10) and (11), one expects the array of the 16 coefficients for the solutes MeOH, EtOH, PrOH, and BuOH to be symmetrical. In fact, the largest deviation from symmetry is for the pair EtOH—BuOH which differ by 37 cal-mole^{-1}-molal^{-1}, no more than expected from the experimental uncertainty.

Another kind of test of these data may be made by comparison with the results of a series of measurements of the heats of mixing of the sodium alkylsulfonates RSO$_3$Na (x) with EtOH (y) which was carried out in the same way as the series described above. The resulting h_{xy} values are plotted as a function of chain length in Fig. 2. They were derived by the analysis in Sec. 4C which is applicable when species x is an electrolyte, except that now h_{xy} is a sum of ionic contributions, $h_{+y} + h_{-y}$ in the case of a 1-1 electrolyte. We study only the sodium salts, so h_{+y} is the same in each case. Thus in Fig. 2 the variation of h_{xy} with chain length must be attributed to h_{-y}, where $-y$ is the RSO_3^- ion. The EtOH row of Table III is also plotted in Fig. 2 for comparison. The significance of the linear portions of these graphs has already been noted at the end of Sec. 1. Thus the simplest explanation of the linear $(N > 3)$ portions of these graphs is that they are due to additive methylene-group contributions. If they are due to additive methylene-group contributions, the linear portions are necessarily parallel, as observed.

The discrepancy with the coefficients derived from Lange's data, also given in Table III, is many times the known sources of error in either his or our experiments. Unfortunately, with our calorimeter it is not possible to get useful results for ΔH for the process in 4B if $m_{x'}$ is much less than 0.4 m, so there is a possibility, which we cannot completely rule out by our experiments, that our neglect of the higher terms in Eqs. (10) and (11) is unjustified. On the other hand, the greatest $|h_{xxx}|$ estimated from Lange's data is 25 cal-mole^{-1}-molal^{-2} (for EtOH), and this is much too small to account for the difference in the two sets of h_{xx} data. Furthermore, the coefficient h_{xx} which may be extracted from the data of Kenttamaa et al.[34] for aqueous t-BuOH agrees with our result within the combined accuracy of the two sets of data, as shown in Table III.

6. EXCESS ENTHALPIES AND DERIVED GURNEY PARAMETERS

To calculate the excess energy of a model system we use the pair correlation functions of the model which fits the free-energy data (Sec. 3) together with[22] the temperature coefficients of the Gurney free-energy parameters

$$S_{xy} \equiv -\partial A_{xy}/\partial T \tag{12}$$

or the actual input for our computational procedure[22]

$$E_{xy} \equiv A_{xy} + TS_{xy} \tag{13}$$

The excess energy per liter of the model system may be written in the form

$$E^{ex}(c_x, c_y) = e_{xx}c_x^2 + 2e_{xy}c_x c_y + \cdots \tag{14}$$

in complete analogy with Eq. (7), where now c_x is the molarity of solute x. As explained below, the McMillan–Mayer coefficients e_{xx}, e_{xy} are nearly equal to the corresponding Lewis–Randall coefficients h_{xx}, h_{xy}. Then we can estimate E^{ex} from the coefficients in Table III.

For example, for a model for aqueous propanol to agree with experiment at $c_x = 0.4$ M, we must have $E^{ex} = 225 \times (0.4)^2 = 36$ cal-liter^{-1}. A model calculation with $E_{xx} = 0$ gives $E^{ex} = 28.8$ cal-liter^{-1}. A calculation made with

$E_{xx} = -100$ cal-mole^{-1} gives $E^{ex} = -19.1$ cal-liter^{-1}. We know that E^{ex} varies linearly with E_{xx},[22] so we deduce that the model will fit the enthalpy data if $E_{xx} = 15$ cal-mole^{-1}.

Calculations for mixtures can be done in a similar way except that, in the absence of data from which the A_{xy} parameters can be derived, we use values estimated by the rule

$$A_{xy} = \tfrac{1}{2}(A_{xx} + A_{yy}) \tag{15}$$

Then the correlation functions for a mixture of desired composition can be calculated and the energy of the mixture fitted to experiment by varying the E_{xy} parameter while E_{xx} and E_{yy} are fixed at values determined by calculations for solutions of the separate solutes.

The results do not depend upon the concentrations at which the model computations are made, at least up to $c_x + c_y = 0.8$ M, except that the round-off errors are serious if the concentrations are too low. Thus in the models the contributions of terms in h_{xxx}, h_{xyy}, and other higher terms are negligible when the total concentration is less than 0.8 M. In this respect the model is consistent with our own experiments but not with Lange's.[31]

The relation of the McMillan–Mayer to the Lewis–Randall coefficients may be derived from the general theory of the conversion.[35] For example, we find

$$e_{xx} = h_{xx} + \alpha RT^2(-a + \phi_E^0/\alpha V_w^0) \tag{16}$$

where α is the thermal expansion coefficient of the solvent, a appears in Eq. (4), V_w^0 appears in Eq. (6), and ϕ_E^0 is the limiting partial molal expansibility of solute x.[26, 36] These corrections amount to only a few cal-mole^{-1}-molal^{-1} but were made before calculating the E_{xx} Gurney coefficients. The results are not affected significantly by the corrections except for MeOH. Because the effect of the corrections is so small and because the quantity corresponding to a is not known experimentally for mixtures and would have to be estimated, the difference between h_{xy} and e_{xy} was neglected for the cases in which x and y are different species, although the equation corresponding to (16) is readily derived.

The energy-interaction parameters determined in this study are given in Table IV. It is remarkable that E_{xy} gets bigger with increasing chain length even for the largest solute particles studied. Thus there is no sign of the leveling-off found for A_{xx} which, as discussed in Sec. 3, we really expect if the model is adequate. The present behavior might be attributed to the alkyl chains not being long enough to dominate the effects of the –OH groups. The data in Fig. 2 make this explanation plausible, but it is not clear that it is consistent with the A_{xx} results.

It seems more likely that the anomalous behavior of the coefficients in Table IV indicates the need for studying models with explicit orientation

Table IV. Gurney Energy Coefficients E_{xy}, cal-mole^{-1}

	MeOH	EtOH	n-PrOH	n-BuOH	t-BuOH
MeOH	-8	-13	-10	-5	-5
EtOH		-10	-4	6	-1
n-PrOH			15	33	29
n-BuOH				68	56
t-BuOH					35

dependence, as discussed in Sec. 2. For such a model the cluster-expansion expression for h_{xy} would be proportional to

$$\int d^3 r \int d\theta [\partial(u_{xy}(r, \theta)/T)/\partial(1/T)] \exp(-u_{xy}(r, \theta)/kT)$$

so that the appropriate average over orientations actually is different from the one specified in Eq. (3). It seems necessary to carry through computations with the more detailed model to see whether the difference in orientation averages is the source of the inconsistency we find on comparing the A_{xx} parameters with the E_{xy} parameters.

Another aspect of interest is the magnitude of E_{xy} in Table IV. In the models studied here the increment in cosphere volume when a $-CH_2-$ group is added to an alkyl chain is about 70 Å^3, which is equivalent to about 2.3 water molecules. The enthalpy change for the process in which a mole of $-CH_2-$ groups (in alkyl chains) is transferred from water to the gas phase is about 800 cal.[17] Assuming that this is all due to change in the local thermodynamic state of the water in the cospheres of the methylene group, with the cosphere geometry assumed here, this gives about 350 cal-mole^{-1} of water in the cospheres. Therefore the very simplest interpretation of E_{xy} for the overlap of cospheres on alkyl chains leads one to expect $E_{xy} \simeq 350$ cal-mole^{-1}. This is much larger than even the largest entry in Table IV. It seems of interest to consider two changes in the model which might remove this inconsistency.

Representing n-BuOH as a chain (Fig. 3) rather than as a sphere will lead to a somewhat larger cosphere volume per methylene group which would change the calculation in the preceding paragraph in such a way as to reduce the discrepancy, provided that the E_{xy} needed to fit the model to the data did not change much. While this seems likely, one cannot be sure without doing the full computation.

Inclusion of a London dispersion term in the pair potential for the model [Eq. (2)] makes a negative contribution to u_{xy} which is independent of temperature and, therefore, a negative contribution to E^{ex}. If no other change were made in the model, then E^{ex} would be too negative compared with the experimental data for h_{xy}. This would have to be compensated by a more positive value of E_{xy} which then would be in better agreement with the hydration data. Unfortunately, to refine the model in this way one needs a satis-factory theory for the London interaction between solute particles so that there are no additional parameters to be determined by fitting the model to the data.

In the course of this work a few calculations for models with COR potentials of the $e^{-r/R}$ form[11] were made. The resulting Gurney parameters are not very different from those for the model that is mostly studied here. For example, the model with the $e^{-r/R}$ COR term fits the aqueous-ethanol data with $A_{xx} = -78$ and $E_{xx} = 9$ cal-mole^{-1} compared to -102 and -10, respec-tively, for the r^{-9} COR potential model. Apparently the conclusions from the present study would be preserved if the entire work were repeated with the other model.

7. EXCESS VOLUMES AND DERIVED GURNEY PARAMETERS

Beginning with the excess volume measurements on aqueous solutions of the pure alcohols reported by Friedman and Scheraga,[26] finding the coefficient v_{xx} that corresponds to h_{xx} in Eq. (7) and comparing with the results of model calculations in complete analogy with the procedure described in Sec. 6, we determined the parameters

$$V_{xx} = \partial A_{xx}/\partial P \qquad (17)$$

where P is the pressure on the pure solvent in which two solute particles of species x are located. The results are

$$V_{xx} = -2.2, -6.0, -4.5, \text{ and } -4.1 \ \mu\text{l-mole}^{-1}$$

for MeOH, EtOH, PrOH, and n-BuOH, respectively.

The sign of V_{xx} is just what one expects for cospheres on alkyl groups if the local molar volume of the cosphere water is somewhat greater than normal, as expected if it has more, or stronger, hydrogen bonds per molecule. The magnitude of V_{xx} may be compared with the volume change when ice melts, namely, $-1620 \ \mu\text{l-mole}^{-1}$. Compared to this figure the above values of V_{xx} are exceedingly small! It is difficult to compare V_{xx} with a solvation volume effect, in analogy to the comparison of E_{xx} with the hydration enthalpy of a methylene group, because of the delicacy of allowing for packing effects.

The constancy of V_{xx} values reported here is expected if the main contri-bution is due to overlap of cospheres on alkyl groups and if these cospheres have similar properties in the series methyl to butyl. However, in view of the results in Sec. 6 one must suspect that the simple results here result from the accidental cancellation of various complicated effects.

8. DISCUSSION

In earlier sections the Gurney free-energy parameters A_{xy} and energy parameters E_{xy} were derived for models which fit the data, and several aspects of the significance of the results were discussed. Two additional aspects are treated in this section.

It is interesting to compare the Gurney A_{xy} and E_{xy} parameters with the corresponding results reported by Kozak, Knight, and Kauzmann.[13] The comparison can only be made for solutions of single alcohols, the class of systems common to both their study and ours. They made the following analysis of the coefficient a' [cf. Eq. (5)] which for solute species x is related to the solute–solute pair potential u_{xx} by the equation

$$a' = -\tfrac{1}{2}K \int_0^\infty [\exp(-u_{xx}(r)/kT) - 1] 4\pi r^2 \, dr \qquad (18)$$

where the conversion factor K is 6.023×10^{-4} if r is expressed in angstroms and, in Eq. (5), c is molarity. Their analysis assumes, in the terminology of Eq. (2), that there is a hard-sphere COR term:

$$\begin{aligned} \text{COR}_{xx}(r) &= \infty \quad \text{if } r < d_{xx} \\ &= 0 \quad \text{if } d_{xx} < r \end{aligned} \qquad (19)$$

where d_{xx} is the equivalent hard-sphere diameter. Then a' is given by

$$a' = 4v_x + a_x'' \qquad (20)$$

where v_x is the volume of a mole of the hard-sphere cores expressed in liters and a_x'' is the remainder of the integral in (18), which is

$$a_x'' = -\tfrac{1}{2}K \int_{d_{xx}}^\infty [\exp(-u_{xx}(r)/kT) - 1] 4\pi r^2 \, dr \qquad (21)$$

Kozak et al. point out that the repulsive term $4v_x$ is only slightly changed if the cores are ellipsoidal with axial ratios as great as 5:1. They evaluated v_x by assuming it is the same as V^0, the partial molal volume of x at infinite dilution. This procedure gives about the same repulsive term in Eq. (20) as we would get using Eq. (18) with only the COR term together with the second set of radius parameters in Table I, except that the tail on the r^{-9} repulsive potential makes the integral somewhat larger than for the hard-sphere core.

It follows that a_x'' in our model would be given by the integral in Eq. (21) but with just the Gurney term GUR_{xx} of our model. Therefore, there is a direct relation between the coefficient a_x'' reported by Kozak et al. and the coefficient A_{xx} reported here; clearly, however, the latter is the more reduced coefficient since the model calculation accounts for the solute sizes through the dependence of the possible volume of cosphere overlap upon the radii r_x^*. Thus it is quite consistent that from Fig. 3 of Kozak et al. we find

$$a_x'' = -30 - 80N_x \qquad (22)$$

where N_x is the number of carbon atoms in normal alcohol x, while in the present study A_{xx} tends to approach a negative asymptote as N_x increases.

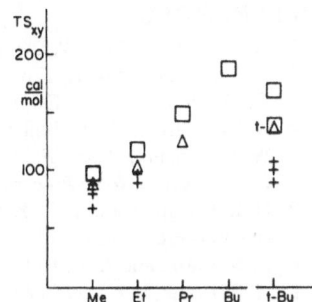

Fig. 5. Values of TS_{xy} for models which fit the data. S_{xy} is the Gurney entropy parameter. + MeOH. ‡ EtOH. Δ n-PrOH. $\square$ n-BuOH. $t\square$ t-BuOH.

The additional detail apparent in the A_{xx} data compared to the a_x'' data must of course reflect the particular model assumptions of this study. If these are sufficiently realistic, then the conclusion of Kozak et al. that a'' becomes more negative with increasing chain length may be interpreted simply: It is mainly due to the increasing overlap volume for two solute particles in contact rather than to changes in the cosphere material.

Kozak et al.[13] also note that the data together with the assumed properties of the hard-sphere core lead to the conclusion that a_x'' becomes more negative with increasing temperature. This corresponds, in the present model, to positive S_{xx}.

Values of S_{xy} obtained in the present work are shown in Fig. 5. The sign agrees with the expectation from the Kozak study.[13] The dependence of S_{xy} upon the length of the alkyl chains reflects the corresponding dependence of E_{xy} which was discussed in Sec. 6. As pointed out there, this trend may well indicate the need for studying more elaborate models.

In the study of models for aqueous tetraalkylammonium halides,[12] it was pointed out that we expect to have

$$TS_{xy} \simeq -2A_{xy} \tag{23}$$

if x and y are particles with large alkyl chains, as a consequence of Lumry's law, which was referred to in the introduction. It may be concluded from the new results and the earlier studies referred to in Sec. 3 that we have $A_{xy} \simeq -140$ cal-mole^{-1} if x and y are both particles with large alkyl groups. Then, if Eq. (23) is right, the data in Fig. 5 should asymptotically approach 280 cal-mole^{-1} as the chains are lengthened. This is not inconsistent with the results shown there.

In summary, we may say that while not all the variables in hydrophobic bonding are adequately controlled using the simple model studied here, it does seem that the model provides a useful step toward placing our knowledge of hydrophobic bonding on a quantitative basis.

REFERENCES

1. H. S. Frank and M. W. Evans, *J. Chem. Phys.* **13**, 507 (1945).
2. J. D. Bernal and R. H. Fowler, *J. Chem. Phys.* **1**, 515 (1933).
3. D. D. Eley and M. G. Evans, *Trans. Faraday Soc.* **34**, 1093 (1938).
4. A. Rahman and F. H. Stillinger, *J. Chem. Phys.* **55**, 3336 (1971).
5. W. Kauzmann, *Advan. Protein Chem.* **14**, 1 (1959).
6. R. W. Gurney, *Ionic Processes in Solution* (Dover Publications, Inc., New York, 1962).
7. H. L. Friedman and C. V. Krishnan, in *Water, A Comprehensive Treatise*, F. Franks, ed. (Plenum Press, New York, 1973), Vol. 3.
8. H. S. Frank and A. L. Robinson, *J. Chem. Phys.* **8**, 933 (1940).
9. H. S. Frank, in *Chemical Physics of Ionic Solutions*, B. E. Conway and R. G. Barradas, eds. (John Wiley and Sons, New York, 1965); H. S. Frank, *Z. Physik. Chem. (Leipzig)* **228**, 364 (1965).
10. H. L. Friedman, in *Modern Aspects of Electrochemistry*, J. O'M. Bockris and B. E. Conway, eds. (Plenum Press, New York, 1971), Vol. 6.
11. H. L. Friedman, C. V. Krishnan, and C. Jolicoeur, *Ann. N.Y. Acad. Sci.* **204**, 79 (1973).
12. P. S. Ramathan, C. V. Krishnan, and H. L. Friedman, *J. Solution Chem.* **1**, 237 (1972).
13. J. J. Kozak, W. S. Knight, and W. Kauzmann, *J. Chem. Phys.* **48**, 675 (1968).
14. H. S. Frank, *J. Chem. Phys.* **13**, 493 (1945).
15. R. Lumry and S. Rajender, *Biopolymers* **9**, 1125 (1970).
16. J. A. V. Butler, *Trans. Faraday Soc.* **33**, 235 (1937).
17. C. V. Krishnan and H. L. Friedman, *J. Solution Chem.* **2**, 37 (1973).
18. C. V. Krishnan and H. L. Friedman, *J. Phys. Chem.* **74**, 3900 (1970).
19. D. S. Reid and F. Franks, Proceedings of the 1st International Conference on Calorimetry and Thermodynamics, Warsaw, 1969, p. 891.
20. E. M. Arnett, W. B. Kover, and J. V. Carter, *J. Am. Chem. Soc.* **91**, 4028 (1969); J. Konicek and I. Wadsoe, *Acta Chem. Scand.* **25**, 1541 (1971).
21. W. Y. Wen and H. G. Hertz, *J. Solution Chem.* **1**, 17 (1972).
22. P. S. Ramanathan and H. L. Friedman, *J. Chem. Phys.* **54**, 1086 (1971).
23. H. L. Friedman and P. S. Ramanathan, *J. Phys. Chem.* **74**, 3756 (1970).
24. W. S. Knight, doctoral dissertation, Princeton, 1962.
25. H. L. Friedman, *J. Solution Chem.* **1**, 387 (1972).
26. M. E. Friedman and H. A. Scheraga, *J. Phys. Chem.* **69**, 3795 (1965).
27. W. Y. Wen and J. H. Hung, *J. Phys. Chem.* **74**, 170 (1970).
28. C. V. Krishnan and H. L. Friedman, to be submitted.
29. R. W. Kreis and R. H. Wood, *J. Phys. Chem.* **75**, 2319 (1971).
30. S. Cabani, G. Conti, and L. Lepori, *Trans. Faraday Soc.* **67**, 1943 (1970).
31. E. Lange and H. G. Margraf, *Z. Elektrochem.* **54**, 73 (1950); W. Dimmling and E. Lange, *Z. Elektrochem.* **55**, 322 (1951); E. Lange and K. Moehring, *Z. Elektrochem.* **57**, 660 (1953).
32. H. L. Friedman and Y. C. Wu, *Rev. Sci. Instr.* **32**, 1236 (1965).
33. C. V. Krishnan and H. L. Friedman, *J. Phys. Chem.* **73**, 1572 (1969).
34. J. Kenttamaa, E. Tommida, and M. Mertti, *Ann. Acad. Sci. Fennicae*, Ser A2, No. 93, 3 (1959).
35. H. L. Friedman, *J. Solution Chem.* **1**, 413 (1972).
36. F. Franks and H. T. Smith, *Trans. Faraday Soc.* **64**, 2962 (1968).

DISCUSSION[6]

Professor R. H. Stokes (*University of New England, Armidale, Australia*). I found that one of the most interesting conclusions of your calculations is

---

[6] See footnote 2 on page 119.

that the Gurney A_{ij} coefficients are negative in nearly all cases, indicating attractive forces when the cospheres overlap. I would just like to point out that in my opinion a reason for this is that a negative term is required in your theory in order to compensate for a very positive term in the form of the covolume term. Your treatment includes a hard-sphere covolume term, and I strongly suspect [R. H. Stokes, *J. Chem. Phys.* **56**, 3382 (1972)] that this should be deleted with the result that the other terms in your equation would not be necessary.

Professor Friedman (*State University of New York, Stony Brook, New York*). I certainly would agree that if I left out something, then the result would be changed. I do not have a clear idea myself in terms of Hamiltonian models what it is one ought to do. Our study of isotope mixtures [*J. Solution Chem.* **1**, 387 (1972)] was partially stimulated by the same questions. Also, because it seems we are mostly getting negative Gurney A_{ij} coefficients and there must be a uniform conclusion or a uniform explanation, the obvious uniform explanation is some general liquid structure effect—not necessarily water structure, but having to do with packing of particles. I do not know how to do anything more with it. I wish I did.

Professor Stokes. My proposal is to take your complete calculation for the primitive model, which in effect evaluates your coulomb terms plus a hard-sphere co-volume term, subtract from it the corresponding term for uncharged spherical particles, and use the result for a coulomb term.

Professor Friedman. But that is not what one measures. Typically, one measures the osmotic coefficient of a solution. It is not the osmotic coefficient of a solution minus that of some reference system, for which your proposal would be appropriate.

Professor Stokes. My proposal is to take the behavior of simple uncharged systems as the norm rather than the behavior of a perfect gas as the norm, which is essentially what you are doing.

Professor H. S. Frank (*University of Pittsburgh*). You have a problem there, Professor Stokes, because simple uncharged molecules in water follow their own idiosyncracies and do not behave as one would want them to.

Dr. F. H. Stillinger (*Bell Laboratories, New Jersey*). I think I am a little disturbed about the fact that the solute cospheres in water are treated the same regardless of the species. One of the things that always struck me about water is its very versatile and imaginative nature—hence my concern with the fact that this statistical-mechanical technology that you use seems to be relatively unversatile and unimaginative. In fact, I would like to see you discuss just electrolytes or just nonionic solute–solute interactions. When put together, I wonder if something isn't really lacking due to structural detail which surely must be present in the solvent but is neglected in your model. Have I misunderstood your strategy?

Professor Friedman. Let me just consider alkali metal ion–argon interaction in water. (Obviously, with a solute such as urea there is another problem because of the orientation effects.) It seems to us that there should be a difference, reflecting what you just said, in the Gurney parameters for the ion–argon model fitted to the salting-out data and for the ion–ion model fitted to the osmotic-coefficient data for the electrolytes themselves. We did not assume the similarity of the Gurney coefficients; we asked what the comparison is going to be like. They come out to be rather similar and in the same range, -100 to 0 cal-mole^{-1} of water displaced.

Dr. Stillinger. Can one reasonably imagine that an ion in water cannot tell an argon atom from a chloride or potassium ion? Surely the ions exert an enormous force and torque on water molecules. The state of local aggregation is thus surely different for uncharged vs charged solutes.

Professor Friedman. I am not arguing that, but the question is: if so, why doesn't it show up in the model interaction potential that one uses to get the thermodynamic properties to fit those of the real system?

Dr. Stillinger. The obvious answer to that is that one doesn't have anything like uniqueness for models that fit the osmotic coefficients.

Professor Friedman. That is most likely true. Now we come to the next point. In order to put in models which are more detailed and show more versatile behavior, one needs some information ahead of time because only one parameter per pair of solute species needs to be adjusted to fit the thermodynamic data. I don't dare to leave the implication that one is seeing anything but the average over monstrous effects, which is generally true when one looks at a change in a thermodynamic function. An interesting proposal that gets to the heart of things is to get more detailed information about solute–solute correlation functions in real solutions. Obviously, given information like the H_2O^{17} self-correlation function [*J. Solution Chem.* **1**, 387 (1972)], but for some other solute in water, one would really be able to arrive at a form for the interaction potential which is more realistic. However, in the meantime, it seems to me to be a fair question to ask whether there is anything misleading in the study of models with an oversimplified representation of the effect of the solvent or whether it is useful here in order to give some overall description of what is going on.

Structure in Aqueous Solutions of Nonpolar Solutes from the Standpoint of Scaled-Particle Theory[1]

Frank H. Stillinger[2]

Received October 30, 1972

Underlying assumptions have been examined in scaled-particle theory for the case of a rigid-sphere solute in liquid water. As a result, it has been possible to improve upon Pierotti's corresponding analysis in a way that explicitly incorporates measured surface tensions and radial-distribution functions for pure water. It is pointed out along the way that potential energy nonadditivity should create an orientational bias for molecules in the liquid–vapor interface that is peculiar to water. Some specific conclusions have been drawn about the solvation mode for the nonpolar rigid-sphere solute.

KEY WORDS: Hydrogen bonding; molecular correlation; potential non-additivity; scaled-particle theory; solubility; surface tension; water.

1. INTRODUCTION

The scaled-particle theory of classical fluids offers a powerful conceptual and computational framework within which to examine molecular order and thermodynamic properties. This method was originally devised to describe only the rigid-sphere model without attractive forces.[1,2] Nevertheless, its scope has since been increased to include models for a wide class of real substances.[3,4] Furthermore, the underlying theory has been substantially strengthened and deepened in comparison to its early version.[5-8]

The initial attempts to apply scaled-particle theory to liquid water were disappointing. Both the surface tension and the isothermal compressibility, along the saturation line from 0° to 100°C, were predicted to be too low and to have improper temperature variations.[9] In view of the strong, directional, and nonadditive interactions that operate in water to produce extensive hydrogen bonding, this failure seems hardly surprising.

---

[1] This paper is substituted for the talk given at the symposium, "The Physical Chemistry of Aqueous Systems," held at the University of Pittsburgh, Pittsburgh, Pennsylvania, June 12–14, 1972, in honor of the 70th birthday of Professor H. S. Frank.

[2] Bell Laboratories, Murray Hill, New Jersey 07974.

One of the key quantities in the scaled-particle theory is $W(\lambda)$, the amount of reversible, isothermal work necessary to create a spherical cavity of radius λa in the fluid of interest, whose interior is devoid of molecular centers. Following the usual convention, we use a here to denote a convenient fixed molecular length, and λ varies in the range $0 \leqslant \lambda < \infty$. In order for a nonpolar spherical solute molecule to dissolve in a liquid, it must at least have available to it the requisite cavity; as a consequence, $W(\lambda)$ for the appropriate size λa becomes an important contribution to the solubility of that nonpolar solute.[10]

Pierotti[11] has specifically applied the scaled-particle theory to description of aqueous solutions of nonpolar gases. Somewhat surprisingly, he finds that it is possible to predict heats, entropies, and molar heat capacities of solution merely under the assumption that the water molecules arrange themselves spatially in the pure liquid as would rigid spheres of appropriate size. In view of the current understanding about the interactions between water molecules,[12-16] this apparent success must be somewhat fortuitous. Certainly, the molecular structure in water revealed by x-ray scattering experiments[17] seems quite different from that appropriate to rigid spheres. It is therefore the purpose of this paper to reexamine one aspect of application of scaled-particle theory to aqueous solutions of spherical, nonpolar solutes, with the aim of restoring chemical detail.

For completeness, Sec. 2 provides an outline of the main ingredients in the scaled-particle theory as well as some numerical results for Pierotti's specific method of application. Section 3 discusses the character of the planar water interface that obtains in the $\lambda \to \infty$ limit and points out some features of molecular arrangements in the interface that are peculiar to water.

We offer in Sec. 4 a more detailed version of scaled-particle theory for water as a solvent than has heretofore been available. This version incorporates both measured surface tensions and radial-distribution functions for the pure liquid. The results are interpreted in Sec. 5 in a way that accords with the special nature of intermolecular forces in water.

2. PRELIMINARY RESULTS

Consider a set of N molecules confined to the interior of a region with volume V. Thermal equilibrium at absolute temperature T will be assumed. For the moment, at least, no special restrictions need to be imposed about the internal structure of the molecules or about the way in which they interact with one another.

We wish to examine the way that this N-molecule solvent responds to the insertion of a special type of solute particle. This solute particle has the property that it interacts with the solvent molecules only in being unable to get closer than distance λa to the center of each. Thus each solvent molecule has associated with it an exclusion sphere of radius λa, and the center of the solute particle must forever remain outside of all of these exclusion spheres. Equivalently, solvent molecules are excluded from a sphere of radius λa surrounding the solute particle.

The probability $P(\lambda)$ that a randomly chosen position in the pure solvent lies outside of all exclusion spheres is equal to the fraction of V uncovered by those spheres. This probability can in turn be related[1] to the cavity-creation work $W(\lambda)$ mentioned previously:

$$P(\lambda) = \exp\left[-W(\lambda)/kT\right] \tag{1}$$

where k is Boltzmann's constant. If λ is zero, $P = 1$ and $W = 0$; by contrast, $P(\lambda)$ will be very small for large positive λ, and $W(\lambda)$ will be large and positive since many solvent particles would normally have to be moved out of the way to create an uncovered location.

For arbitrary λ, the density of solvent molecule centers, at the surface of the empty radius-λa sphere S_λ surrounding the solute, is traditionally denoted by $\rho G(\lambda)$, where

$$\rho = N/V \tag{2}$$

By considering the work expended during increase of the solute size from 0 to λa, it may be shown[1] that

$$W(\lambda)/kT = 4\pi\rho a^3 \int_0^\lambda (\lambda')^2 \, G(\lambda') \, d\lambda' \tag{3}$$

Another fundamental relation in the scaled-particle theory results from expressing $P(\lambda)$, the probability that S_λ is empty, in terms of molecular correlation functions $g^{(n)}$ for molecular centers in the pure solvent. One thus has the following identity:[1]

$$P(\lambda) = 1 + \sum_{n=1}^\infty \left[(-\rho)^n/n!\right] \int_{S_\lambda} d\mathbf{r}_1 \cdots \int_{S_\lambda} d\mathbf{r}_n \, g^{(n)}(\mathbf{r}_1 \cdots \mathbf{r}_n) \tag{4}$$

The terms in this series will all vanish for orders n exceeding the maximum number of solvent molecule centers that can be packed into sphere S_λ.

Let a be the distance of closest approach of two solvent molecules to each other. Then when $0 \leqslant \lambda \leqslant \frac{1}{2}$, all terms in (4) beyond $n = 1$ vanish. Since $g^{(1)} \equiv 1$, we then have in this initial λ range

$$P(\lambda) = 1 - (4\pi\rho a^3/3)\,\lambda^3$$
$$W(\lambda)/kT = -\ln\left[1 - (4\pi\rho a^3/3)\,\lambda^3\right] \tag{5}$$
$$G(\lambda) = \left[1 - (4\pi\rho a^3/3)\,\lambda^3\right]^{-1}$$

As λ begins to exceed $\frac{1}{2}$, two solvent centers can fit into S_λ, so the $n = 2$ term in series (4) begins to contribute. Nevertheless, each of $P(\lambda)$, $W(\lambda)$, and $G(\lambda)$ remain continuous and differentiable at $\lambda = \frac{1}{2}$, and only the last of these three functions can suffer a simple discontinuity in its second λ derivative there.[1]

As $\lambda \to \infty$, $W(\lambda)$ becomes dominated by work against the external pressure p and against the surface tension γ for the cavity–solute interface.[3] Thus we have

$$W(\lambda) = (4\pi pa^3/3)\,\lambda^3 + (4\pi\gamma_\infty a^2)\,\lambda^2 - (16\pi\gamma_\infty \delta a)\,\lambda + O(1) \qquad (6)$$

here γ_∞ stands for surface tension in the planar interface limit, and δ provides the leading curvature dependence for the mechanical tension γ:[18]

$$\gamma \sim \gamma_\infty[1 - (2\delta/\lambda a)] \qquad (7)$$

The integral connection (3) between W and G allows us to conclude that the latter has the following large-λ behavior:

$$G(\lambda) = (p/\rho kT) + (2\gamma_\infty/\rho akT)/\lambda - (4\gamma_\infty \delta/\rho a^2 kT)/\lambda^2 + \cdots \qquad (8)$$

Insofar as G is concerned, the essence of Pierotti's calculation is to consider the three terms explicitly shown in Eq. (8) to be alone an adequate approximation for all $\lambda > \tfrac{1}{2}$. The pressure p is chosen according to experimental circumstances, and the solvent contact distance a is assigned the value 2.75 Å (at least below 70°C). The requirement that the $\lambda > \tfrac{1}{2}$ approximation to G must continuously and differentiably connect to the exact expression (5) at $\lambda = \tfrac{1}{2}$ then allows one to calculate γ_∞ and δ. The results are the following:

$$\gamma_\infty = \frac{3ykT}{\pi a^2}\left\{\frac{1}{1-y} + \frac{3}{2}\frac{y}{(1-y)^2} - \frac{p}{\rho kT}\right\} \qquad (9)$$

$$\delta = \frac{a}{8}\left\{1 + \frac{3y}{2 + y - 2(1-y)^2(p/\rho kT)}\right\} \qquad (10)$$

where $y = \pi\rho a^3/6$.

Table I lists some values computed for γ_∞ and δ by Eqs. (9) and (10) at selected points along the saturation curve for water. The table also includes the measured liquid–vapor interfacial tension for comparison, as well as the dimensionless compressibility factors $p/\rho kT$.

Once γ_∞ and δ have been evaluated, $G(\lambda)$ may then be obtained. Figure 1 shows the resulting $G(\lambda)$ at 25°C (and the corresponding pressure for the vaporization curve). Its most distinctive feature is the maximum at $\lambda a = 2.009$ Å. Similar maxima occur for other temperatures, always at

$$(\lambda a)_{\max} = 4\delta \qquad (11)$$

in this Pierotti approximation. Evidently the predicted G's are rather insensitive to temperature below 100°C.

The most significant point to realize about this approximation for $G(\lambda)$ is that the only explicit information it requires about the molecular structure of water is a, the distance of closest approach. Thus water could as well as

---

[3] The Gibbs dividing surface for which γ is appropriate is the geometric surface of S_λ.

Table I. Surface Tension (γ_∞) and Curvature Parameter (δ) Calculated for Liquid Water at Its Saturated Vapor Pressure[a] Using the Pierotti Approximation[b]

t (°C)	ρ (10^{24} cm^{-3})	$p/\rho kT$	γ_{lv}(expt) (dyn-cm^{-1})	γ_∞ [Eq. (2.9)] (dyn-cm^{-1})	δ [Eq. (2.10)] (Å)
4	0.033443	6.3533×10^{-6}	75.07	51.44	0.5026
25	0.033346	2.3068×10^{-5}	72.01	54.97	0.5022
50	0.033043	8.3683×10^{-5}	67.93	58.35	0.5010
75	0.032599	2.4606×10^{-4}	63.49	60.96	0.4992
100	0.032043	6.1391×10^{-4}	58.78	62.86	0.4970
200	0.028917	8.2325×10^{-3}	37.81	63.82	0.4845
300	0.023818	4.5597×10^{-2}	14.39	52.18	0.4648

[a] Measured values for ρ, p, and γ_{lv} have been taken from E. Schmidt, *Properties of Water and Steam in SI Units* (Springer-Verlag, New York, 1969).
[b] The molecular size a has been assumed to remain constant at 2.75 Å.

not have consisted of rigid spheres (with long-range attractive forces to stabilize the liquid), and *no* interaction anisotropy.

The shape of the $G(\lambda)$ curve is intimately related to the occurrence of "contact pairs" of solvent molecules at the solute's exclusion cavity S_λ. As Fig. 2 shows, we consider two infinitesimal volume elements dv_1 and dv_2 in contact with (but exterior to) the exclusion sphere and separated by angle θ measured from the center of that sphere. The probability that both dv_1 and dv_2 are simultaneously occupied by solvent molecule centers then may be written

$$\rho^2 G^{(2)}(\lambda, \lambda, \theta)\, dv_1\, dv_2 \tag{12}$$

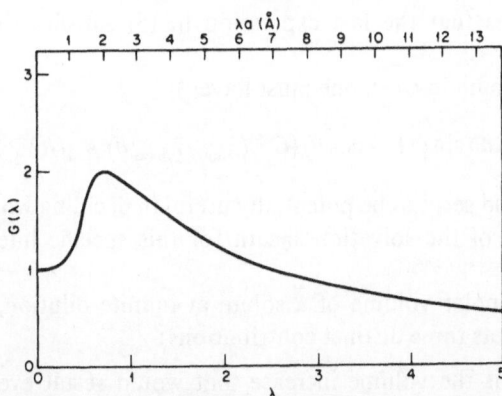

Fig. 1. $G(\lambda)$ for liquid water, calculated in the Pierotti approximation. The temperature is 25°C, and the pressure is that of the vapor-pressure curve. The maximum occurs at $\lambda a = 2.009$ Å.

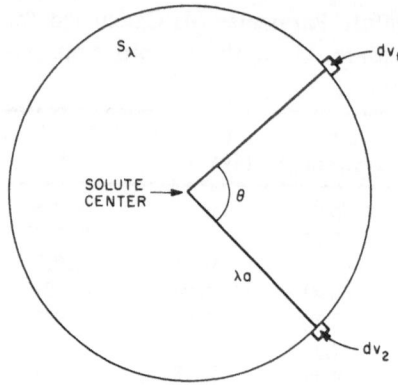

Fig. 2. Arrangement of differential volume elements dv_1 and dv_2 on the solute exclusion sphere, used in definition of the contact-pair correlation function $G^{(2)}(\lambda, \lambda, \theta)$.

thereby introducing a contact-pair analog $G^{(2)}$ of the singlet quantity G. For very large λ and nonzero θ, we naturally expect to have the reduction

$$G^{(2)}(\lambda, \lambda, \theta) \sim [G(\lambda)]^2 \tag{13}$$

that results from statistical independence of dv_1 and dv_2.

A general relationship has been derived[7] which links G and $G^{(2)}$:

$$\partial G(\lambda)/\partial \lambda = -2\pi \rho a^3 \lambda^2 \int_0^{\pi} d\theta \sin \theta (1 - \cos \theta)\{G^{(2)}(\lambda, \lambda, \theta) - [G(\lambda)]^2\} \tag{14}$$

Thus the rate of change of G with λ depends specifically on the deviations of $G^{(2)}$ from the asymptote (13); those deviations in turn sensitively reflect the arrangement patterns preferred by solvent molecules around the solute.

When $0 \leqslant \lambda < \frac{1}{2}$, at most one solvent molecule can be in contact with S_λ, so $G^{(2)}$ must vanish for all angles θ. Equation (2.14) then reduces to

$$\partial G(\lambda)/\partial \lambda = 4\pi \rho a^3 \lambda^2 [G(\lambda)]^2 \qquad (0 \leqslant \lambda < \tfrac{1}{2}) \tag{15}$$

One easily verifies that the last expression in (5) satisfies this differential equation.

At the maximum in $G(\lambda)$, one must have

$$0 = \int_0^{\pi} d\theta \sin \theta (1 - \cos \theta)\{G^{(2)}(\lambda_{\max}, \lambda_{\max}, \theta) - [G(\lambda_{\max})]^2\} \tag{16}$$

This identity would seem to be potentially useful in deciding what must be the geometric nature of the solvation sheath for this specific interesting solute size.

The partial molar volume of a solute at infinite dilution, $\bar{v}^{(0)}$, roughly speaking, represents three distinct contributions:

1. The first is the volume increase that would result even for a point particle (not interacting with the solvent molecules) on account of its kinetic contribution to the pressure. The magnitude of this part will be proportional to the pure solvent's isothermal compressibility.

2. The second is just the geometric volume occupied by the solute particle. In the general case, there may be some ambiguity about the precise magnitude of this volume, but for the present model solute it is clearly $\tau(\lambda) = 4\pi a^3 \lambda^3/3$.

3. Finally, one expects that solvent molecule rearrangement to accommodate the solute particle will normally result in a change in packing efficiency for those solvent molecules.[4] This will also contribute to $\bar{v}^{(0)}$.

It is known[7] that $\bar{v}^{(0)}$ for our model solute may be written in terms of $G(\lambda)$:

$$\bar{v}^{(0)}(\lambda) = (kT/\rho)(\partial\rho/\partial p)_T + \tau(\lambda) + 4\pi a^3 kT(\partial\rho/\partial p)_T \int_0^\lambda d\lambda'(\lambda')^2$$
$$\times (\partial/\partial\rho)[\rho G(\lambda') - (p/kT)] \tag{17}$$

The three terms added together in the right member, are respectively, the three contributions (1), (2), and (3) just mentioned.

Entropy provides a fundamental thermodynamic parameter for measuring structure in the solution. From the standpoint of the statistical theory of the present nonpolar solute model, the most revealing comparison occurs between entropy for the pure solvent plus noninteracting point solute (all in volume V) and entropy for the same number of water molecules plus repulsive solute molecule in volume $V + \tau(\lambda)$. With this convention, the volume accessible to the centers of solvent molecules is unchanged, and the resulting entropy variation measures only the result of restructuring around the inserted exclusion sphere. The Helmholtz free energy associated with expansion of the solute sphere (λ increasing from zero to the required final value) and the change in system volume is

$$\Delta_1 A(\lambda) = W(\lambda) - p\tau(\lambda) \tag{18}$$

the corresponding entropy change will be

$$\Delta_1 S = -(\partial W/\partial T)_V + \tau(\lambda)(\partial p/\partial T)_V \tag{19}$$

Ordinarily, solution thermodynamic properties are experimentally observed at constant p. The resulting system volume increment then would be $\bar{v}^{(0)}(\lambda)$ rather than $\tau(\lambda)$. Thus the more conventional solution process quantities are

$$\Delta_2 A(\lambda) = W(\lambda) - p\bar{v}^{(0)}(\lambda) \tag{20}$$

and

$$\Delta_2 S = -(\partial W/\partial T)_V + (\partial p\bar{v}^{(0)}/dT)_V \tag{21}$$

For water under ordinary temperature and pressure conditions, the second terms in the right members of Eqs. (19) and (21) are negligibly small, so that solution entropy can be explained satisfactorily in terms of W alone.

---

[4] In principle, this phenomenon could be assessed quantitatively by computing the mean Voronoi (nearest-neighbor) polyhedron volume for solvent molecules with and without the solute present.

3. PLANAR INTERFACE

As λ approaches infinity, the surface of S_λ locally takes on the appearance of an impenetrable planar wall, so far as the neighboring water molecules are concerned. We now turn attention specifically to the question of how these water molecules are arranged in the immediate vicinity of this limiting flat wall.

Heretofore we have not had to commit ourselves about the position of the "center" of a water molecule with respect to its nuclear framework. Indeed the general exact results of the preceding section are invariant to the choice of "center" (provided we still have $a > 0$). It is characteristic, furthermore, of the Pierotti approximation that specification of the "center" position is unnecessary. Now, however, we must recognize that it is this "center" which encounters and is repelled by the surface of S_λ.

We thus choose to identify the position of the oxygen nucleus in each water molecule as the "center." This choice is suggested by the periodic arrangement of oxygen nuclei in ice (in contrast to the disorder that characterizes proton positions) and by the fact that the pair-interaction potential between water and the simple solute Ne is nearly spherically symmetric about the oxygen.[19] With this convention, one recognizes that the protons of the water molecules can penetrate S_λ to an extent consistent with the intramolecular bond length ($\simeq 1$ Å).

Since there are no forces of attraction between water molecules and the surface of S_λ, we cannot expect that the surrounding liquid will "wet" the surface. This fact is borne out by the small value of the molecular density actually in contact with the flat wall:

$$\rho G(\infty) = \rho(p/\rho kT) \tag{22}$$

Table I shows that the compressibility factor $p/\rho kT$ is only about 2×10^{-5} at room temperature, so the contact density is only 2×10^{-5} of that in bulk water. In effect, then, the very large sphere S_λ must be immediately surrounded by a thin film of water vapor.[5] As one proceeds outward from the S_λ surface toward the interior of the bulk-water phase, the average density must rise (probably monotonically at low temperatures) to ρ, its large-distance limit.

While the external pressure p is at or near the saturated vapor pressure for the given temperature, there is essentially no driving force within the system to eliminate the vapor film surrounding S_λ. The film thus can become rather thick on the molecular scale, and its character can properly be assessed in terms of known facts about bulk water vapor. The zone of transition between low vapor density in the film and high liquid density farther out is therefore determined primarily by the same factors[20] that determine the normal liquid–vapor interface structure at that temperature. The surface tension γ_∞ then should be very close to the measured liquid–vapor surface tension γ_{lv}

---

[5] At room temperature, the water-vapor second virial coefficient is negative, so in fact the water molecule density (22) is slightly *less* than that of the saturated vapor.

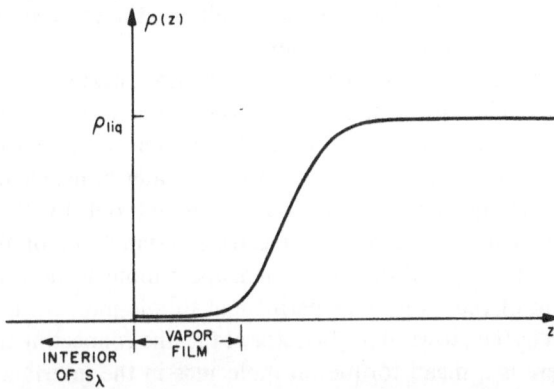

Fig. 3. Water molecule density $\rho(z)$ vs normal distance z from the surface of S_λ. This schematic diagram refers specifically to the $\lambda \to \infty$ limit for which the impenetrable surface appears locally to be flat. The vapor film results from the nonwettability of the surface.

(some values are listed in Table I). Figure 3 indicates schematically the water-molecule density distribution expected to obtain near the flat boundary.

As the external pressure is isothermally increased, the vapor film will be squeezed out. The pressures required to effect significant changes, though, are rather large in ordinary terms. If the film were to be expelled to the extent that the surface density rose to half that in the bulk water (i.e., $p/\rho kT = \frac{1}{2}$), about 700 bars would be required at 25°C. Small pressure increments above the saturation vapor pressure would mainly have the effect of moving an unperturbed liquid–vapor interface inward, and the γ_∞ that is relevant to the large-λ asymptotic expansion of $G(\lambda)$ [Eq. (8)] remains equal substantially to the measurable interfacial tension γ_{lv}. But at very high external pressures the surface density profile will have been crushed into a totally different form, and the required γ_∞ must come from some other source.

The parameter δ [also appearing in the asymptotic series (8)] has the dimension of length. This length (more precisely 2δ) measures the apparent distance inward from the surface of S_λ at which the surface tension of the spherical interface seems to act mechanically. At very low pressure increments there can be no doubt that -2δ ought to measure the vapor film width. Of course, under extremely large p it is unclear what δ measures; although we expect it to remain comparable to molecular size, even its sign is uncertain.

These considerations cast some doubt on the accuracy of the approximate water calculation outlined in the preceding section. Table I shows that the computed surface tension has the wrong temperature variation. Furthermore, the computed values for δ have the wrong sign and are certainly too small. Though we cannot be certain what its value strictly ought to be, it seems clear that -2δ should be no less than measured interfacial widths. Experiments by Kinosita and Yokota,[21] using an ellipsometric technique, suggest that this width is about 8 Å at 25°C. The Pierotti calculation therefore seems seriously to misrepresent δ near saturation pressure.

Having established that the low-pressure interface next to the flat repelling surface is closely related to the free liquid surface, it is relevant to inquire how water molecules are oriented within both. The preferred orientations depend on the molecular structure and on the nature of water-molecule interactions.

One orienting agency was previously pointed out by Stillinger and Ben-Naim,[22] which stems from the electrical asymmetry of the separate water molecules. The sign of the axial quadrupole moment indicates that the effective position of the molecule's permanent dipole moment is forward of the oxygen nucleus (i.e., toward the bisector of the line connecting the protons). As a result, there is a mean torque on molecules in the interface tending to orient their dipoles toward the bulk liquid. Consequently, the mean electrostatic potential increases upon passage through the interface from the vapor side to the liquid side, since molecules at the surface tend to immerse their protons and expose their lone-pair electrons.

The peculiar character of interaction nonadditivity in water is also capable of biasing the interfacial-region orientational distribution. The predominant nonadditive component to the potential in a water molecule assembly is probably three-molecule nonadditivity, which depends strongly on the hydrogen-bond pattern.[15] Any hydrogen-bond network with perfect fourfold coordination has an invariant number of three types of molecular neighbor trimers:

(a) Double donor trimer—one molecule simultaneously donating its protons in hydrogen bonds to two other molecules;

(b) Double acceptor trimer—one molecule simultaneously acting as the acceptor for two protons donated by distinct neighbors;

(c) Sequential trimer—a central molecule simultaneously accepts a proton from one neighbor and donates one of its own to a second neighbor.

The relative network occurrence frequencies for these three neighbor types are respectively $1:1:4$. The three-molecule potential energy nonadditivity $V^{(3)}$ is positive for the first two types but negative for the last type. Figure 4 shows specific examples of each of the three types. For counting purposes in an extended network, these trimers may be regarded as "belonging to" the middle one of the three molecules.

Although liquid water might properly be described as a random, three-dimensional, hydrogen-bond network,[23] it surely cannot have invariant fourfold coordination. Instead, some of the hydrogen bonds must be broken and others severely strained in length and direction. Nevertheless, a significant fraction of the molecules (in cold water especially) should be four-coordinated, though possibly with somewhat distorted hydrogen bonds.

In a rough way, we can think of the free liquid surface at low temperature as having been formed by passing a mathematical surface through the bulk

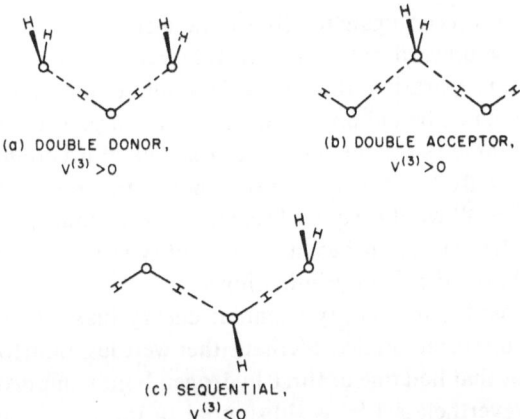

(a) DOUBLE DONOR,
$V^{(3)} > 0$

(b) DOUBLE ACCEPTOR,
$V^{(3)} > 0$

(c) SEQUENTIAL,
$V^{(3)} < 0$

Fig. 4. The three distinct types of water-molecule trimers, involving two linear hydrogen bonds (dotted lines). These all occur in hydrogen-bond networks with fourfold tetrahedral coordination. The sign of the potential nonadditive component, $V^{(3)}$, is shown for each.

liquid, snipping all bonds that cross the surface, and then separating the two halves of the bulk phase. Figure 5 shows two water molecules that become residents of the outermost surface layer after the cutting and separation process. Both were four-coordinated to begin with, and both have two of the hydrogen bonds snipped.

One can readily count how many trimers of each type "belonging to" the resultant surface molecule have been disrupted by formation of the surface. Figure 5 shows the counts for its two examples. In view of the $V^{(3)}$ signs shown in Fig. 4, it is clearly more costly in energy, other things being equal, to cut apart configuration 5(b) than 5(a). Accordingly, we would expect to find doubly coordinated surface molecules of type 5(a) energetically preferred in a real surface over those of type 5(b).

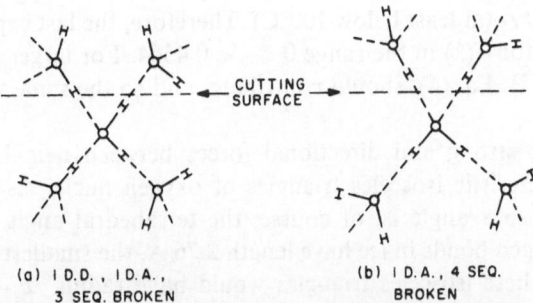

(a) I D.D., I D.A.,
3 SEQ. BROKEN

(b) I D.A., 4 SEQ.
BROKEN

Fig. 5. Two types of four-coordinated water molecules which become two-coordinated surface-layer molecules. The "cutting surface" is the mathematical surface at which bonds are snipped to produce fresh liquid surfaces. The numbers of trimers of each type disrupted by surface formation are shown for the two initial configurations.

A configuration conjugate to 5(b) should also be considered, having both protons pointing upward across the cutting surface to neighbors above it. After cutting and separation, these protons would point out of the underlying liquid and, of course, be unbonded. We know from preceding considerations of electrical asymmetry for the molecules that this third configuration "5(c)" is itself energetically less favorable even than 5(b). We note here for completeness that "5(c)" would require breakage of one double donor trimer and four sequential trimers, so surface molecules of type 5(a) are still the preferred species on the basis of $V^{(3)}$ discrimination.

It is not possible to identify a similar energy bias, based on $V^{(3)}$ signs, which operates on surface molecules that either were less than four-coordinated to begin with, or that had one or three hydrogen bonds snipped during surface preparation. Nevertheless $V^{(3)}$ is substantial in magnitude relative to kT at room temperature,[15] and the argument just posed is apparently relevant to a significant fraction of the molecules. Therefore, this phenomenon ought to comprise a major interfacial orienting effect whose presence should be acknowledged both in scaled-particle theories of water and in study of the free liquid surface. One must keep in mind, for the remainder of this paper, that the measurable surface tension γ_{lv} is numerically affected by this non-additivity phenomenon.

4. REVISED G FOR WATER

We now undertake to improve the calculation of $G(\lambda)$ for the rigid-sphere solute in pure water. Both the liquid–vapor surface tension and the radial distribution function for pure water will be used as input data. We shall continue to set $a = 2.75$ Å to be consistent with the preceding length scale, though no direct structural significance will now be implied by that choice.

The most accurate determination to date of the oxygen–oxygen pair correlation functions $g^{(2)}(r)$ in liquid water has been carried out by Narten and Levy.[24]6 Their results show that virtually no pairs of oxygen nuclei occur closer than 2.40 Å (at least below 100°C). Therefore, the last expression in (5) will be correct for $G(\lambda)$ in the range $0 \leqslant \lambda \leqslant 0.4364$. For larger λ, at least the pair term in $P(\lambda)$, Eq. (4), should contribute, and so the same would be true of $G(\lambda)$.

In ice, the strong and directional forces between neighbor molecules produce characteristic isosceles triangles of oxygen nuclei, as illustrated in Fig. 6(a). The apex angle is, of course, the tetrahedral angle $\theta_t = 109°28'$; since the hydrogen bonds in ice have length 2.76 Å, the smallest sphere which could enclose these isosceles triangles would have radius $\lambda a = 2.25$ Å. For ice then, nothing beyond the pair ($n = 2$) terms in $P(\lambda)$ and $G(\lambda)$ would be

---

6 The author is grateful to Dr. Narten for supplying a numerical tabulation of the function $g^{(2)}(r)$, which has been used in the present paper.

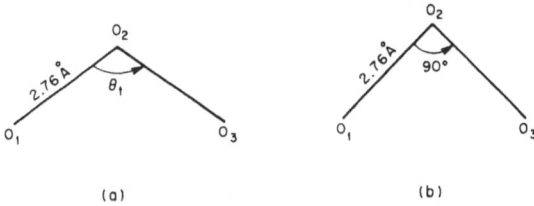

(a)   (b)

Fig. 6. Triads of oxygen nuclei. In (a) the arrangement shown corresponds to bonded neighbors in ice, with $\theta_t = 109°28'$. The distorted triad (b) is used to estimate the radius of the smallest sphere which could reasonably be expected to circumscribe triads in the cold liquid.

required, provided that λa did not exceed 2.25 Å or, equivalently, that λ did not exceed 0.8182.

The hydrogen-bond pattern present in ice surely undergoes major distortion upon melting. Still, the coordination number in the liquid remains low, and it seems abundantly clear that the tendency toward tetrahedral bonding is a dominant feature in the liquid.[25] Consequently, it seems to be a reasonable assumption for cold liquid water that triads of oxygen nuclei are seldom distorted into a more compact arrangement than would result from reduction of θ_t to 90°. The correspondingly distorted configuration, shown in Fig. 6(b), will now just fit into a sphere with radius

$$\lambda a = 1.95 \text{ Å} \tag{23}$$

which occurs at $\lambda = 0.7091$. Although some oxygen-nucleus triads in the liquid may have apex angles less than 90°, it seems likely that the necessarily weakened bonds will increase in length, thereby still obeying the estimate (23).

In order to specify $G(\lambda)$ beyond the limit (23) explicitly in terms of molecular correlation functions, knowledge of $g^{(3)}$, $g^{(4)}, \ldots$ would be required. That knowledge, of course, is unavailable.[7] Instead, we can rely on the conventional Laurent series format for $G(\lambda)$, Eq. (8), suitably truncated. The continuity and differentiability of $G(\lambda)$ at point (23) subsequently can be used to fix unknown parameters.

The summary for $G(\lambda)$ thus is the following:

$$G(\lambda) = [1 - (4\pi/3)\rho a^3 \lambda^3]^{-1} \qquad (0 \leqslant \lambda a \leqslant 1.20 \text{ Å})$$

$$G(\lambda) = \frac{1 + (\pi\rho a^3/\lambda) \int_0^{2\lambda} dt\, g^{(2)}(t)\, t^2(t - 2\lambda)}{1 - (4\pi/3)\rho a^3 \lambda^3 + (\pi\rho a^3)^2 \int_0^{2\lambda} dt\, g^{(2)}(t)\, t^2(\tfrac{1}{6}t^3 - 2\lambda^2 t + \tfrac{8}{3}\lambda^3)}$$
$$(1.20 \text{ Å} \leqslant \lambda a \leqslant 1.95 \text{ Å})$$

$$G(\lambda) = (p/\rho kT) + (2\gamma_{lv}/\rho a kT\lambda) + (G_2/\lambda^2) + (G_4/\lambda^4) \qquad (1.95 \text{ Å} \leqslant \lambda a < \infty) \tag{24}$$

---

[7] For simple liquids, the Kirkwood superposition approximation might suffice to estimate $g^{(3)}$, $g^{(4)}, \ldots$ in terms of $g^{(2)}$. That approximation is likely to be very poor for liquid water, however, and will be avoided here.

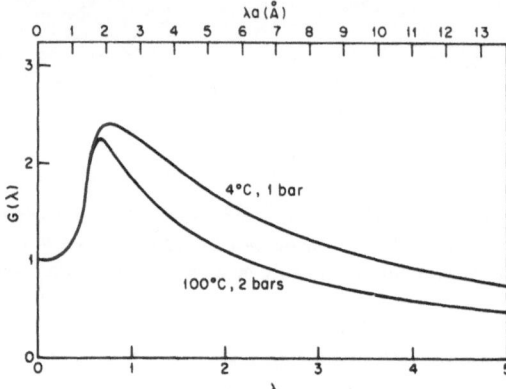

Fig. 7. Contact correlation function $G(\lambda)$ calculated according to the method of Sec. 4.

The second of these functional forms results after carrying out the $n = 2$ integration in Eq. (4) and then using Eqs. (1) and (3) to yield the explicit G expression. The λ^{-3} term is missing in the last of the three forms in (24), as required by the general theory,[7] so that $W(\lambda)$ will be free of contributions proportional to $\ln\lambda$. The quantities G_2 and G_4 are adjustable parameters whose values effect smooth connection at $\lambda a = 1.95$ Å.

The function $G(\lambda)$ has been calculated according to this procedure for the two sets of conditions: (1) 4°C, 1 bar; and (2) 100°C, 2 bars.[8] The results are displayed in Fig. 7. After comparing these curves with that in Fig. 1, we see that the present more accurate procedure tends to give $G(\lambda)$ substantially larger maxima (though at roughly the same λ value) than the Pierotti hard-sphere approximation. Furthermore, the Pierotti approximation is far less sensitive to temperature (it depends essentially on the number density alone, which is nearly constant) than the more detailed result.

5. CONCLUSIONS

The fact that the revised $G(\lambda)$ calculation leads to larger maxima is relatively easy to understand. Unlike the Pierotti hard-sphere approximation, it accounts explicitly for the strong and directional hydrogen-bonding forces in water, not only through the pair correlation function $g^{(2)}$ that it utilizes but also in the selection of the λ range toward which triplets first contribute. As the exclusion sphere S_λ expands, it is forced to stretch and tear the hydrogen-bond network in its neighborhood. While this process occurs, the remaining hydrogen bonds probably reach around S_λ in a tightly drawn net, which surely enhances G.

[8] The slight overpressures are invoked to prevent the vapor film from widening as $\lambda \to \infty$, as discussed in Sec. 3. If this measure were not adopted, the asymptotic development (8) for $G(\lambda)$ would contain terms that were not merely integer powers of λ.

At 4°C, the $G(\lambda)$ maximum shown in Fig. 7 has the value

$$G(\lambda_{max}) = 2.3959$$

$$a\lambda_{max} = 2.1075 \text{ Å} \tag{25}$$

The inward stress, or "pressure," exerted by the water molecules at the surface of S_λ subsequently may be evaluated:

$$p_{max} = \rho kTG(\lambda_{max})$$

$$= 3.0657 \text{ kbar} \tag{26}$$

This presumably measures the extent of network stretch.

Next, we can use the general expression (14) [in the special case shown in (16)], to reveal some information about solvation of S_λ at the maximal size λ_{max}. It seems reasonable to suppose that the contact-pair correlation function $G^{(2)}$ is simply proportional to $g^{(2)}$ at the appropriate straight-line distance:

$$G^{(2)}(\lambda, \lambda, \theta) \cong A_0[G(\lambda)]^2 g^{(2)}[r(\lambda, \theta)] \tag{27}$$

where

$$r(\lambda, \theta) = a\lambda[1 - \cos\theta]^{1/2} \tag{28}$$

By substituting this approximation for $G^{(2)}$ into Eq. (16), a unique determination of the multiplier A_0 results:

$$A_0 = 2\left\{ \int_0^\pi d\theta \sin\theta (1 - \cos\theta) g^{(2)}[r(\lambda_{max}, \theta)] \right\}^{-1}$$

$$= 4a^4 \lambda_{max}^4 \left\{ \int_0^{2a\lambda_{max}} dr\, r^3 g^{(2)}(r) \right\}^{-1} \tag{29}$$

The value implied by Eq. (29) for A at 4°C is

$$A_0 = 1.048 \tag{30}$$

Figure 8 shows a plot of the resulting function $G^{(2)}(\lambda_{max}, \lambda_{max}, \theta)$ vs angle. Similar curves could be obtained for $\lambda \neq \lambda_{max}$ from Eq. (14), but the results would be trivial for small λ; the results would furthermore be unreliable at large λ because estimate (27) would then become inappropriate.

It is important to reflect upon the local solvent structures which might contribute to the function $G^{(2)}(\lambda, \lambda, \theta)$. To be sure, a wide variety of hydrogen-bond network fragments must be present in liquid water,[23] but statistically they seem to present important common features. Figure 9 shows a specific example of the way that the network can surround S_λ when the size parameter is approximately λ_{max}. This particular cage consists of 12 water molecules arranged into four pentagons and two hexagons. Only the oxygen positions

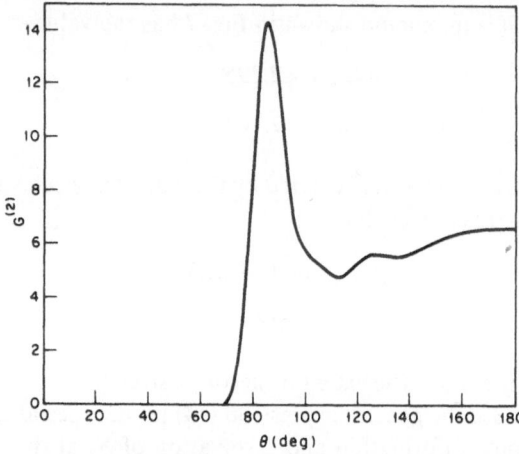

Fig. 8. Surface-pair correlation function $G^{(2)}(\lambda_{max}, \lambda_{max}, \theta)$ for water at 4°C and 1 bar. Figure 2 provides the relevant geometry, with $\lambda a = 2.1075$ Å.

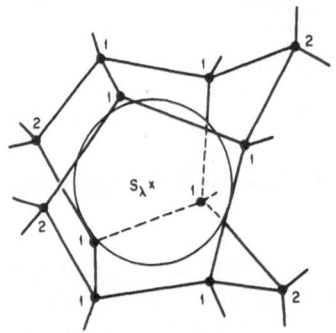

Fig. 9. "Random" water-molecule cage enclosing the rigid-sphere solute, whose exclusion sphere is denoted by S_λ. The oxygen nuclei are shown as dark circles, and hydrogen bonds as dark lines. Protons can be distributed asymmetrically along the bonds in a variety of canonical ways. Oxygen nuclei "1" are closer to the solute than those of type "2."

are explicitly shown (as dark circles). The protons can be assigned to the hydrogen bonds (dark lines) in a variety of ways that is analogous to the variety underlying the configurational disorder in ice.[26]

Notice that the eight water molecules nearest to the center of S_λ, which have been indicated by "1" in the figure, are all oriented so that one of the four tetrahedral bond directions points radially outward, i.e., away from the center of S_λ. For each of these eight, then, the remaining three bond directions straddle S_λ. As a result, the dipole-moment direction for these solvent molecules cannot point either toward, or directly away from, the hard-sphere solute. The four more remote solvent molecules denoted by "2" in Fig. 9, however, can point their dipole-moment vectors either inward or outward along the radial direction.

Model building shows that this orientational bias for nearest-neighbor solvent molecules is very marked and extends to S_λ radii well in excess of $a\lambda_{max} \cong 2.1$ Å. Evidently the geometric requirement that the nearest solvents form a predominantly convex solvation "cage" forces them to adopt the straddling mode. The cages present in aqueous clathrates[27] exhibit precisely this characteristic too, though they represent only a small fraction of the total possible cage geometries. Probably the largest convex cage that could reasonably be expected to form on energetic grounds would be the one whose bonds have the pattern displayed by soccer balls: Sixty water molecules form 12 pentagons and 20 hexagons, and each of these molecules straddles the cavity interior.[9]

The water molecule bonding situation depicted in Fig. 9 is unrealistic in that no severely stretched, twisted, or broken hydrogen bonds are shown. The true situation is therefore more complicated, but the orientational bias just adduced for first-layer solvent molecules must still be a substantial phenomenon since extensive hydrogen bonding exists. For the statistical tendency to be present, of course, we need not demand that the random solvation cages be perfectly bonded. In any event, the orientational bias certainly becomes less pronounced as temperature increases.

The situation which leads to the small-cavity solvation effects on water molecule orientation is not relevant to large λ. When λa exceeds the range 3–4 Å, the processes leading to formation of the free liquid surface (at low pressure) described in Sec. 3 would probably dominate. One should keep in mind, however, that many interesting solutes have sizes below this range, so that convex cage statistics is indeed relevant to their solutions.

As a useful "thought experiment," we can imagine placing electrostatic charge uniformly on our hard-sphere solute particle to convert it into a monatomic ion. The strong electric fields that result would have the effect of twisting first-layer solvation molecules out of "straddling" configurations into those with radial directions for dipole moments. This structural rearrangement likely plays an important role in the thermodynamics of ionic solvation.[28]

It would be very desirable to augment the conclusions in this paper with independent evidence. The molecular-dynamics technique has recently been adapted to the study of pure water.[29] In principle, it could be modified to incorporate rigid-sphere solutes of arbitrary size. Not only would this permit the $G(\lambda)$ curves in Fig. 7 to be checked quantitatively, but the geometry of solvation cages could also be observed. It would then be interesting to see if the approximation, on which the function $G^{(2)}$ shown in Fig. 8 was calculated, was in fact accurate. In the long run, there is hope that the combination of such stylistically independent procedures could lead to understanding solvation subtleties such as hydrophobic bonding in the biochemical regime.[30]

ACKNOWLEDGMENT

The author has enjoyed a protracted and constructive dialogue with Professor Henry S. Frank on the nature of water and aqueous solutions. This

---

[9] Assuming that the hydrogen bonds forming this maximal cage have length 2.8 Å, its radius would be about 6.4 Å.

interaction has helped the former to maintain in his work a respectable balance between the sterile intricacy of formal theory and the seductive simplicity of poetic "explanation."

REFERENCES

1. H. Reiss, H. L. Frisch, and J. L. Lebowitz, *J. Chem. Phys.* **31**, 369 (1959).
2. E. Helfand, H. L. Frisch, and J. L. Lebowitz, *J. Chem. Phys.* **34**, 1037 (1961).
3. E. Helfand, H. Reiss, and H. L. Frisch, *J. Chem. Phys.* **33**, 1379 (1960).
4. H. Reiss, *Advan. Chem. Phys.* **IX**, 1–84 (1965).
5. D. M. Tully-Smith and H. Reiss, *J. Chem. Phys.* **53**, 4015 (1970).
6. H. Reiss and D. M. Tully-Smith, *J. Chem. Phys.* **55**, 1674 (1971).
7. F. H. Stillinger and M. A. Cotter, *J. Chem. Phys.* **55**, 3449 (1971).
8. M. A. Cotter and F. H. Stillinger, *J. Chem. Phys.* **57**, 3356 (1972).
9. H. Reiss, *Advan. Chem. Phys.* **IX**, 78 (1965), Table XVI.
10. H. Reiss, H. L. Frisch, E. Helfand, and J. L. Lebowitz, *J. Chem. Phys.* **32**, 119 (1960).
11. R. A. Pierotti, *J. Phys. Chem.* **69**, 281 (1965).
12. K. Morokuma and L. Pederson, *J. Chem. Phys.* **48**, 3275 (1968).
13. P. A. Kollman and L. C. Allen, *J. Chem. Phys.* **51**, 3286 (1969).
14. J. Del Bene and J. A. Pople, *J. Chem. Phys.* **52**, 4858 (1970).
15. D. Hankins, J. W. Moskowitz, and F. H. Stillinger, *J. Chem. Phys.* **53**, 4544 (1970).
16. G. H. F. Diercksen, *Theoret. Chim. Acta (Berlin)* **21**, 335 (1971).
17. A. H. Narten and H. A. Levy, *J. Chem. Phys.* **55**, 2263 (1971).
18. F. P. Buff, *J. Chem. Phys.* **19**, 1591 (1951).
19. M. Losonczy, J. W. Moskowitz, and F. H. Stillinger, to be published.
20. B. Widom, *J. Chem. Phys.* **43**, 3892 (1965).
21. K. Kinosita and H. Yokota, *J. Phys. Soc. Japan* **20**, 1086 (1965).
22. F. H. Stillinger and A. Ben-Naim, *J. Chem. Phys.* **47**, 4431 (1967).
23. J. D. Bernal, *Proc. Roy. Soc. (London)* **A280**, 299 (1964); J. D. Bernal, in *Liquids: Structure, Properties, Solid Interactions*, T. J. Hughel, ed. (Elsevier Publishing Co., New York, 1965), p. 25.
24. A. H. Narten and H. A. Levy, *J. Chem. Phys.* **55**, 2263 (1971).
25. A. H. Narten and H. A. Levy, *Science* **165**, 447 (1969).
26. L. Pauling, *J. Am. Chem. Soc.* **57**, 2680 (1935).
27. L. Pauling, *The Nature of the Chemical Bond* (Cornell University Press, Ithaca, 1960), p. 469.
28. H. S. Frank and W.-Y. Wen, *Disc. Faraday Soc.* **24**, 133 (1957).
29. A. Rahman and F. H. Stillinger, *J. Chem. Phys.* **55**, 3336 (1971); F. H. Stillinger and A. Rahman, *J. Chem. Phys.* **57**, 1281 (1972).
30. W. Kauzmann, *Advan. Protein Chem.* **14**, 1 (1959).

EDITOR'S NOTE

This paper was substituted for the talk originally given by Dr. Stillinger at the H. S. Frank Symposium. Consequently, no Discussion section is available.

Raman Spectra from Partially Deuterated Water and Ice VI to 10.1 kbar at 28°C[1]

G. E. Walrafen[2]

Received July 20, 1972

High-pressure argon-ion laser-Raman spectra (4880 Å excitation) have been obtained from partially deuterated water and ice VI (20 volume % D_2O) in the OD and OH stretching regions to pressures of 10.1 kbar at 28°C. The Raman spectra from ice VI are the first to be reported at room temperature, and they are similar to the liquid spectra obtained at 9.7 kbar. Raman shifts corresponding to contour intensity maxima were observed to change with pressure rise in the OD and OH stretching regions from $\Delta\bar{\nu} \doteq 2513$–2490 cm^{-1} and $\Delta\bar{\nu} = 3402$–3380 cm^{-1}, respectively, for pressures from 1 bar to 10.1 kbar (ice VI). In addition, a shoulder observed at 1 bar on the OD contour near $\Delta\bar{\nu} = 2650$ cm^{-1} became less distinct and was visually absent for pressures from 6.4 to 10.1 kbar, although a shoulder on the OH contour at about $\Delta\bar{\nu} = 3250$ cm^{-1} intensified gradually for pressures to 9.7 kbar, and abruptly upon freezing at 10.1 kbar. The small effects of pressure on the OD component percentages obtained from computer analysis indicate that hydrogen-bond breakage is not a significant effect of pressure rise, and a downward change in the position of the OD stretching component having the largest Raman shift indicates that the nonhydrogen-bonded OD units or broken O–D···O bonds that exist at 1 bar are probably transformed by close packing due to compression into weak O–D···O bonds that are angularly deformed. In addition, intensification of the OH component at $\Delta\bar{\nu} = 3220$ cm^{-1} upon freezing or upon pressurizing the liquid to 9.7 kbar is indicated by the computer analyses, and an increase in intermolecular coupling is thus favored, as opposed to enhancement of Fermi resonance, because the positions of components at $\Delta\bar{\nu} = 3220$ cm^{-1} and $\Delta\bar{\nu} = 3405$ cm^{-1} are nearly independent of pressure. The computer results also strengthen previous evidence indicating that the OD component which occurs at about $\Delta\bar{\nu} = 2654$ cm^{-1} at 1 bar arises from broken O–D···O bonds, when it is understood that the severely deformed O–D···O bonds of ice VI give rise to intensity at a Raman shift of $\Delta\bar{\nu} = 2617$ cm^{-1}, a difference of 37 cm^{-1} in the direction of stronger hydrogen-bonding.

KEY WORDS: High-pressure; laser-Raman; spectra; structure; water; ice VI; HDO; 10 kbar.

1. INTRODUCTION

High-pressure argon-ion laser-Raman spectra from solutions of HDO in H_2O (10 and 20 mole % D_2O) were obtained previously at 25°C to pressures

---

[1] This paper was presented at the symposium, "The Physical Chemistry of Aqueous Systems," held at the University of Pittsburgh, Pittsburgh, Pennsylvania, June 12–14, 1972, in honor of the 70th birthday of Professor H. S. Frank.

[2] Bell Laboratories, Murray Hill, New Jersey 07974.

of 3790 bar[1] and 7200 bar.[2, 3] The upper pressure was limited by the strength of the Raman cell and the pressure obtainable from a hand pump in one study,[1] and by cell and valve leakage in the other.[2] The leakage problems were overcome in the present work, and Raman spectra are now reported to pressures of 10.1 kbar for a 20 volume % solution of D_2O in H_2O. In particular, the spectrum from partially deuterated ice VI is reported here for the first time at room temperature.

Raman spectra from H_2O and D_2O ice VI have been reported previously at 1 bar and 77°K.[4] However, the Raman spectra were obtained by use of mercury excitation methods, and the OD and OH stretching intensities were low. In addition, the OD and OH stretching contours were obscured by intense mercury-arc lines. Accordingly, accurate contour shapes were not obtained, and, further, the Raman spectra from ice VI at 1 bar are not comparable with those from liquid water because of the very low temperatures involved. Such limitations, however, do not apply to the present high-pressure results, and thus useful structural information is obtained from comparisons of the ice VI spectra with Raman spectra from the liquid at high pressures. The high-pressure laser-Raman results from water and ice VI are now presented.

2. EXPERIMENTAL

The experimental high-pressure laser-Raman back-scattering arrangement employed with the Cary model-81 Raman spectrophotometer is shown schematically in Fig. 1. The in-line high-pressure Raman cell was positioned such that its center B was located at the focus of the image slicer. The focal length A-B is about 48 cm. The argon-ion laser beam (0.5–0.8 W at 4880 Å) was reflected from the very small mirror M_1 (~3 mm diameter) along the axis

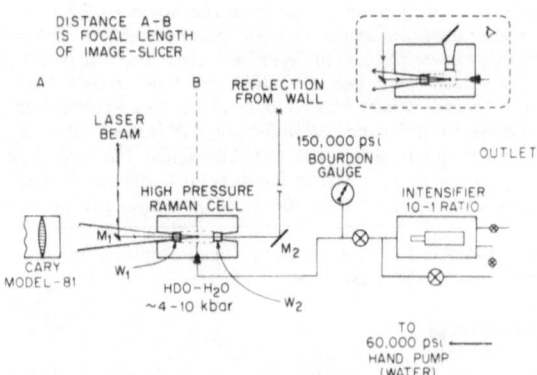

Fig. 1. Schematic diagram illustrating the high-pressure argon-ion laser-Raman back-scattering apparatus employed with the Cary model-81 Raman spectrophotometer. The 90° Raman cell shown in the inset was employed for simultaneously observing ice VI under laser illumination and obtaining the Raman spectrum.

of the Raman cell through the sapphire windows W_1 and W_2 (0.75 in. diameter, 0.75 in. thick). The beam was then reflected by the large mirror M_2 to the wall for observation. The back-scattered Raman radiation, with the exception of the small amount blocked by mirror M_1, was collected by the image slicer. The electric vector of the laser radiation was rotated by a half-wave plate until the Raman intensity at $\Delta\bar{\nu} = 3250$ cm^{-1} was maximal. This rotation corrected the small rotation produced by the sapphire. However, a polaroid analyzer and a wedge scrambler were not employed because low Raman intensities resulted from the unfavorable cell geometry. Some loss of intensity thus occurred in the OH stretching contour near $\Delta\bar{\nu} = 3250$ cm^{-1}, but this was unavoidable. The Raman cells and high-pressure intensifier (10:1 ratio) were supplied by Autoclave Engineers, Inc. The 150,000 psi calibrated Bourdon gauge was supplied by Pressure Products (Astra type). The gauge could be used to 165,000 psi.

In work with the in-line Raman cell, the presence of ice VI was determined by (1) the pressure reading, (2) some difficulty in actuating the intensifier, and (3) a continual change in shape, followed by a great increase in the diameter, and finally by a marked decrease in the intensity of the laser reflection from the wall. The absence of partial freezing was suggested by (1) complete cessation of small pressure drops due to *contraction* upon freezing (corrected by further actuation of the intensifier), (2) stability of the pressure above 10 kbar for a period of 0.5 to 1.5 h after initial freezing, (3) agreements between Raman spectra from three separate freezings, and (4) the failure of the pressure reading to fall for a few seconds after suddenly opening the system to ambient pressure after a 1.5-h freezing period.

In work with the 90° Raman cell (see inset of Fig. 1), unequivocal evidence for complete freezing to ice VI was obtained by simultaneously viewing the ice VI in the cell while obtaining a spectrum. The intensifier, the high-pressure line, and the valves and connectors were heated to about 40°C. The pressure in the line and intensifier was then increased to about 11 kbar, but it was observed to drop as ice VI formed in the cooler Raman cell. As freezing occurred the diameter of the laser beam was found to increase when observed at 90°, and a distinct beam was unobservable when all of the liquid had frozen. The ice VI, however, was still well-illuminated. Cracks and imperfections in the ice could be seen readily. Further, motion of small particles was not observed upon complete freezing, whereas such motion began immediately when the pressure was decreased below 10 kbar.

The Raman spectra from ice VI obtained with the in-line and 90° cells are shown subsequently to be very similar. The contour height near $\Delta\bar{\nu} = 3250$ cm^{-1} was slightly larger under conditions definitely known to correspond to complete freezing. However, the observed differences are probably within experimental error when the high degree of polarization at $\Delta\bar{\nu} = 3250$ cm^{-1} is considered.

The D_2O employed in this work (99.8 atom %) was supplied by the British

Oxygen Co., Ltd. A 20 volume % solution of D_2O in H_2O was used instead of more dilute solutions because low Raman intensities resulted from the use of the high-pressure cell.

Decompositions of Raman OD and OH stretching contours were performed with a du Pont 310 Curve Resolver using Gaussian component shapes.

3. RESULTS

3.1. Direct Spectral Observations

Raman spectra from 20 volume % D_2O in H_2O are shown in Fig. 2 for pressures from 1 bar to 10.1 kbar at 28°C. The principal conclusions obtained from direct examinations of the spectra (without computer decomposition) are as follows: (1) the spectral effects of pressure are generally not large, (2) the shoulder on the OD stretching contour near $\Delta\bar{\nu} \approx 2650$ cm^{-1} at 1 bar becomes less distinct and is visually absent for pressures of about 6.4 kbar and above, and (3) the shoulder on the OH stretching contour near $\Delta\bar{\nu} \approx 3250$ cm^{-1} increases gradually in relative intensity to pressures of 9.7 kbar, and abruptly upon freezing to ice VI at 10.1 kbar.

Raman frequency shifts corresponding to intensity maxima from the OH and OD stretching contours are shown in Fig. 3. From 1 bar to 10.1 kbar the $\Delta\bar{\nu}$ values vary from 3402–3380 cm^{-1} for the OH (a) and from 2513–2490 cm^{-1} for the OD stretching contour (b).

Gaussian OH stretching components centered at average $\Delta\bar{\nu}$ values of 3125, 3220, and 3405 cm^{-1} are indicated in the following section. Ratios of contour heights were thus determined at the $\Delta\bar{\nu}$ values of the component centers, viz., H_{3125}/H_{3405} and H_{3220}/H_{3405} [Fig. 3(c)]. The percentage of the component centered at an average $\Delta\bar{\nu}$ value of 3405 cm^{-1} from 1 bar to 10.1 kbar (the percentage determined from integrated Gaussian component and total integrated contour intensities) was found to be nearly independent of pressure. Accordingly, the contour height ratios can provide an accurate measure of the change in contour shape with pressure, particularly for the lower $\Delta\bar{\nu}$ values. From Fig. 3(c) it is evident that a gradual increase in the ratio H_{3220}/H_{3405} occurs from 1 bar to 9.7 kbar, followed by an abrupt increase upon freezing at 10.1 kbar. The abrupt intensification is visually evident from Fig. 2. In addition, an increase in the ratio H_{3125}/H_{3405} also occurs, but this ratio does not appear to change discontinuously upon freezing.

3.2. Results from Gaussian Analog Computer Analyses

The OD and OH stretching contours of Fig. 2 were decomposed into four[5] and five[6] Gaussian components, respectively, by means of an analog computer (see spectra of Fig. 2 corresponding to 1 bar and 10.1 kbar). In the case of the OD stretching contour, the four components are centered (within

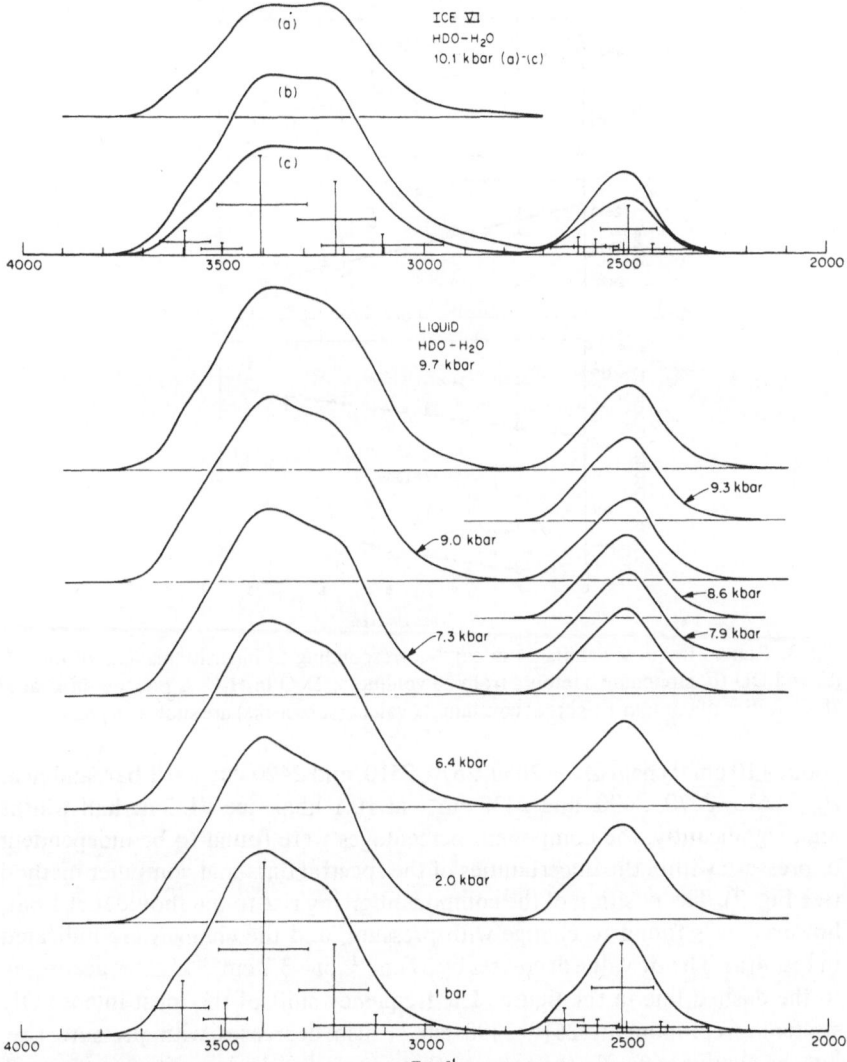

Fig. 2. High-pressure argon-ion laser-Raman spectra (4880 Å) obtained by back-scattering from a 20 volume % solution of D_2O in H_2O with a Cary model-81 Raman spectrophotometer. The spectra were transferred to horizontal baselines by measuring contour heights at increments of 12.5 cm^{-1}, and they were then normalized to equal heights at $\Delta\bar{\nu} = 3400$ cm^{-1} in the case of the liquid spectra, as well as for spectrum (b) from ice VI. Spectrum (a) from ice VI was obtained by back-scattering from the 90° cell while simultaneously viewing ice VI. All of the remaining spectra were obtained with the in-line Raman cell. Spectrum (a) was obtained with a time constant of 0.5 sec, a slit width of 20 cm^{-1} (double slit), and a scanning rate of 2.5 cm^{-1}-sec^{-1}. Spectra corresponding to 1 bar, and to 2.0, 6.4, 7.3, and 9.0 kbar, as well as spectrum (c), were obtained with a time constant of 0.1 sec, a slit width of 20 cm^{-1} (double slit), and a scanning rate of 1 cm^{-1}-sec^{-1}; the remaining spectra were obtained with a time constant of 0.25 sec. All spectra were obtained at 28°C. Positions and half-widths of Gaussian components are shown by bars in two spectra.

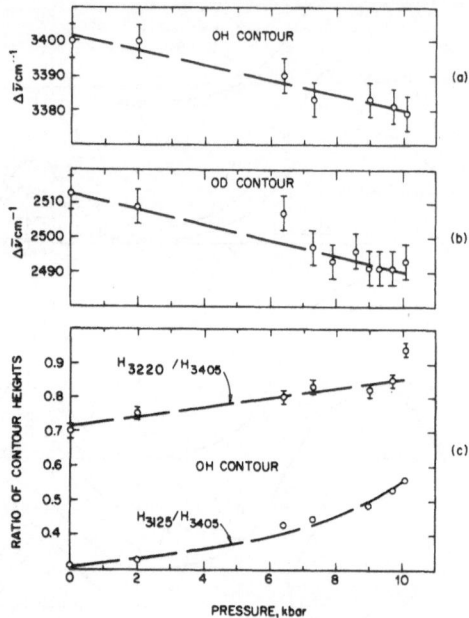

Fig. 3. Raman frequency shifts $\Delta\bar{\nu}$ in cm^{-1} corresponding to intensity maxima of the OH (a) and OD (b) stretching contours from 20 volume % D_2O in H_2O vs pressure in kbar at 28°C. Ratios of contour heights at constant $\Delta\bar{\nu}$ values (subscripts) are shown vs pressure (c).

about ±10 cm^{-1}) near $\Delta\bar{\nu}$ = 2650, 2570, 2510, and 2420 cm^{-1} at 1 bar, and near $\Delta\bar{\nu}$ = 2615, 2570, 2490, and 2430 cm^{-1} at 10.1 kbar, ice VI. The half-widths and, significantly, the component percentages were found to be independent of pressure within the uncertainties of the spectral data and computer method (see Fig. 2). The position of the component giving rise to the shoulder at 1 bar, however, was found to change with pressure, and the changes are indicated in Fig. 4(a). The $\Delta\bar{\nu}$ value decreases by 37 cm^{-1}, or ~3.7 cm^{-1}-kbar^{-1}, according to the dashed line in the figure. The frequency shift of the most intense OD stretching component (2510–2490 cm^{-1}) also decreases with pressure rise, but the change is nearly the same as that shown for the intensity maximum in Fig. 3(b), i.e., about 2 cm^{-1}-kbar^{-1}, and the $\Delta\bar{\nu}$ values are not plotted separately.

The five Gaussian components employed in the computer analyses of the OH stretching contour were found to show negligible changes in Raman frequency shift with pressure, and the half-widths of the components are also nearly independent of pressure (see Fig. 2). The average $\Delta\bar{\nu}$ values and uncertainties corresponding to the component centers from 1 bar to 10.1 kbar are 3600 ± 10, 3505 ± 10, 3405 ± 10, 3220 ± 10, and 3125 ± 25 cm^{-1}. Further, the percentages of two of the five components were found to indicate marked changes with pressure [Fig. 4(b) and (c)]. The percentage of the $\Delta\bar{\nu}$ = 3220 cm^{-1} component was found to increase from about 22% to 25% from 1 bar to 9.7

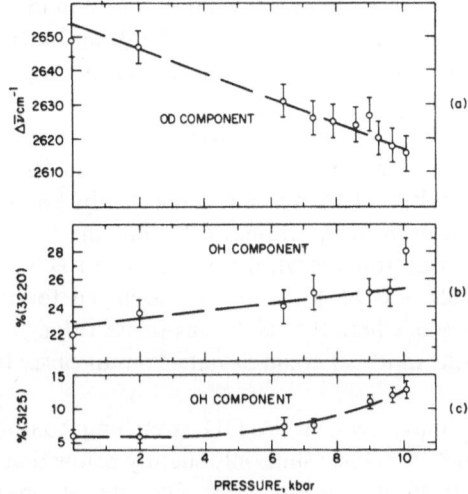

Fig. 4. Frequency shifts (a) and component percentages (b and c) corresponding, respectively, to four-Gaussian OD and five-Gaussian OH analog computer decompositions of Raman stretching contours from 20 volume % D_2O in H_2O vs pressure in kbar at 28°C.

kbar, and then to increase abruptly to 28% upon freezing at 10.1 kbar [the value of 28% refers to spectrum (c) of Fig. 2]. The increase in the 3220 cm^{-1} component percentage is similar to that noted previously for the contour height ratio H_{3220}/H_{3405} [cf. Fig. 4(b) with Fig. 3(c)]. In addition, the percentage of the 3125 cm^{-1} component increases from 6% to 15% from 1 bar to 10.1 kbar, and this effect is also similar to that noted previously for the ratio H_{3125}/H_{3405} [cf. Fig. 4(c) with Fig. 3(c)]. The abrupt increase in the 3220 cm^{-1} component percentage is particularly noteworthy and is discussed subsequently in relation to intermolecular coupling. The percentages of the remaining three of the five components centered near $\Delta\bar{\nu} = 3600$, 3505, and 3405 cm^{-1}, however, were found not to show significant changes with pressure.

4. DISCUSSION

From contour analyses of the OD contours it was concluded that the most marked and significant spectral effect of pressure in the OD stretching region is the change in the position of the Gaussian component from $\Delta\bar{\nu} = 2654$ cm^{-1} at 1 bar to $\Delta\bar{\nu} = 2617$ cm^{-1} at 10.1 kbar [Fig. 4(a)]. The percentage and half-width of the 2654–2617 cm^{-1} component, however, were found to be constant from 1 bar to 10.1 kbar within the accuracy of the Raman data and analog computer method. (Similar conclusions have been reached in regard to the position, percentage, and half-width of the OD component in two previous high-pressure Raman studies.[1,2]) Accordingly, because the percentage of the 2654–2617 cm^{-1} component does not change appreciably with pressure rise, it is suggested that the fraction of OD units involved in the interaction giving rise to the decrease in the $\Delta\bar{\nu}$ value is roughly constant. Further, it is concluded that hydrogen-bond breakage is not involved and that another mechanism is indicated instead.

The structure of ice VI has been investigated by means of x-ray diffraction.[7] The high density at $-175°C$, 1.31 g-cm^{-3},[8] results from interpenetration of two frameworks that are not interconnected by bonds. Four-coordination is preserved in each of the frameworks, but the O–O–O angles depart markedly in some cases from the tetrahedral value of $109.5°$, *viz.*, $76°$ and $128°$. If the value of $(105° + 109.5°)/2 = 107°$ is employed for the average intramolecular HOH angle, where $105°$ is the gas-phase value,[8] it is possible that some O–H$\cdots$O units in ice VI could deviate from linearity by as much as $11°$ to $16°$ at $-175°C$.

The intensity maximum in the OD stretching contour of the ice VI spectrum occurs at a frequency shift only slightly below that of the maximum from the OD spectrum obtained at 1 bar. The intensity maximum of the ice VI spectrum, therefore, almost certainly corresponds to strongly hydrogen-bonded OD groups. The component of the ice VI spectrum at $\Delta\bar{\nu} = 2617$ cm^{-1}, on the other hand, must correspond to weaker hydrogen bonds, and it is reasonable to assign the 2617 cm^{-1} component to angularly deformed O–D$\cdots$O bonds in view of the x-ray evidence described. Further, from numerous Raman studies of HDO–H$_2$O solutions,[6] there is now no serious doubt that the OD component centered near $\Delta\bar{\nu} = 2650$ cm^{-1} arises from nonhydrogen-bonded OD units in the liquid, and this assignment has recently received support from molecular dynamics methods which indicate the presence of dangling OH groups.[9] Therefore, it is reasonable to conclude that the nonhydrogen-bonded OD units or broken O–D$\cdots$O bonds that exist at 1 bar are transformed by close packing due to pressure into weak O–D$\cdots$O bonds that deviate markedly from linearity; i.e., the observed decrease in $\Delta\bar{\nu}$ arises from the formation of weak, bent hydrogen bonds.

It could be argued, of course, that the decrease observed in the $\Delta\bar{\nu}$ value of the OD component arises simply from increasing the density, e.g., to about 1.24 g-cm^{-3} at 10 kbar,[10] because the Raman $\Delta\bar{\nu}$ value of the OD stretching component is known to decrease when the density is increased from 0.65 to 0.90 g-cm^{-3} at $400°C$.[3] Nevertheless, a mechanism would still be required to explain the decreases in $\Delta\bar{\nu}$, and the most reasonable mechanism involves hydrogen-bond bending, as opposed to hydrogen-bond breakage.

Computer analyses of the OH stretching contour were accomplished with five Gaussian components (see Fig. 2). The percentages of two of the five components centered near $\Delta\bar{\nu} = 3220$ cm^{-1} and $\Delta\bar{\nu} = 3125$ cm^{-1} indicated marked increases with pressure rise, but the positions and half-widths of all five components, as well as the percentages of the three components centered near 3405, 3505, and 3600 cm^{-1}, were not significantly influenced by pressure.

The gradual intensification of the 3220 cm^{-1} component evident in

Fig. 4(b) from 1 bar to 9.7 kbar and the abrupt intensification upon freezing at 10.1 kbar indicate that the 3220 cm^{-1} component must arise primarily from an effect other than Fermi resonance, *viz.*, from intermolecular coupling.[6] The 3220 cm^{-1} component has recently been assigned to Fermi resonance,[11] but that assignment is highly questionable. If Fermi resonance were primarily responsible for the 3220 cm^{-1} component, it is unlikely that the relative intensity of that component would rise when the positions of the 3220 and 3405 cm^{-1} components remain nearly constant. However, the relative intensity could rise at constant Raman frequency shift if the intermolecular coupling increases with increasing pressure. From Fig. 4(b) it is evident that the percentage of the component at $\Delta\bar{\nu} = 3220$ cm^{-1} increases nearly as much upon freezing at 10.1 kbar as upon pressurizing the liquid from 1 bar to 9.7 kbar. The long-range order resulting from the formation of lattices in ice VI would greatly enhance the intermolecular coupling—thus the relative intensities of coupling components, as indicated. A gradual intensification of the component at $\Delta\bar{\nu} = 3125$ cm^{-1} with increasing pressure is also indicated in Fig. 4(c). This component could also arise from the effects of intermolecular coupling.

Finally, it is now widely accepted that the OD stretching component at $\Delta\bar{\nu} = 2650$ cm^{-1} arises from broken O–D$\cdots$O bonds or nonhydrogen-bonded OD units in the liquid, but the present work provides additional support for this conclusion. The intensity maxima in the OD stretching contours of Fig. 2 almost certainly correspond to strong O–D$\cdots$O bonds. Thus the shift of the OD stretching component from $\Delta\bar{\nu} = 2654$ cm^{-1} at 1 bar to $\Delta\bar{\nu} = 2617$ cm^{-1} at 10.1 kbar is in the direction of stronger hydrogen-bonding. However, the OD component from ice VI at $\Delta\bar{\nu} = 2617$ cm^{-1} reasonably corresponds to weak, angularly distorted O–D$\cdots$O bonds. Accordingly, it seems reasonable to conclude that the larger $\Delta\bar{\nu}$ value of 2654 cm^{-1} at 1 bar corresponds to broken O–D$\cdots$O bonds or nonhydrogen-bonded OD units in the liquid.

REFERENCES

1. G. E. Walrafen, *J. Chem. Phys.* **55**, 768 (1971).
2. G. E. Walrafen, *J. Chem. Phys.* **55**, 5137 (1971).
3. E. U. Franck and H. Lindner, doctoral dissertation of the latter, University of Karlsruhe, 1970.
4. J. P. Marckmann and E. Whalley, *J. Chem. Phys.* **41**, 1450 (1964).
5. G. E. Walrafen and L. A. Blatz, *J. Chem. Phys.* **56**, 4216 (1972).
6. G. E. Walrafen, in *Water: A Comprehensive Treatise*. Vol. I. *Physics and Physical Chemistry of Water*, F. Franks, ed. (Plenum Press, New York, 1972).
7. B. Kamb, *Science* **150**, 205 (1965).
8. D. Eisenberg and W. Kauzmann, *The Structure and Properties of Water* (Clarendon, Oxford, 1969).
9. A. Rahman and F. H. Stillinger, *J. Chem. Phys.* **55**, 3336 (1971), F. H. Stillinger and A. Rahman, *J. Chem. Phys.*, in press.

10. Estimated from extrapolation of data reported by T. Grindley and J. E. Lind, *J. Chem. Phys.* **54**, 3983 (1971).
11. W. F. Murphy and H. J. Bernstein, *J. Phys. Chem.* **76**, 1147 (1972).

DISCUSSION

Professor E. U. Franck (*University of Karlsruhe, West Germany*). The temperature at which these measurements were made, namely 28°C, is still in the range where the viscosity has the abnormal negative pressure coefficient. One way to explain this abnormality would be to assume a breaking of hydrogen bonds (at pressures up to 10 kbar, but particularly up to 2 kbar). Could you comment on this?

Dr. Walrafen (*Bell Laboratories, Murray Hill, New Jersey*). The negative pressure coefficients in the viscosity of water below about 30°C were first brought to my attention by Professor Frank in a private discussion about 10 years ago. Of course, the present Raman data were obtained at about 28°C where the viscosity minimum is at best very shallow, but it is nevertheless tempting to look for some sort of intensity minimum with pressure even at 28°C. However, the high-pressure data have been obtained as carefully as possible with the presently available techniques, and they do not reveal a direct relationship to the viscosity variations as far as can be determined. Accordingly, a suitable explanation of the viscosity minimum may be more involved than previously supposed. Or, further Raman studies of the pressure dependence of the *intermolecular* intensities at temperatures from 0° to 30°C may provide a more complete understanding that will elucidate the viscosity changes.

Professor H. S. Frank (*University of Pittsburgh*). Of possible relevance to this remark are the data, which Walrafen did not show, of Professor Franck's 400° curve, namely, the one in which he reduced the density down to a tenth of a g-cm^{-3}, where the band at 2720 cm^{-1} which corresponds to the free molecules began to appear because the density was so low. I have a feeling that, unless we remember that even the broken-bonded bands you are discussing are far from being at a frequency corresponding to an undisturbed vibration, we are likely to be misled. I think that even if the broken bond is still broken, the degree of departure of its frequency from that of an undisturbed vibration must increase very considerably when the density is increased. This means that some of the comparisons we have been trying to make may be a bit of a trap.

Dr. Walrafen. If Dr. Lindner's thesis (he is a former student of Prof. E. U. Franck) is examined carefully, it is evident that he also obtained Raman spectra from HDO at constant (very high) temperatures and at densities from 0.65 to 0.9 g-cm^{-3}. At a constant temperature of 400°C, the Raman frequency shift corresponding to maximum contour intensity was observed to decrease at high pressures, as it does at 28°C.

Professor Frank. Yes, under conditions where, in the clearly hydrogen-bonded bands, change in density makes a very pronounced change, there are interactions (which are not hydrogen-bonding) which also show up spectroscopically in the same sense.

Dr. Walrafen. At a pressure of 1 atm, the mixture model is quite adequate for treating the structure of water. However, if the broken hydrogen bonds, that is, dangling O–H units, re-form at 10 kbar such that the $O-H\cdots O$ angles deviate markedly from 180° and/or the O—O distances change, then some modification of the mixture model, or even another more complicated model, might be required.

Professor Frank. What I want to ask at one bar depends on what bar I'm at.

Professor K. S. Pitzer (*University of California, Berkeley*). It seems that Dr. Stillinger's graphs are very helpful in having a realistic understanding of this. You really have a continuum of mobile geometries; some simplification is obtained due to a relative predominance of a small number of "structures" which can be discussed qualitatively. One can correlate everything in between in a multidimensional way, and this is really seconded by Professor Frank. We have to remember that the so-called "broken bonds" here are still close enough to other atoms (which have relatively strong electrical fields) that they are not H_2O or HDO in the vacuum. We will get, therefore, a sort of first approximation in which the sensible thing to do is to picture two or three idealized "structures" which represent, of course, maxima—not very sharp maxima, but still maxima of Dr. Stillinger's diagrams—and then eventually a high approximation is obtained which can integrate across the whole curve.

Dr. Walrafen. It may be inadvisable at the present time to rely so very heavily on theoretical calculations, particularly in view of the remarkable effects obtained from stimulated Raman scattering. The stimulated Raman spectra provide very clear evidence for mutually exclusive stimulation of the hydrogen-bonded or nonhydrogen-bonded stretching components, that is, the stimulated maxima agree favorably with the frequency shifts corresponding to the spontaneous Raman (Gaussian) components. Of course, the spontaneous Raman components are very broad, and this breadth is related to Professor Pitzer's comment, i.e., to a range of structures within a single class; but the mutually exclusive stimulated Raman scattering that is observed indicates that the hydrogen-bonded–nonhydrogen-bonded description is a very good one. Further, the idea that the broken hydrogen bonds engage in other highly cohesive interactions has been stressed several times in my publications. Indeed, I strongly agree with Professor Pitzer with regard to the presence of strong electrical fields, and I wish that this point were more generally appreciated.

Professor P. A. Giguère (*Université Laval, Canada*). This is not a question but, if I may be allowed, a brief report on some of our current research, which happens to be closely related to what we have just heard. It has to do with the laser-Raman spectra of pure liquid hydrogen peroxide. There are two main reasons why the O–H stretching bands in that compound should be less complex than those in water. First, the internal coupling between the two O–H stretching vibrations is very much smaller than in water. For instance the symmetric and antisymmetric frequencies are separated by less than 10 cm^{-1} in the free H_2O_2 molecule compared with 100 cm^{-1} in water. Then, there can be no appreciable Fermi resonance with the overtone of the OH bending modes (near 2800 cm^{-1}).

Accordingly the Raman band at 3400 cm^{-1} is not as wide as that in water: half-width of ~270 cm^{-1} vs ~400 cm^{-1}. Its contour is nearly symmetric with a *soupçon* of a shoulder, but... on the low-frequency side. Life is seldom simple! With liquid hydrogen peroxide, we are still in the age of innocence inasmuch as we have not yet proposed models to argue about. For one thing, there is no appreciable scattering in the 3600 cm^{-1} region often assigned to so-called free OH groups.

Dr. Walrafen. Did you conduct studies over a range of temperatures?

Professor Giguère. Yes we did, over the range from 40° to −50°. You can supercool pure liquid hydrogen peroxide that much if you are very careful. Our intensity measurements are not too reliable, but the band intensity seems to increase on cooling. The contour becomes more symmetrical, and the frequency maximum is shifted, even more so than in water.

Dr. Walrafen. Would you expect different hydrogen-bonded configurations at different sites, and would you also expect to find spectroscopic evidence of their existence?

Professor Giguère. Certainly there is a greater number of possibilities than in H_2O because of the two extra proton acceptor sites (A) on the oxygen atoms. For instance, one can visualize as many as 16 different D_2A_j species against 4 in water. However, I doubt very much that one could differentiate between them from the spectra.

Dr. Walrafen. Were your spectra obtained from a solid?

Professor Giguère. I was talking about liquid H_2O_2 and D_2O_2. The hybrid species HDO_2 shows bands of uncoupled OH and OD groups resembling very much those in water and heavy water.

Professor G. S. Kell (*National Research Council, Ottawa*). For the picture I have of water, there are more interstitial molecules at higher than at low pressures and the results you were showing at the end of your talk are in opposite directions. As we go toward ice VI, the oxygen–oxygen separation is almost the same as in the low-pressure liquid. So, I would assume that the change in density in the liquid at increased pressure is, in large part, brought about by deformation of angles and by water molecules tending to occupy interstitial positions which will be unfavorable energetically at low pressures because of the considerable energy of hydrogen bonding.

Dr. Walrafen. In reply to Dr. Kell's first question as to why there are two stimulated Raman peaks in each of the hydrogen-bonded and nonhydrogen-bonded regions, I would say that the answer is that the HDO concentration was nearly maximal in the two cases described, and thus additional hydrogen-bonded and nonhydrogen-bonded coupling components due to HDO–HDO interactions are present. With regard to the second question, my understanding is that the high density of ice VI arises from the existence of two tetragonally distorted interpenetrating tetrahedral lattices that are completely unbonded to each other, i.e., there are no hydrogen bonds *between* the interpenetrating lattices, although other strongly cohesive forces are operative. Further, if the average *intramolecular* angle in ice VI is about $107°$, the $O—H \cdots O$ units might be expected to deviate from $180°$ by as much as $\sim 11°$ to $\sim 16°$. Hence, I would expect that angular distortions of the $O—H \cdots O$ and $O—D \cdots O$ bonds in liquid $HDO–H_2O$ mixtures would be of a similar magnitude. One might further expect that the concentration of holes, or the density fluctuations, would be decreased at 9.7 kbar, the highest pressure at which the $HDO–H_2O$ mixture remained liquid.

Solvation Equilibria in Very Concentrated Electrolyte Solutions[1,2]

R. H. Stokes[3] and R. A. Robinson[4]

Received September 20, 1972

A stepwise hydration-equilibrium model for aqueous electrolytes is developed and shown to give a good quantitative account of osmotic coefficients of strong, highly soluble electrolytes up to concentrations of 20–30 m. The main parameters needed are two equilibrium constants describing the stepwise hydration and the number of hydration sites. Choice of ion-size parameter has only a minor effect.

KEY WORDS: Calcium bromide; calcium chloride; hydration of ions; hydrochloric acid; lithium bromide; lithium chloride; osmotic coefficient; perchloric acid; potassium hydroxide; sodium chloride; sodium hydroxide; solvation of ions.

1. INTRODUCTION

Some 25 years ago we proposed a model[1] for electrolyte solutions up to a few molar in concentration in which the Debye–Hückel treatment of ion–ion interactions was combined with the idea that the species in solution are hydrated ions. Subsequently, Glueckauf[2] showed that the model gave physically more satisfactory hydration numbers on the assumption that the hydrated species and the free-water molecules mix according to Flory–Huggins or volume-fraction statistics rather than to the mole-fraction statistics which we assumed. More recently, Friedman and collaborators[3] have treated the "primitive model" of electrolytes by the solution of the integral equations of

---

[1] This paper was presented at the symposium, "The Physical Chemistry of Aqueous Systems," held at the University of Pittsburgh, Pittsburgh, Pennsylvania, June 12–14, 1972, in honor of the 70th birthday of Professor H. S. Frank.

[2] Contribution from the Diffusion Research Unit, Research School of Physical Sciences, Australian National University, Canberra, A.C.T., Australia and the University of New England, Armidale, N.S.W., Australia.

[3] On leave at the Australian National University from the University of New England.

[4] Present address: Department of Chemistry, University of Florida, Gainesville, Florida 32601.

modern statistical-mechanical theories of dense fluids. They have obtained much higher values for osmotic and activity coefficients than are found experimentally and have concluded that hydration models (which explain why the experimental results are higher than the Debye–Hückel prediction) are therefore inappropriate. However, it has been shown[4] by applying their arguments to an ideal solution that their high results are due to a covolume term which cannot appropriately be included in comparisons with actual solutions.

Our treatment of the problem, and Glueckauf's, both included the over-simplification of assigning to the ions a fixed hydration number independent of concentration. This inevitably results in failure of the equations at high concentrations. For example, if the lithium ion is assumed to exist as tetra-hydrate, all the water in the solution is bound at a molality of $55.51/4$, i.e., $13.9\ m$, and the water activity should then be zero. In fact, lithium chloride is soluble beyond $20\ m$, and the water activity in the $20\ m$ solution is still 0.11. Clearly, one needs a model in which the hydration phenomenon is treated as an equilibrium of various stages of hydration. The essential validity of this idea is strongly supported by important mass-spectrometric studies of the enthalpy and free-energy changes for stepwise hydration processes in the gas phase, reported in recent years by Kebarle[5] and associates. We have had some success in accounting for the thermodynamic behavior of single and mixed sugar solutions[6] using a model of this sort, and our early work[1] included a demonstration that very concentrated electrolytes could be treated by means of a B.E.T.-type approach in which water molecules are "adsorbed" on ionic sites.

The present paper provides a treatment of electrolytes of the highly hydrated type which is satisfactory up to the highest accessible concentrations.

2. DEVELOPMENT OF EQUATIONS

We follow Glueçkauf in assuming that the ions in the various stages of hydration mix with the water according to volume-fraction statistics. We consider a solution of volume V containing, in equilibrium, the following species: n_A moles of free solvent, n_0 moles of anhydrous cations, n_1 moles of singly hydrated cations, n_i moles of i-hydrated cations, and n_D moles of anions.

This solution would have been prepared by mixing c moles of anhydrous solute with the necessary amount of water to give a final volume of 1 liter. During this preparative process, large enthalpy and entropy changes will have occurred, but we are not concerned with these in our hypothesis about the mixing statistics of the species present at equilibrium. For this, we are considering the mixing of the appropriate amounts of the various hydrates, free solvent, and anion. That it is valid to calculate *activities* of the various species present at equilibrium from such an argument can be deduced from a rigorous statistical-mechanical treatment by Saroléa-Mathot[7]; it turns out to be a consequence of the fact that the Gibbs free energy of the solution is a minimum when the concentrations of the various species have their equilibrium values.

Using the pure liquid of each species as the standard state, the activity a_j of species j is given by volume-fraction statistics, *in the absence of interionic effects*, as

$$\ln a_j = \ln [(n_j \, \bar{V}_j)/V] + 1 - (\bar{V}_j/V) \sum_{k=A, \, D, \, i} n_k \qquad (1)$$

The summation in the last term of Eq. (1) extends over *all* the species present, both solvent and solute, i.e., it is the total number of moles of distinct entities present at equilibrium in the volume V. The $\bar{V}$'s are best regarded as partial molar volumes, and to avoid the later frequent appearance of factors of 1000 or 0.001, we shall regard them as being expressed in liters per mole, while concentrations c are in moles per liter.

$$\sum c_k \bar{V}_k = 1 \qquad (2a)$$

Then

$$h = \sum_{i=0, \, 1, \, \ldots} i c_i \Big/ \sum_{i=0, \, 1, \, \ldots} c_i \qquad (2b)$$

where $c_k = n_k/V$ and h denotes the average hydration number. We shall also assume that the various hydrates have partial molar volumes linearly related to the degree of hydration:

$$\bar{V}_i = \bar{V}_0 + i \bar{V}_A \qquad (3)$$

If we denote the stoichiometric molar concentration by c (unsubscripted) and give v_+ and v_- their usual significance (number of ions per "molecule" of electrolyte),

$$c_D = v_- c \qquad \sum_i c_i = v_+ c \qquad v = v_+ + v_-$$

Eq. (2a) becomes

$$c_A \bar{V}_A + c v_+ (\bar{V}_0 + h \bar{V}_A) + c v_- \bar{V}_D = 1 \qquad (4)$$

This may be written in terms of the conventional partial molar volume of the solute, $\bar{V}_B = v_+ \bar{V}_0 + v_- \bar{V}_D$, as

$$c_A \bar{V}_A + c(\bar{V}_B + h \bar{V}_A) = 1 \qquad (5)$$

The solvent activity, still in the temporary absence of interionic forces, is given by Eqs. (1) and (5) upon elimination of c_A as

$$\ln a_A(\text{uncharged}) = \ln [1 - c(\bar{V}_B + h \bar{V}_A)] + c[\bar{V}_B + (h - v) \, \bar{V}_A] \qquad (6)$$

If we start with "uncharged" (but hydrated) ions in a solvent of activity given by Eq. (6) and "switch on" the interionic forces, the chemical potential of the solvent will change by an amount $\mu_A(\text{el})$. While it is true that the Debye theory undertakes the evaluation of $\mu(\text{el})$ only in rather dilute solutions,[8] the recent treatment by Waisman and Lebowitz,[9] in which the statistical-mechanical equation of state for charged hard spheres is solved, does not appear to be subject to this restriction. There remains, of course, a large uncertainty about the appropriate dielectric constant, but the present model should absorb most of the effects of the change in dielectric constant by its consideration of the water molecules bound to ions. The Debye and Lebowitz solutions show only minor differences even at high concentrations, and it has been shown[4] that reasonable adjustments of the "ionic diameter" can lead to agreement between the two approaches. It is important to note that for strongly hydrated solutes the term $\mu_A(\text{el})$ is in any case small compared to $\mu_A(\text{config}) = RT\ln a_A(\text{uncharged})$. Using the Debye result as developed by Fowler and Guggenheim[8],

$$\mu_A(\text{el})/RT = (v_A/24\pi a^3)(\kappa a)^3 S(\kappa a) \tag{7}$$

where a is the ionic diameter, κ the Debye reciprocal length, and v_A the partial volume of the solvent per molecule, we obtain in our notation

$$\ln a_A(\text{el}) = Q\bar{V}_A c^{3/2} S(\kappa a) \tag{8}$$

where

$$Q = \kappa^3/(24\pi N c^{3/2})$$

The function S is given by the Debye theory as

$$S(x) = (3/x^3)[1 + x - 1/(1 + x) - 2\ln(1 + x)] \tag{9a}$$

The corresponding result of Waisman and Lebowitz[9] may be written

$$S'(x) = (6/x^3)[\tfrac{2}{3}(1 + 2x)^{3/2} - \tfrac{2}{3} - x - x(1 + 2x)^{1/2}] \tag{9b}$$

Equations (6) and (8) sum to

$$\ln a_A = \ln[1 - c(\bar{V}_B + h\bar{V}_A)] + c[\bar{V}_B + (h - v)\bar{V}_A] + Q\bar{V}_A c^{3/2} S(\kappa a) \tag{10}$$

The essential difference between Eq. (10) and the expression for $\ln a_A$ which can be derived from Glueckauf's[2] treatment is that h is no longer a constant but is dependent on the water activity a_A. We now develop this relationship.

The attachment of the i-th water molecule to a cation may be written in terms of a thermodynamic equilibrium constant K_i:

$$K_i = a_i/(a_{i-1} \cdot a_A) \tag{11}$$

with, according to Eq. (1),

$$\ln a_i = \ln(c_i \bar{V}_i) + 1 - \bar{V}_i \Sigma c_k + \ln a_i(\text{el}) \tag{12}$$

Writing $a_i = A \cdot c_i y_i$, where y_i is a molar-scale activity coefficient so defined as to approach unity at zero concentration of all solute species, we can evaluate the constant A from this limiting case, for we then have

$$\lim(c \to 0)\, c_k = c_A^\circ = 1/\bar{V}_A^\circ \tag{13}$$

$$\lim(c \to 0)\, \bar{V}_i = \bar{V}_i^\circ = \bar{V}_0^\circ + i\bar{V}_A^\circ \tag{14}$$

where superscript zeros denote zero-concentration values. Hence $\ln A = \ln \bar{V}_i^0 + 1 - \bar{V}_i^0/\bar{V}_A^0$ and

$$\ln y_i = \ln (\bar{V}_i/\bar{V}_i^\circ) - V_i \Sigma c_k + (V_i^\circ/V_A^\circ) + \ln y_i(\text{el}) \tag{15}$$

Then from Eq. (11),

$$\ln K_i = \ln [c_i/(c_{i-1} \cdot a_A)] + \ln (y_i/y_{i-1}) \tag{16}$$

Denoting the activity coefficient factor by Y and using Eqs. (3) and (15),

$$\ln Y = \ln (y_i/y_{i-1}) = \ln [(\bar{V}_i \cdot \bar{V}_{i-1}^\circ)/(\bar{V}_i^\circ \cdot \bar{V}_{i-1})] + 1 - V_A \Sigma c_k \tag{17}$$

Obviously the first term on the right of Eq. (17) is extremely close to zero; we treat it as exactly zero. We have also assumed cancellation of the interionic activity coefficient contributions of species i and $i-1$.

Equations (2) and (5) yield with Eq. (17)

$$\ln Y = c(\bar{V}_B + h\bar{V}_A - v\bar{V}_A) \tag{18}$$

which is independent of i. Hence all the stepwise equilibria at a given concentration have their K_i's modified by the same activity coefficient factor Y:

$$c_i/c_{i-1} = (K_i/Y) a_A \tag{19}$$

To make further progress we need some simplifying assumptions about the successive K_i values. The work of Kebarle et al.[5] clearly shows that we must expect K_1 to be large, K_2 smaller, and so on. However, we cannot simply take over the gas-phase hydration results; the K's will be greatly different in the liquid phase. For instance, competition between the fields of neighboring ions will tend to reduce the strength of binding. We shall therefore use a model in which

$$K_1 = K$$
$$K_2 = kK$$
$$K_i = k^{i-1} K$$

The standard free-energy change for attachment of a water molecule thus becomes less negative by an amount $RT \ln k$ for each successive step.

It follows that the concentration of the i-hydrate is related to that of the unhydrated ion by

$$c_i = c_0 \cdot (K/Y)^i \cdot k^{i(i-1)/2} a_A^i \tag{20}$$

The number of moles of bound water per liter of solution is

$$v_+ ch = c_0 \sum_{i=0}^{n} i(K/Y)^i k^{i(i-1)/2} a_A^i \tag{21}$$

where n is the upper limit of significant binding sites. The total concentration of cations is

$$cv_+ = c_0 \sum_{i=0}^{n} (K/Y)^i k^{i(i-1)/2} a_A^i \tag{22}$$

so that

$$h = \sum_{l=0}^{n} i(K/Y)^i k^{i(i-1)/2} a_A^l \bigg/ \sum_{l=0}^{n} (K/Y)^i k^{i(i-1)/2} a_A^l \qquad (23)$$

(The series in Eq. (22) can be put in the form $\sum x^n e^{-an^2}$, which is related to Jacobi's theta functions. We have not, however, found any way of exploiting this relationship.)

Finally, we have for the water activity the implicit equation (24) relating the water activity to the concentration, K and k:

$$\ln a_A = \ln(1 - c\bar{V}_h) + c(\bar{V}_h - v\bar{V}_A) + Q\bar{V}_A c^{3/2} S(\kappa a) \qquad (24)$$

where

$$\bar{V}_h = \bar{V}_B + h\bar{V}_A \qquad (25)$$

and h is given by Eq. (23) in which Y is given by Eq. (18). The numerical constant $Q = 0.7849$ for 1:1 electrolytes in water at 25°C as solvent; otherwise see Eq. (7).

3. SOLUTE ACTIVITY COEFFICIENTS

The conventional mean molar activity coefficient of the solute is most readily obtained from the proposition that the chemical potential of the electrolyte is the sum of those of the anhydrous ($i = 0$) cation species and the anion.
Hence

$$v \ln(cy_\pm) = v_+ \ln(c_0 y_0) + v_- \ln(cy_D) + A' \qquad (26)$$

Substituting for c_0/c from Eq. (22) and considering the limit of zero concentration to evaluate A' gives

$$\ln y_\pm = (1/v)(v_+ \ln y_0 + v_- \ln y_D) + (1/v)\ln(\Sigma_0/\Sigma) \qquad (27)$$

where Σ is the summation in Eq. (22) and Σ_0 is its value in the limit $a_A = 1$ and $Y = 1$. This yields on simplification

$$\ln y_\pm = c\bar{V}_A[r(r + h - v)/v] + \ln y_\pm^{(el)} + (1/v)\ln(\Sigma_0/\Sigma) \qquad (28)$$

where $r = \bar{V}_B/\bar{V}_A$.

The corresponding molality-scale expression is

$$\ln \gamma_\pm = [0.018mr(r + h - v)/(1 + 0.018mr)v] - \ln(1 + 0.018mr)$$
$$+ (1/v)\ln(\Sigma_0/\Sigma) + \ln \gamma_\pm^{(el)} \qquad (29)$$

We can see the relation of Eq. (29) to Glueckauf's earlier result in which h was taken as a constant as follows. Equation (23) can be written

$$h = \partial \ln \Sigma / \partial \ln a_A \qquad (30)$$

so that constant h implies

$$\ln(\Sigma_0/\Sigma) = h \ln a_A \qquad (31)$$

Substituting Eq. (31) in Eq. (29), we obtain Eq. (7) of ref. 10, which was shown to be equivalent to Glueckauf's result when the complete Debye–Hückel expression for $\ln a_A(\text{el})$ is included.

We have not used Eq. (29) for comparisons of theoretical and experimental results because Eq. (24) provides a more direct calculation of the osmotic coefficient which is usually the experimentally determined quantity.

4. EVALUATION OF THE PARAMETERS

Strictly speaking, Eqs. (23–24) involve four parameters, the equilibrium constants K and k, the ion size parameter a, and the upper limit n assigned to the number of hydration sites. For typical strongly hydrated and highly soluble 1:1 electrolytes, however, we can assume a value of a of about 4×10^{-8} cm. The choice of a is more critical if we are seeking an exact fit below, say, $2\ m$, but it has little effect at really high concentrations. Also, for the values of K and k involved, the series (23) converges so rapidly that even the choice $n = \infty$ gives good results at high concentrations where a_A is small. Truncation of the series by setting $n = 4$ or 5 for 1:1 electrolytes, or $n = 9$ for 2:1 electrolytes, which are physically reasonable values for the number of molecules in a complete hydration shell, gives rather better results at intermediate concentrations. The choice of K and k is, however, quite critical, though once a K near the best value has been found, it can be varied by several percent while a compensating change in k is made without materially altering the quality of fit.

Because of the implicit occurrence of a_A [in h via Eq. (23)] on the right of Eq. (24), an iterative computer method provides the only practical way of fitting a given set of activity and partial volume data to find best values for K and k. The method used will vary according to the nature of the equation-solving and error-minimizing subroutines available, and so it is not detailed here. Essentially it depends on choosing reasonable a and n values, then selecting a value of K and finding the k required to fit the most concentrated solution point. With these values, the standard deviation of the whole set of data from the equations is calculated. This is repeated with other choices of K until the standard deviation is minimized. Though this method puts a lot of weight on the highest point, it is more economical of computer time than a full optimization of both K and k.

Figure 1 shows the quality of fit obtained for nine electrolytes. The quantity plotted is the molal osmotic coefficient $\phi = -(55.51/vm)\ln a_A$. The parameters used are given in Table I.

In all cases we have insisted on fitting the whole range of the available reliable data. In the cases of the two acids, this may involve neglect of a

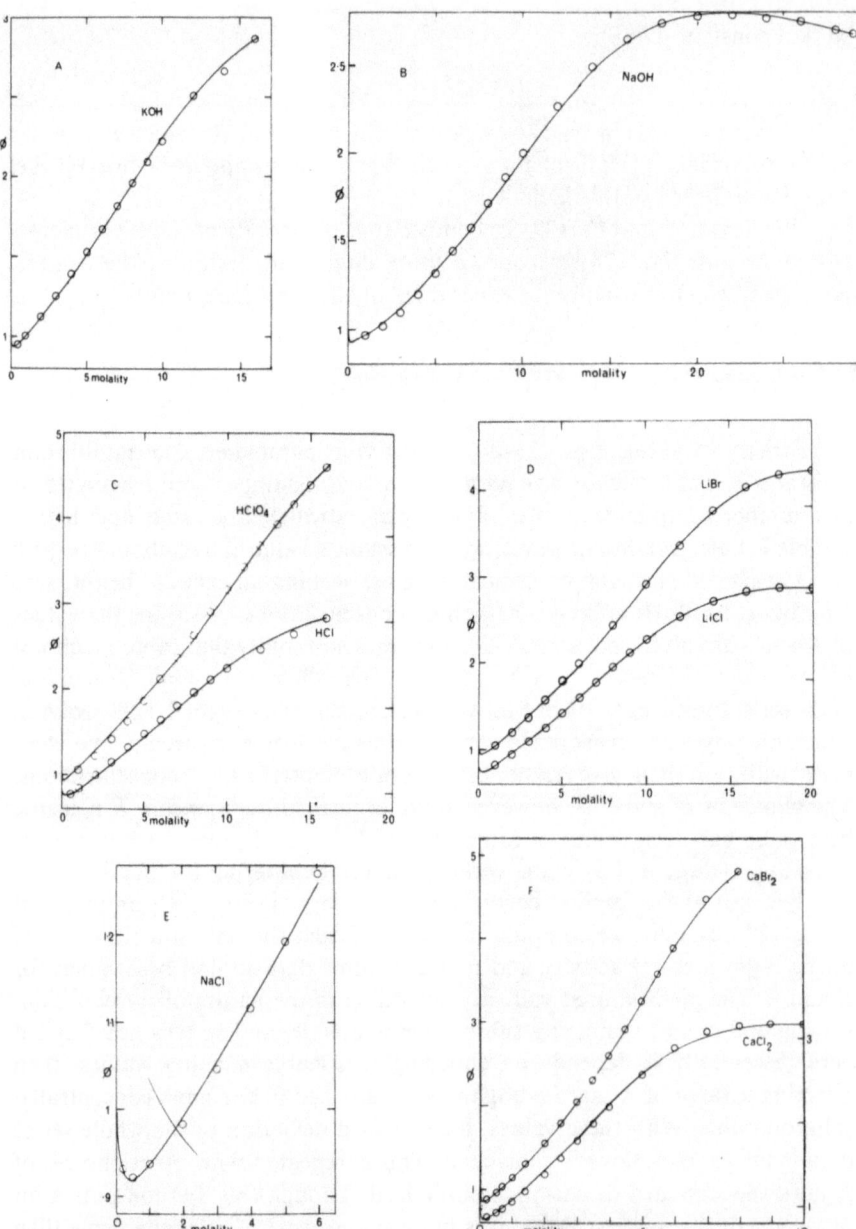

Fig. 1. Osmotic coefficients of electrolytes in water at 25°C. (A) Potassium hydroxide; (B) sodium hydroxide; (C) hydrochloric acid and perchloric acid; (D) lithium chloride and lithium bromide; (E) sodium chloride; (F) calcium chloride and calcium bromide. Curves are calculated using the parameters in Table I. Circles show experimental values. The standard deviations given in Table I are those of the points shown in the curves. Note that in C, D, and F the ϕ values for hydrochloric acid, lithium chloride, and calcium chloride are displaced downwards by 0.2, as shown by the right-hand ordinates, in order to avoid overlap.

Table I. Parameters of Eqs. (23) and (24) for Aqueous Solutions at 25°C

Solute	a, Å	n	K	k	σ^a	Max. Molality
KOH	4	5	150.2	0.303	0.018	16
NaOH	4	4	77.6	0.375	0.018	29
HCl[b]	4	5	88.9	0.403	0.041	16
HClO₄	4	5	2527	0.192	0.044	16
LiCl	4	5	81.6	0.414	0.015	20
LiBr	4	5	492	0.290	0.017	20
NaCl	3.5	3	61.1	0.368	0.008	6
CaCl₂	4	9	48.7	0.678	0.041	10
CaBr₂	4	9	804.6	0.595	0.014	8

[a] Standard deviation of ϕ.
[b] More recent calculations show that increasing n to 6 gives a much better fit for HCl with $K = 135$, $k = 0.338$, $\sigma = 0.019$.

significant amount of the molecular form at high concentrations. There is not much point in testing the theory in cases where solubility limits one to solutions below 5 or 6 m, as here one can get results of comparable accuracy with a fixed value of h, i.e., by Glueckauf's original equation. Sodium chloride has, however, been included in Table I, though it was necessary to fix the value of k by arbitrary choice to ensure convergence of the calculations.

The present treatment assumes that all the hydration sites are on the cation. We could equally well have chosen the anion, but when there is some

Table II. Detailed Results for LiCl at 25°C ($a = 4$ Å, $n = 5$, $K = 81.6$, and $k = 0.414$)

m	c	$\bar{V}_A$ cm³-mole⁻¹	$\bar{V}_B$ cm³-mole⁻¹	h	$\Delta\phi_{el}{}^a$	ϕ_{calc}	ϕ_{expt}
1	0.979	18.06	19.15	4.577	−0.104	1.025	1.018
2	1.921	18.04	20.01	4.533	−0.102	1.165	1.142
3	2.826	18.01	20.62	4.482	−0.098	1.316	1.286
4	3.696	17.98	21.03	4.422	−0.093	1.474	1.449
5	4.532	17.96	21.28	4.352	−0.090	1.638	1.619
6	5.335	17.95	21.43	4.271	−0.087	1.806	1.791
7	6.107	17.94	21.50	4.178	−0.083	1.977	1.965
8	6.851	17.94	21.52	4.071	−0.080	2.146	2.143
9	7.567	19.94	21.50	3.953	−0.077	2.310	2.310
10	8.258	17.95	21.46	3.823	−0.074	2.464	2.464
12	9.570	17.96	21.38	3.540	−0.070	2.725	2.730
14	10.80	17.98	21.33	3.249	−0.066	2.910	2.915
16	11.95	17.97	21.35	2.973	−0.062	3.019	3.023
18	13.03	17.93	21.47	2.722	−0.059	3.065	3.057
20	14.03	17.86	21.69	2.503	−0.056	3.063	3.063

[a] $\Delta\phi_{el}$ is the contribution to ϕ from the interionic forces given by Eq. (8).

hydration of both ions, as is probably the case with NaOH and KOH, one should really have separate h series and equilibrium constants for each. This would provide too many adjustable parameters for a useful theory.

Table II gives fuller details for one case, LiCl at 25°C. The c, $\bar{V}_A$, and $\bar{V}_B$ values for round m were derived from density data given in the International Critical Tables.[11] The values of h are those simultaneously satisfying Eqs. (23) and (24) with the given values of K and k. The Debye–Hückel term $\Delta\phi_{el}$ is included to demonstrate its minor role at high concentrations. Very similar results are obtained if the Waisman–Lebowitz expression for $\Delta\phi_{el}$ is used.

5. DISCUSSION

Figure 2 shows how the distribution of the various stages of hydration changes with concentration for the case of lithium chloride. The ordinate is the fraction (c_i/c_{total}) calculated from the data in Table II.

Several points about Table I may be noted.

(a) For most of the 1:1 electrolytes, k has a value near 0.35, so that the standard free-energy change for successive equilibrium steps is less negative by approximately RT per mole.

(b) Bromides have considerably larger K values than the corresponding chlorides. Bearing in mind that K and k are largely determined by the high-concentration region, where water molecules on a cation must also be very close to anions, one can see that the larger the anion the less it will tend to oppose the attraction of the cation for water molecules.

(c) This effect can also be seen with the pair HCl and HClO₄. Here, however, the real situation is probably complicated by the presence of the molecular acids at high concentrations, which is not considered in the present model. This may also account for the relatively poor fit obtained in the case of the acids.

Figure 1 shows that the equations are capable of describing quantitatively the whole range of concentration which includes osmotic coefficients of almost 5. (In terms of activity coefficients, it is worth noting that $\gamma_\pm$ for LiBr at 20 m is 485.) It is of especial interest that the tendency of ϕ to rise less

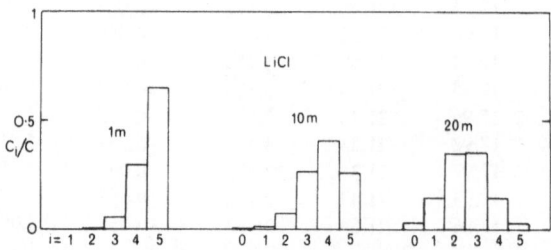

Fig. 2. Distribution of hydration stages for lithium chloride at 1, 10, and 20 mole-kg⁻¹.

rapidly at high molalities is correctly reproduced, even to the actual maximum observed with sodium hydroxide.

NOTATION

A solvent

B solute (electrolyte as a whole)

D anion

K equilibrium constant for first hydration step

Q numerical coefficient of Eq. (8): $Qc^{3/2} = \kappa^3/24\pi N$

S, S' functions defined by Eqs. (9a) and (9b)

V volume of solution (liter)

$\bar{V}_j$ partial molar volume of species j (liter-mole^{-1})

Y activity coefficient ratio [Eq. (17)]

a ion-size parameter

a_j activity of species j

c concentration (mole-liter^{-1})

h average hydration number

i number of water molecules bound to a particular cation $(0 \leqslant i \leqslant n)$

k ratio of successive equilibrium constants for hydration

k in summations, denotes sum over all species in solution

m molality

n number of hydration sites

n_j number of moles of species j

r $\bar{V}_B/\bar{V}_A$

y molar-scale activity coefficient

v_A molecular volume of solvent

$\gamma_\pm$ mean molal activity coefficient

κ Debye–Hückel characteristic reciprocal length

v, v_+, v_- stoichiometric numbers of ions per "molecule" of electrolyte

ACKNOWLEDGMENT

R. H. S. thanks the Director and staff of the Research School of Physics, Australian National University, for facilities provided during his study leave there.

REFERENCES

1. R. H. Stokes and R. A. Robinson, *J. Am. Chem. Soc.* **70**, 1870 (1948).
2. E. Glueckauf, *Trans. Faraday Soc.* **51**, 1235 (1955).
3. J. C. Rasaiah and H. L. Friedman, *J. Chem. Phys.* **48**, 2742 (1968); **50**, 3965 (1969).
4. R. H. Stokes, *J. Chem. Phys.* **56**, 3382 (1972). (Note that in Eq. (5) of this paper the coefficient of the term $(1 + 2x)^{3/2}$ should have been written 2/3, not 2. The correct result was used in calculating data for Fig. 1 of the paper.)

5. M. Arshadi, R. Yamdagni, and P. Kebarle, *J. Phys. Chem.* **74**, 1475 (1970).
6. R. H. Stokes and R. A. Robinson, *J. Phys. Chem.* **70**, 2126 (1966).
7. L. Saroléa-Mathot, *Trans. Faraday Soc.* **49**, 8 (1953).
8. R. H. Fowler and E. A. Guggenheim, *Statistical Thermodynamics* (Cambridge University Press, 1949), Chap. IX.
9. E. Waisman and J. L. Lebowitz, *J. Chem. Phys.* **52**, 4307 (1970).
10. R. H. Stokes and R. A. Robinson, *Trans. Faraday Soc.* **53**, 301 (1957).
11. *International Critical Tables* (McGraw-Hill, New York, 1928, Vol. III.

DISCUSSION

Dr. F. Franks (*Unilever Research Laboratories, Bedford, England*). The fits produced by Professor Stokes are certainly impressive. It is, however, generally true that free energies are very insensitive to the model used for fitting experimental data. The temperature and pressure derivatives provide more stringent tests for the relevance of a given model. Has Professor Stokes compared the experimental heats and volumes of dilution with those computed from his equation?

Professor Stokes (*Australian National University*). I haven't done a great deal on this yet. One can adapt this theory, of course, to describe the heats of dilution, but this involves making some sort of assumption about the way in which the heat value per step will change also. To keep it as simple as possible, I assumed one heat for the first step, and a constant decrease in heat for all subsequent steps, and further assumed that the entropy terms were the same for all steps, so that the heat change per step decreased by $RT\ln k$. In this way, I could get a pretty satisfactory account of the lithium chloride and bromide data. I could not do any good at all for sodium hydroxide for which the heats at room temperature are very odd—they turn down, become negative, then go very positive again. That sort of complex behavior does not seem to be well accounted for at all. I think it is due in physical terms to the idea that one is getting some ion association at modest concentrations and later on the stripping off of hydration is contributing more to the heat of dilution. But with lithium chloride and bromide, I could get the correct behavior with a sensible value for the heat parameter up to 18 m or so (the maximum concentration for which heat data are available), but with errors up to 5%. However, considering how large the heats are (they go up to about 2 kcal-mole^{-1}), this is quite encouraging. For sodium chloride at room temperature, the heat of dilution is of the opposite sign to the Debye term. I cannot do anything with that unless I make a very unlikely assumption about the sign of the heat in this process. I would expect these additions of water molecules would have to be exothermic. To account for the sodium chloride results, they would have to be made endothermic and an improbably high entropy term would have to be included to get any attachment of the water molecules at all. I cannot do much with HCl either. The answer is that we have only partial success with the heat data unless a number of rather vague assumptions are made which only amount to curve fitting.

Dr. W. L. Marshall (*Oak Ridge National Laboratory*). I agree with Professor Franks that the fits look extremely impressive. I would like to know whether your parameters represent the maximum of solvation at the limit of infinite dilution.

Professor Stokes. No, the model represents the maximum number that can be put on the sites available, which may not all be filled even at infinite dilution.

Dr. Marshall. You haven't assumed any type of association. Could you say something about that?

Professor Stokes. It is correct that we assumed no ion association. I think probably there should be since there is ion association in many of these systems. The heat data indicate it particularly, as for instance the sodium chloride data. For sodium chloride the heat of dilution is negative at room temperature; as one raises the temperature to 100°C, it becomes highly positive—exactly as the lithium chloride result at room temperature. Sodium chloride is probably completely ionized at 100°C, but not at room temperature. If there is some ion pairing, it would have a fairly substantial heat effect. I could put this in, of course, but it would just be another parameter, and I think this sort of treatment only means something if one keeps the parameters to a sensible minimum.

Professor R. M. Fuoss (*Yale University*). At those concentrations, it would be impossible to assign a unique partner to any given ion; therefore, ion association theory is irrelevant.

Professor Stokes. That is probably correct, but it is in the intermediate range where the trouble arises, e.g., the heat of dilution of sodium hydroxide is negative in the region 2 to 4 m but becomes positive again at higher concentrations.

Professor H. L. Friedman (*State University of New York, Stony Brook*). Your equations would give the characteristic behavior of Harned systems (common-ion mixtures), but only with the sum of the Harned coefficients equal to zero. I think another coefficient is needed for ion pairing because there must be something in the mixture that is not in the end solutions.

Professor Stokes. If the water activity changes, all those equilibria are altered, and a complicated behavior results. I have not tried to do this yet, but, in principle, by upsetting those equilibria, one can make all sorts of things happen.

Professor Friedman. But one cannot find a new interaction in the mixture that is not present in the end solutions, according to the equations you propose.

Professor Stokes. No, I suppose one would not.

Dr. F. H. Stillinger (*Bell Laboratories, New Jersey*). I don't understand why the anions were not allowed to become hydrated. The interaction between water molecules and anions is hardly negligible. I thought, in the case of hydroxide ions, that this interaction would be particularly powerful in view of the possibility of hydrogen-bonded formation.

Professor Stokes. I think with sodium hydroxide the hydration is mainly on the hydroxide. We have formally put the hydration on the cation, but one could put it on the anion and still get the same answer.

Dr. Stillinger. But, the point is, you were able to fit the osmotic-coefficient curves for extremely high concentrations very well by making what I consider to be an unphysical or unchemical approximation.

Professor Stokes. I can say instead that the hydration is all on the anion, and it would not alter any of the calculations.

Professor H. S. Frank (*University of Pittsburgh*). No, but in that case you are in trouble, because with sodium chloride you have said it is all on the sodium, and unless sodium is very different in sodium hydroxide, you might have a total hydration number of 10.

Dr. Stillinger. Certainly one cannot alter the inherent volume-fraction statistics too drastically. Just the total number of unbonded molecules present is relevant. You have done volume-fraction statistics, though perhaps not identified as such, but the sign in your version appears in a totally prejudiced way as that of ions of just one type.

Professor Stokes. When one works out the expression for the water activity, the other species only come in as sums of concentrations or sums of concentrations times volumes, and it really wouldn't matter how the hydration was assigned between anion and cation.

Professor Fuoss. Does that mean if one assigns some solvation to both ions, the only change is that there are fewer on the sodiums and more on the chlorides?

Professor Stokes. Essentially yes. I do not really physically regard these water molecules as strict partners of one ion. It is rather a matter of sites in the concentrated solution where water molecules can be suitably fitted into the electric fields then prevailing.

Professor K. S. Pitzer (*University of California, Berkeley*). It seems to me that a water molecule in most concentrated solutions is going to have a positive ion on one side and a negative ion on the other. It is really almost a semantic question to which ion one assigns it. I would like to emphasize that one of the most important things said here relates to what is on the blackboard. Namely, the comments that in these concentrated solutions the Debye–Hückel type of coulomb term is still only a relatively small but significant perturbation is one key point. Then one can think about electrolytes more or less as one thinks about nonelectrolytes. Let me make one modest suggestion. In nonelectrolyte solutions we talked a great deal about these same sorts of ideas and models, and we ought to look back at those discussions and see which general theory was most successful and that can be transferred to the concentrated electrolyte problem. That is a line for future progress.

Dr. G. J. Safford (*Union Carbide Corporation, New York*). In recent neutron spectroscopic studies we have been investigating concentrated ionic solutions. At these high concentrations, the exchange times of individual molecules exceed the neutron time scale. Hence, the remnant diffusional broadening left appears to experimentally correlate with motions of hydrated ions. In particular, if the anion basicity or the proton acceptor strength of the anion is increased, it is found that a very rapid decrease occurs in the self-diffusion coefficient associated with motions of hydrated cations as well as a distortion of the frequencies characteristic of cation–water complexes. It thus appears possible that one does not have to consider just simple anion–cation direct or indirect pairing, but there actually may be a bridging or cooperative effect in the system where both anions and water molecules are shared between cations.

Professor Stokes. I think that is highly likely and supports the idea that water molecules are perhaps better regarded as being on the sites than on the ions.

Professor J. B. Hyne (*University of Calgary, Canada*). I was particularly interested in the exceptionally high value of K for perchloric acid which you said you could not explain. The HCl value of 88.9 and $HClO_4$ at 2527 would normally lead one to conclude that the difference was due to changing from chloride to perchlorate and not in the proton. The affinity of a perchloric acid entity for the first molecule of water must then be exceedingly high. If that is the case, and there would appear to be no other explanation for it at the moment, then one should perhaps reexamine the role of the perchlorate ion in determining water structure.

Professor Stokes. The equilibrium constants given are those which fit the data for HCl and $HClO_4$, but I do not feel satisfied with the equations for these acids. One should probably modify the equation to start with H_3O^+ rather than the bare proton, which is what the equations in their present form imply.

Professor Pitzer. Is part of the picture that you are making a transfer of the protons from the chloride to water in one case and from the perchlorate to water in the other?

Professor Stokes. The value of K changes by a factor of about 30, which is about 3 or 4 times RT in the free energy.

Professor Pitzer. Yes. If you think of the difference in pulling protons off of the perchlorate as compared to a chloride that could easily be different by quite an amount.

Dr. L. D. Hansen (*Brigham Young University*). I hate to bring a theory down out of the clouds, but if one considers for a moment the descriptive chemistry of perchloric acid and HCl, I think the results are perfectly reasonable. Perchloric acid exists as a monohydrate and cannot be removed from that monohydrate without destroying it, while HCl is well known as an anhydrous covalent molecule. It is well known that HCl and water have different affinities than perchloric acid. I have a few comments I would like to make about the temperature derivative and the discussion about the heats of dilution. You claimed that putting water on the ion should be exothermic, but, as I view it, the water molecule is being taken off the ion as more concentrated solutions are reached and being replaced with an anion. Consequently, both endothermic and exothermic heats can be explained.

Professor Stokes. The process is involved in very concentrated solutions. I have not followed the heat calculations very far. I have only done them for a few salts, and I get the feeling that I should use exothermic enthalpy values for the attachment of each water molecule, but perhaps I shouldn't.

Dr. Hansen. The other comment I was going to make was about the entropies of these same processes. They will change as one goes through the series of adding water molecules, simply because the symmetry around the metal ion, or whatever it is, will change in the process. This is a calculable thing if we make a guess at what the symmetry is.

Professor Stokes. It is a calculable thing in dilute solutions where one has freedom to rotate, and so on. I am not so sure how calculable it is at the high concentrations we are talking about.

Dr. Walrafen. Are you saying one could not obtain $HClO_4$ concentrations above those corresponding·to the monohydrates?

Professor Stokes. You can, but it is dangerous..

Dr. Walrafen. I agree, but French scientists have conducted studies of $HClO_4$–Cl_2O_7 mixtures.

Dr. Hansen. It is true one can get there, but the common stopping point— even if one wants to live just a little bit dangerously—is the monohydrate of perchloric acid, whereas HCl is very easy to get anhydrous.

Dr. M. Falk (*National Research Council, Halifax, Canada*). I would like to add that the "monohydrate" of perchloric acid is really hydroxium perchlorate, $H_3O^+ClO_4^-$. It is an ionic compound.

Professor D. G. Miller (*University of California, Livermore*). About 20 years ago, Huggins [*J. Phys. Chem.* **52**, 248 (1948)] published a paper about the statistics to use with solutions. The conclusion was that mole-fraction statistics are approximately right for dilute solutions of mixed spheres even with different sized spheres. Volume-fraction statistics are more appropriate to flexible chains. Shouldn't one then use mole-fraction statistics?

Professor Stokes. Yes, that was more or less a guess. I have looked recently at this question. Using the model system of hard spheres of different sizes for which Lebowitz has given the equation of state, I calculated the entropy of mixing. If there is any difference in size of spherical particles, the entropy of mixing is always more positive than the ideal mole-fraction-statistics value, and it may in special cases agree with the volume-fraction-statistics value; but if it does, it is an accident. It always goes in that direction. However, whether it agrees exactly with the volume-fraction statistics depends on what sort of initial densities one assumes for the two unmixed hard-sphere liquids.

Professor J. Padova (*Israel Atomic Energy Commission, Yavne, Israel*).

I would like to say that the fitting looks impressive, but I have been wondering about one point. The only parameter that appears in the last relation is actually the h of the hydrated ion. If one looks upon compressibility as giving the solvation number at infinite dilution, as well as the change of the solvation number with concentration, it would seem that one could use that relation to give the equilibrium constant. The concentration dependence of the hydration number has been shown to give, in most of the cases, the hydration number of the crystallo-hydrates in various solvents.

Professor Stokes. Yes, but there are several solid hydrates of lithium chloride, for instance, and I wonder which one is aimed for.

Professor Padova. The solvation number of the solid hydrates should not be used but the primary solvation number at infinite dilution and its change with concentration that is given by either compressibility or partial molal volume measurements to calculate the osmotic coefficient in water.

Professor Stokes. The question with the hydration number is whether different methods of measuring give the same average. It is not clear what the relation is between this sort of thermodynamic average hydration number and one determined from compressibility.

Professor Padova. I think you stated that the final hydration number is given by the fit of the activity coefficients which usually agrees with that given by either compressibility or partial volume measurements.

Professor Stokes. The question is whether one can make this rigorous division into primary and secondary, and so on, and especially in the concentration ranges we are discussing.

Dr. Stillinger. In calculating the coulombic parts of the osmotic-coefficient quantity, one is forced to commit oneself to a specific value for the dielectric constant. Can you tell me what you have done in these very concentrated solutions?

Professor Stokes. I have simply taken the bulk dielectric constant of water from the start. I know this is not good, but if there is any solvent intervening between the ions, the hydrated ions can be regarded as an entity in themselves, and the field starts from the surface of the hydrated ion. I know that this dielectric-constant problem is present. There are beginning to be some figures for dielectric constants at very high concentrations—I have seen some for lithium chloride up to about 10 m. They have not fallen as much as expected—it is still somewhere around 30. It does seem to me that a good deal of that drop must be due to the fact that one is placing into the solution the ions which have low dielectric constant anyway. The internal dielectric constant of the ion is influencing that bulk value.

Dr. Stillinger. Yes, but in the most concentrated solutions (lithium halides) you have claimed by hindsight that essentially all of the water is bound up in hydration sites. I do not care whether they are bound to cations or anions. The point is, they are bound and, presumably, not sufficiently free to rotate in an exciting electric field to give rise to anything like a dielectric constant of 75 or 80. The rather small values quoted for the coulombic contribution to these thermodynamic functions that have appeared on your slide therefore may well be misleading by a factor of 3 or 4.

Professor Stokes. That is quite possible. Even if they were four times larger, they still would not be large compared to the observed ϕ's, however. I agree one might be able to fit with a different ion size by using a more realistic dielectric constant. The ion size I used is only a rough average, of course. It will be changing with concentration too. I agree that the approximation there is a bit crude.

Professor J. H. Gibbs (*Brown University*). I gather that these residual water molecules are assumed not to change in nature as the concentration of the ion is changed.

Professor Stokes. That is characteristic of this sort of chemical-equilibrium picture. One makes a too-rigid division between bound and unbound molecules.

Professor Gibbs. If the water molecules are associated with each other, as surely they are, presumably the correction for this is absorbed in one of your parameters, provided the nature of that association is not changing as one changed the concentration of ions. If there is a rather tightly organized tetrahedral network in water, as many believe, surely as the concentrations go up that network must be enormously changed, a result that ought to show up in a failure of your calculation.

Professor Stokes. One might reasonably expect so, but the astonishing thing about these very concentrated solutions is, of course, that they are liquids. They are very like water (rather more viscous, certainly in most cases), but it is amazing that when there is one ion per water molecule, there will still be what is recognizably an aqueous solution. It is clearly based on water, and it looks like water and behaves like water.

Dr. W. T. Lindsay (*Westinghouse Electric Corporation, Pittsburgh*). It seems to me that a serious test of whether there is a physical basis for this approach would be to look at the temperature dependence of the osmotic coefficients along with the concentration dependence. In the case of sodium chloride solutions, for which data are available to 300°C for the complete concentration range up to saturation, the differences between experimental osmotic coefficients and Debye–Hückel osmotic coefficients (with adjustable ion size parameters) decrease considerably with increasing temperature. It appears that your approach applied to these data would lead to the conclusion that hydration decreases substantially as the temperature increases, a conclusion that seems physically unrealistic.

Professor Stokes. It is physically exactly what one would expect if one obtained a heat evolved on hydration.

Dr. Lindsay. I would think that hydration should tend, if anything, to be even more energetic and effective at the higher temperatures because the dielectric constant of water falls markedly and the temperature–dielectric constant product εT also decreases with increasing temperature.

Professor Frank. When the water–water interaction is decreased, hydration is exceedingly potent in supercritical systems, as Professor Franck will tell us. Waters stick more strongly to ions in spite of the high temperature because water has no competition from other water.

Dr. Lindsay. If one accepts that a much lesser degree of hydration is unrealistic in high-temperature liquid solutions, one then has to invoke ion pairing to account for the fall of osmotic coefficients to levels comparable to Debye–Hückel osmotic coefficients. This would add still another parameter to the several you already have, so that economy of parameters is not an advantage. By contrast, the data from 75° to 300°C can be fitted nicely by the Debye–Hückel expression, with added terms, requiring only three parameters, namely, the ion-size parameter and the coefficients b and c of the terms bm and cm^2. Both b and c become very small at high temperatures.

Professor Stokes. It is not true that they can be fitted at these concentrations, though the Debye–Hückel parameter with a linear term will fit very nicely up to a few molar.

Dr. Lindsay. At 300°C one can fit the data up to saturation (over 10 m) with just the ion-size parameter because the b and c coefficients both become essentially zero. This is a coincidence probably, but it is a fact that Debye–Hückel osmotic coefficients fit the sodium chloride data better over a broader range of concentration the higher the temperature.

Professor R. L. Kay (*Carnegie-Mellon University, Pittsburgh*). I must end this discussion at this point, but I would like to remark that it only shows how little we really know about hydration when two of the world's experts on hydration cannot agree even as to the sign of the temperature dependence.

Ionic Association in Hydrogen-Bonding Solvents[1]

D. Fennell Evans[2]
and Sister Mary A. Matesich[3]

Received August 18, 1972

Studies on ionic association in hydrogen-bonding solvents including water, aliphatic alcohols, fluorinated alcohols, formamide, ethylene glycol, and propanol-acetone mixtures are compared. Data were derived from measurements of conductivity, ultrasonic absorption, and viscosity of electrolyte solutions and diffusion of nonelectrolytes. It is concluded that electrolyte solvents can be divided into three categories according to whether they solvate effectively cations only, anions only, or both cations and anions. Water and most other hydrogen-bonding solvents constitute the third class, a group in which patterns of ionic aggregation are very similar. Fluorinated alcohols belong to the second class, showing significantly different ion-pairing behavior.

KEY WORDS: Ion association; hydrogen-bonded solvents; non-aqueous solvents; conductance; diffusion; ultrasonics; electrolyte solutions; fluorinated alcohols; solvent mixtures; water structure.

1. INTRODUCTION

As Henry Frank pointed out in his pioneering work,[1,2] many of the properties of aqueous solutions can be understood only in terms of the three-dimensional structure of liquid water and the modification of this structure by solutes. This focus on the perturbation of solvent structure as a major factor in the behavior of solutions has stimulated many fruitful investigations of aqueous solutions. The interpretation of this complex behavior is clear in broad outline and has provided much of the general framework for understanding reaction

[1] This paper was presented at the symposium, "The Physical Chemistry of Aqueous Systems," held at the University of Pittsburgh, Pittsburgh, Pennsylvania, June 12–14, 1972, in honor of the 70th birthday of Professor H. S. Frank.
[2] Chemical Engineering Department, Carnegie-Mellon University, Pittsburgh, Pennsylvania 15213.
[3] Department of Chemistry, Ohio Dominican College, Columbus, Ohio 43219.

kinetics in mixed solvents and interactions in biological systems as well as physicochemical properties of aqueous solutions.

Until recently, few systematic studies of electrolyte solutions in other hydrogen-bonding solvents have been carried out. In common with water, these solvents possess both acidic and basic solvation sites. Furthermore, the directional nature of the hydrogen bond leads to some internal structure in these liquids. In water, the favored tetrahedral orientation of hydrogen bonds leads to the formation of three-dimensional networks. In the monohydric alcohols, the presence of only one acidic proton per molecule limits structuring to linear chains. Unsubstituted amides occupy an intermediate position between alcohols and water. Although these molecules are capable of forming multiple hydrogen bonds, a two-dimensional sheet-like structure occurs in the crystal, and the molecular geometry is not conducive to forming tetrahedral, three-dimensional arrays. Polyhydric alcohols like glycerol and ethylene glycol are also capable of multiple hydrogen bonding, and the nature of the intermolecular association in these liquids is not well understood. The existence of such a range of order caused by hydrogen bonding suggests that electrolytes in these solvents may show features similar to those observed in water, although perhaps to a lesser degree.

The presence of both acidic and basic groups in water and the other hydrogen-bonding solvents provides effective solvation for both cations and anions. These protic solvents have long been considered a different class from dipolar aprotic solvents which interact effectively with cations through their lone pairs but have no sharply defined positive sites for interaction with anions. There are strong indications that polyfluorinated alcohols form a third solvent class in which strong solvation of anions by hydrogen bonding occurs but the reduced basicity of oxygen causes very poor solvation of cations. What results is a spectrum from good cation solvents through good anion solvents with water and the hydrogen alcohols showing both characteristics. This is summarized in Table I.

We have carried out a series of studies on the conductivity, ultrasonic absorption, and viscosity of electrolyte solutions and the diffusion of non-electrolytes in these solvents. Although the picture which emerges is not

Table I. Typical Solvents That Exhibit a Spectrum of Interactions with Ions

Strong interactions with cations	Strong interactions with cations and anions	Strong interactions with anions
R_2CO	ROH	CF_3CH_2OH
RCN	$RCONH_2$	$(CF_3)_2CHOH$
RNO_2	H_2O	

complete, a number of interesting observations shed further light on the complex behavior caused by the interplay between solvating properties and structural characteristics in hydrogen-bonding solvents.

2. ION PAIR FORMATION IN PROTIC AND APROTIC SOLVENTS

The behavior of electrolytes in aqueous solution has been discussed in terms of structural effects, involving the perturbation of solvent structure by ions, and in terms of ionic aggregation. Because of the high dielectric constant of water, it is difficult to distinguish between the relative importance of these two factors. One approach is to study ionic association in hydrogen-bonding solvents of lower dielectric constant. A comparison of electrolyte behavior in these solvents to that in water has identified those features which water shares with all hydrogen-bonding solvents, and this is developed in detail below. All association constants were obtained from the concentration dependence of the electrolyte conductance.

One would expect that the pattern of ionic association would be affected not only by solvation but also by structural effects. Simple theories of ionic association are based on the model of charged hard spheres in a dielectric continuum. Since solvation and structuring come about precisely because the solvent is not a dielectric continuum, these effects should result in association constants that differ from theoretical predictions.

The two theories which have been used most extensively are due to Fuoss[3] [Eq. (1)] and Bjerrum[4] [Eq. (2)]:

$$K_F = (4\pi N/3000) \, a_F^3 \exp(e^2/a_F \varepsilon kT) \tag{1}$$

$$K_B = (4\pi N/1000)(e^2/\varepsilon kT)^3 \int_2^d e^t \, t^{-4} \, dt \tag{2}$$

where K_F and K_B refer to the equilibrium constants predicted by the Fuoss and Bjerrum models, respectively, for the ion pairing process of Eq. (3):

$$M^+ + X^- \rightleftharpoons MX \tag{3}$$

Although both of these equations are based on the solvent continuum model, in Eq. (1) it is assumed that all the ion pairs have the same interionic separation. In Eq. (2) ions are counted as paired over a range of distances of approach from some minimum a_B to a limit d which is usually taken as equal to q [Eq. (4)]:

$$d = q = e^2/2\varepsilon kT \tag{4}$$

This distance corresponds to the separation at which the coulomb attraction is balanced by energy of random thermal motion of the ions. Both of these theories enable one to calculate the absolute magnitude of the association constant, its dependence on ion size, and its dependence on solvent dielectric constant.

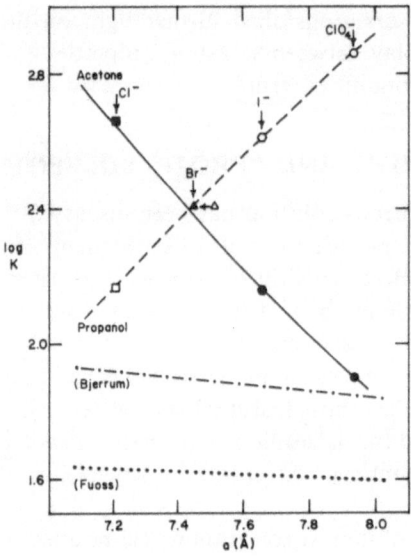

Fig. 1. $\text{Log} K_A$ values for Bu_4N^+ salts in the isodielectric solvents acetone and propanol ($\square$ Cl; $\triangle$ Br; $\diamond$ I; $\circ$ ClO_4) along with predictions of the Bjerrum (dashed line) and Fuoss (dotted line) equations.

Figure 1 compares association constants for tetrabutylammonium chloride, bromide, iodide, and perchlorate in the isodielectric solvents acetone[5] and 1-propanol[6] to those calculated from Eqs. (1) and (2). It is immediately obvious that the experimental values are much larger in magnitude than the calculated values. Furthermore, the very different patterns in the two isodielectric solvents point to significant differences in solvation or structural factors. In acetone the association constant decreases as anionic radius increases, in qualitative accord with theory, whereas in propanol the opposite dependence on anion size is seen. This behavior is generally typical of aprotic and hydrogen-bonding solvents, respectively, and is explored below.

We consider first the dependence of K_A on dielectric constant. Equation (1) predicts that a plot of $\log K_A$ vs $1/\varepsilon$ for any given salt should be a straight line. With Eq. (2) some curvature is predicted, but all solvents should give the same theoretical curve. Such a plot is shown in Fig. 2 for the tetrabutylammonium bromide and iodide in the primary alcohols methanol[7] through pentanol, in the ketones propanone[5] and butanone,[8] and in 2-propanol.[9] Within a good approximation, a straight line is obtained for each salt in EtOH,[6] PrOH, BuOH,[10] and PeOH.[10] However, neither the ketones nor methanol and 2-propanol lie on the same line. Association constants for the two ketones determine a line of essentially the same slope as the alcohol line, indicating that in each series of solvents the association constant changes in the same way with dielectric constant. The deviation exhibited by methanol, $\varepsilon = 32.6$, parallels the behavior in other systems such as mixed solvents as the dielectric constant exceeds 30 and can be accounted for by Eq. (2). 2-Propanol deviates from the primary alcohols in that the association constants are two

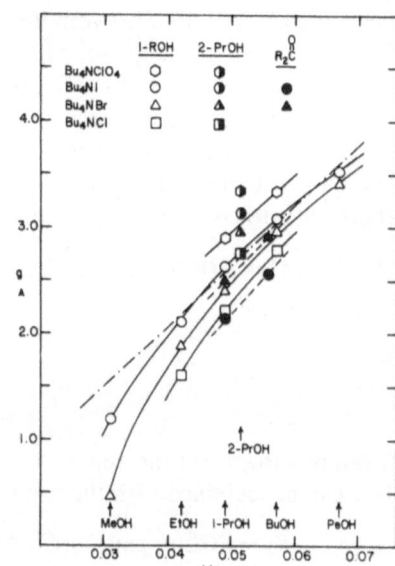

Fig. 2. The variation of $\log K_A$ with $(1/\varepsilon)$ for Bu_4N^+ salts in alcohols and ketones.

to three times larger than one would expect. Thus it would appear that a given salt in a series of similar alcohols or ketones follows the simple exponential law predicted by electrostatics but that a relatively small change in solvent molecular structure leads to deviations.

Although not all the alcohols fall on the same line for $\log K_A$ vs $1/\varepsilon$, all show the same peculiar dependence of K_A on anion size shown in Fig. 1 for propanol. One can suggest that this is due to solvation of the anions through hydrogen bonding. Since smaller anions interact more strongly with OH groups,[11] the extent of solvation would be on the order $Cl^- > Br^- > I^- > ClO_4^-$. Thus chloride would be effectively larger than perchlorate, and, consequently, according to Eq. (1) or Eq. (2), a would be larger and K_A smaller for the chloride salts. However, this cannot account for the fact that Bu_4NClO_4 is 10 times more highly associated in 1-propanol than in acetone. If perchlorate ion is not highly solvated by propanol, it should be effectively the same size in both solvents, and the extent of ion pairing should be the same. Thus, arguments based on differential anion solvation alone are incapable of explaining the observed behavior.

An additional factor which may be considered is whether the effective dielectric constant is in fact the same in acetone and propanol. The dielectric constants of hydrogen-bonding solvents are unusually large given their gas-phase dipole moments.[12] This has been explained in terms of intermolecular hydrogen bonding. However, in the immediate vicinity of an ion, the normal hydrogen bonding between solvent molecules is diminished because of the orientation of alcohol dipoles by the ion. As a result, the local dielectric constant may well be lower than the bulk dielectric constant, and the coulomb attraction between ions is enhanced, causing greater association. This argument accounts for the increased magnitude of the association constants in propanol but not for the peculiar dependence on anions.

This complex behavior cannot then be accounted for by continuum theories or by simple solvation and must take the discrete molecular nature of the solvent and its dynamic interaction with ions into account. The simplest approach is to consider ion pairing as a two-step association process [Eq. (5)]:

$$(ROH)_m M^+ + A(ROH)_n^- \overset{K_{12}}{\rightleftharpoons} M(ROH)A(ROH)_{m+n-1}$$

$$I \hspace{5cm} II$$

$$\overset{K_{23}}{\rightleftharpoons} MA(ROH)_{m+n-1} + ROH$$

$$III \hspace{2cm} (5)$$

It can be shown that the apparent association constant K_{Σ} is given by Eq. (6). This can be identified with the conductance association

$$K_{\Sigma} = \Sigma C(\text{ion pairs})/[(C_{M+})(C_{A-})] = K_{12}\{1 + K_{23}/[ROH]\} \hspace{1cm} (6)$$

constant K_A with the assumption that neither type of ion pair is conducting.

One can estimate the value of K_{12} from Eq. (1) since Eq. (1) results from a consideration of diffusion-controlled formation and dissociation of encounter ion pairs. Species II is the encounter complex of solvated ions I. A judicious choice for a_F of 8 to 12 Å is reasonable for such a species, and the value of K_F calculated for solvents of intermediate dielectric constant is relatively insensitive to the choice of a_F in this range.[10] Setting K_F equal to K_{12} and K_A equal to K_{Σ} makes it possible to calculate the K_{23} values given in Table II.

Also included in Table II are relative acidities of the alcohols. The primary alcohols are 8 to 12 times more acidic than 2-propanol and consequently would be expected to hydrogen-bond more effectively with anions. Thus, less additional stabilization would result from converting a solvated ion pair to a contact ion pair in the primary alcohols. K_{23} values for 2-propanol

Table II. Estimated Values of K_{23} for Tetrabutylammonium Salts from Eq. (6)

Salt	EtOH	1-PrOH	2-PrOH	1-BuOH
Bu$_4$NCl	9.5	36	180	94
Bu$_4$NBr	34	75	240	130
Bu$_4$NI	67	120	350	190
Bu$_4$NClO$_4$	—	240	540	360
Acidity, $K_e{}^a$	0.95	0.5	0.076	0.6

[a] J. Hine and M. Hine, *J. Am. Chem. Soc.* **74**, 5266 (1952).

should then be larger than for the primary alcohols. Since the ease of desolvation of anions is inversely related to their radii and directly related to the magnitude of K_{23}, K_{23} should be on the order $ClO_4^- > I^- > Br^- > Cl^-$, as observed. Thus the multiple-step association process qualitatively accounts for the association behavior in the alcohols and for the larger values of K_A in 2-propanol.

The evidence given above is all thermodynamic in nature. The verification of such a mechanism requires kinetic data which allow the determination of the rate constants interrelating the various species. Such evidence is provided by ultrasonic studies on these systems, as discussed in the next section.

In the course of these investigations we have accumulated a large body of data on electrolyte mobilities in the alcohols. In the past, much of the discussion of such data has been in terms of positive and negative deviations from Stokes' law,[13] as caused by structural, solvation, and dielectric friction effects (Zwanzig equation).[14, 15] In the final analysis, all such discussion assumes implicitly the validity of Stokes' law.

We have accordingly tested the validity of Stokes' law by studying the diffusion of nonelectrolytes in alcohols where solvation and dielectric effects should be absent and structural effects at a minimum. Diffusion constants for tetramethyltin and tetrabutyltin in a series of alcohols[16] have been multiplied by F^2/RT and are shown in Fig. 3 along with the Walden products for the corresponding tetraalkylammonium ions. Also shown as a dashed line is Stokes' law.

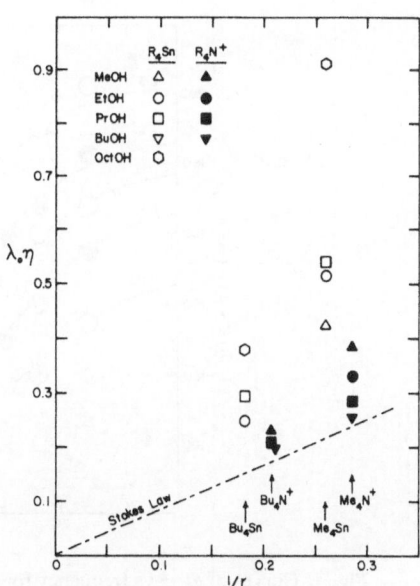

Fig. 3. Comparison of Walden products for tetraalkylammonium ions and tetra-alkyltins in the alcohols. (B. Lamartine and D. F. Evans, unpublished data.)

What is surprising is that the deviations for nonelectrolytes are as extensive as for electrolytes. This unexpected result demands extensive study of the
diffusion of nonelectrolytes in a wide series of solvents.

3. THE MECHANISM OF ION PAIRING

Concentration-dependent changes in the ultrasonic absorption of
electrolyte solutions provide valuable insight into ion-pairing dynamics. Three
situations can be suggested which would lead to different kinds of ultrasonic
behavior: (1) Only one kind of ion pair is formed; (2) solvent-separated ion
pairs form upon encounter and are converted to contact ion pairs at a rate
comparable to that of formation of species II, Eq. (5); (3) the conversion of
solvent-separated ion pairs to contact ion pairs occurs at a rate appreciably
different from the rate of formation of the former. 2-Propanol was chosen as
a solvent for detailed investigation[17] because the relaxation effects were
found to be large in amplitude and centrally located in the frequency range
utilized.

For solutions of Bu_4NI at various concentrations in 2-propanol, plots
of α/f^2 vs f are presented in Fig. 4, where α is the ultrasonic absorption
coefficient and f the frequency. These data are typical of all the systems studied
and are characteristic of a single relaxation process. The solid lines have been
calculated for such a process using Eq. (7):

$$\alpha/f^2 = A/[1 + (f/f_r)^2] + B \tag{7}$$

where f_r is the relaxation frequency, A is the amplitude of the relaxation process,
and B is the high-frequency value of α/f^2.

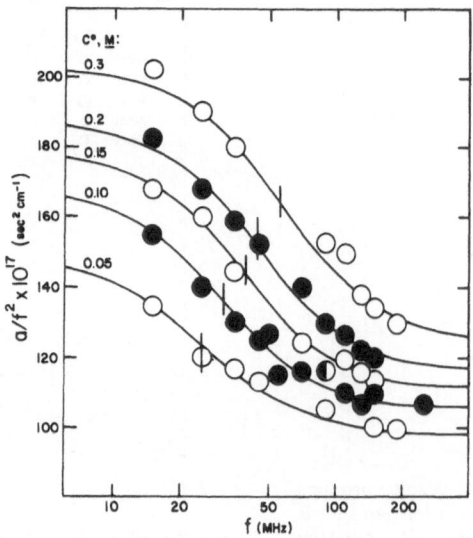

Fig. 4. Curves of α/f^2 vs frequency for Bu_4NI solutions in 2-propanol at 25°C.

The relaxation frequency varies with concentration in a manner which depends on the mechanism responsible for the process. For the systems we have studied, all the data are consistent with the two-step diffusion–desolvation mechanism in which the desolvation step II–III [Eq. (5)] is rapid compared to the diffusional step I–II. For this case, the relaxation frequency is given by Eq. (8):

$$f_r = (1/2\pi)[k_{12}f(c) + k_{21}/(1 + K_{23})] \qquad (8)$$

where k_{12} is the rate of conversion of I to II, k_{21} is the reverse rate, and $f(c)$ is given by $2\sigma C^\circ \gamma_\pm^2[1 + d\ln\gamma_\pm/d\ln\sigma]$, with σ the degree of dissociation and C° the stoichiometric concentration of salt.

A plot of f vs $f(c)$ for Bu_4NI in 2-propanol is shown in Fig. 5, and according to Eq. (8) the slope gives a value of 2.5×10^{10} M^{-1}-sec^{-1} for k_{12} and an intercept near zero. Debye[18] [Eq. (9)] and Eigen[19] [Eq. (10)] have derived equations predicting k_{12} and k_{21} from diffusion coefficients for M^+ and A^-:

$$k_{12} = [4(D_M + D_A)NZ_M Z_A e^2/1000\,kT][\exp(Z_M Z_A e^2/kT\varepsilon a) - 1] \qquad (9)$$

$$k_{21} = [3(D_M + D_A)Z_M Z_A e^2/\varepsilon kTa^3][1 - \exp(-Z_M Z_A e^2/\varepsilon kTa)] \qquad (10)$$

The ratio of k_{12} to k_{21} gives K_{12} in a form identical to Eq. (1).

The predicted values for k_{12} and k_{21} for Bu_4NI in 2-propanol are 1.6×10^{10} M^{-1}-sec^{-1} and 2.7×10^8 sec^{-1}, respectively. The first agrees satisfactorily with the value of 2.5×10^{10} calculated from the slope of Fig. 5 considering the precision of the ultrasonic measurements and the uncertainty in calculating activity coefficients. According to Eq. (7), the intercept should be $k_{21}/(1 + K_{23})2\pi$. Using the predicted value for k_{21} and the value of 350

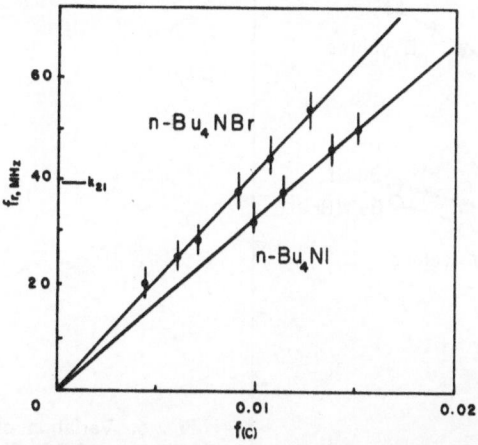

Fig. 5. Dependence of relaxation frequency f_r on concentration function defined in Eq. (8) for Bu_4NI solutions in 2-propanol at 25°C.

for K_{23} taken from Table II, the predicted intercept is 0.13 MHz, in agreement with the zero intercept found in Fig. 5.

Neither of the other mechanisms is consistent with the data. Mechanism 1 would require that the intercept of Fig. 5 equal $k_{12}/2\pi$ or approximately 40 MHz. Not only is the intercept much smaller than that, but so are several of the experimental points. On the other hand, if the diffusion and desolvation steps had comparable rates, a second relaxation frequency should be observed, provided that there are suitable volume and heat changes associated with this step. No such relaxations were found at low frequencies down to 60 kHz using the spherical resonator technique. Brillouin scattering results at high frequencies revealed no excess absorption attributable to a second relaxation process.[20]

Blandamer[21] has suggested that part of the ultrasonic absorption of tetraalkylammonium ions in solution is caused by internal rotation about carbon–carbon bonds, since 3,3-diethylpentane "neat" and in hexane solution shows an absorption in this region with a relaxation frequency near 40 MHz. However, we have several lines of evidence showing that this compound is not a good model for R_4N^+ ions and establishing that the absorption we are observing does not arise from rotation.[20]

The magnitude and concentration dependence of B [Eq. (7)] exhibit a behavior that seems peculiar to electrolyte solutions in hydrogen-bonding solvents. In Fig. 4 it can be seen that the high-frequency limiting value for α/f^2 increases with concentration for Bu_4NI in 2-PrOH. This also occurs for Bu_4NCl, Bu_4NBr, $LiCl$, and Me_4NCl in this solvent. As shown in Fig. 6 for Bu_4NBr and Bu_4NI, the concentration dependence is linear. This effect

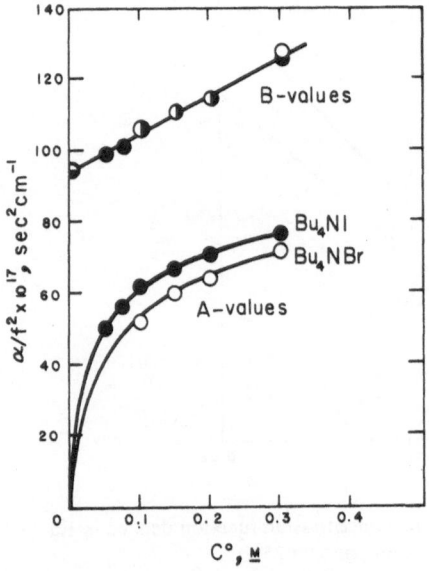

Fig. 6. Variation of ultrasonic parameters A and B for Bu_4NBr (open circles) and Bu_4NI (closed circles) with total concentration.

persists into the high-frequency Brillouin scattering, where the excess absorption of solution over pure solvent remains. These salts show the same behavior in other alcohols. However, for nonelectrolytes such as 2,2-diethylpentane and tributylamine in 2-PrOH, the B values for the solutions are essentially the same as the pure solvent.[20] Furthermore, α/f^2 for Bu_4NBr in acetonitrile and for tetraalkylammonium salts in acetone approaches the value for the pure solvent at high frequencies. However, high B values have also been observed for tetraalkylammonium salts in formamide and for electrolytes in water. High B values thus appear to be exclusively associated with electrolyte solutions in hydrogen-bonding solvents. The origin of this effect is unclear at present, but hydrogen-bonding interactions with anions or modification of the hydrogen-bonded structure of the solvent by the electrolyte are two possibilities. Further investigation of this interesting effect is needed.

In conclusion, the ultrasonic measurements described here clearly confirm the simple two-step mechanism proposed for ionic association in the alcohols. However, some ambiguity remains, as there may be several configurations of solvent-separated ion pairs of comparable energy.[22] When these considerations are carried to the extreme and the size of the solvent molecule is neglected, one has the continuum situation envisioned by Bjerrum. However, the predictions of Bjerrum's theory [Eq. (2)] are too small. At this time it is not possible to merge the quantitative predictions of the continuum and discrete models in any smooth manner.

4. ION PAIRING IN SOLVENT MIXTURES

The radically different ionic-association pattern observed for tetrabutylammonium halides in acetone and propanol, then, seems to be caused by hydrogen-bonding of propanol to anions and the consequences of this interaction on ion-pairing dynamics. A transition from one association pattern to the other must occur as the proportions of the two solvents are changed in mixtures across the composition range. Since a study of this behavior should shed further light on the details of anion solvation, the conductance of these salts was determined in acetone–propanol mixtures.[5]

Although the two solvents are isodielectric, their mixtures are clearly not, as Fig. 7 shows. Also included is a plot of the viscosities. This behavior is typical of mixtures of a hydrogen-bonding solvent with an aprotic solvent and can be rationalized in the following way. The addition of acetone to propanol disrupts self-association in propanol and causes a rapid initial decrease in viscosity. This is paralleled by a drop in dielectric constant. Although pure acetone and 1-propanol are isodielectric, acetone has a dipole moment of 2.9, whereas that of propanol is only 1.7. If the disruption of hydrogen-bonded chains by acetone reduces the enhancement of the dielectric constant by hydrogen bonding, the decrease in dielectric constant of the

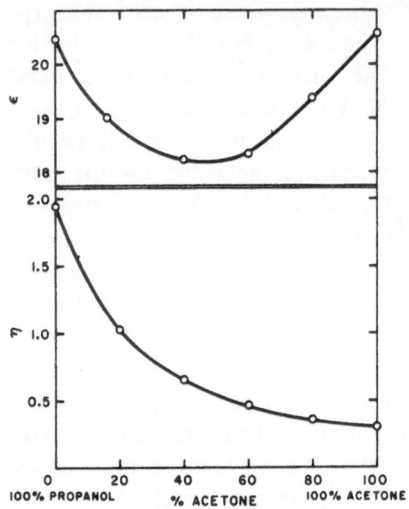

Fig. 7. Viscosity and dielectric constant at 25°C as a function of mole
fraction acetone in 1-propanol.

mixtures up to about 40 mole % acetone is reasonable. Beyond this point the
higher dipole moment of acetone causes the dielectric constant to increase.

As a result of the 10% decrease in dielectric constant at 40% acetone,
K_A would be predicted to increase by a factor of 2. Figure 8 shows that exactly
opposite behavior is observed; K_A for the halides passes through a minimum

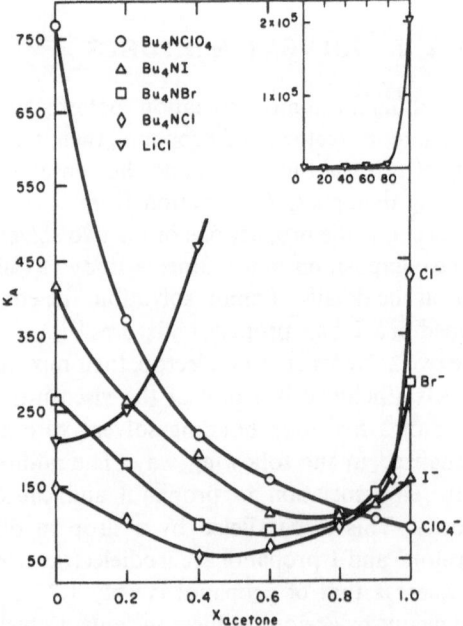

Fig. 8. Association constant at 25°C as a function of mole fraction acetone in 1-propanol.

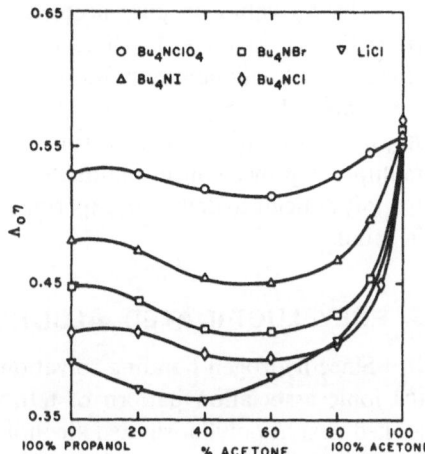

Fig. 9. Walden product at 25°C as a function of mole fraction acetone in 1-propanol.

at this point. Mobility data for these salts shown in Fig. 9 likewise show a minimum near 50% acetone. In pure propanol, the mobilities of the tetra-butylammonium salts are in the order $ClO_4^- > I^- > Br^- > Cl^-$. Heat of transfer data reveal that chloride receives much greater additional stabilization in alcohols than perchlorate,[23] a result attributable to more effective hydrogen bonding to the smaller anion. This is compatible with chloride having the largest hydrodynamic radius of the anions studied, due to increased solvation. In pure acetone, on the other hand, there is virtually no difference in the anionic mobilities. Acetone is a typical aprotic solvent in that it solvates cations strongly but anions only weakly.

The mobility order for the tetrabutylammonium salts observed in pure propanol persists until the mole fraction of acetone reaches approximately 80%. This suggests that the same hydrodynamic entities present in pure propanol exist through this composition range. The salts show nearly parallel behavior in the small variations in Walden product up to 80% acetone, suggesting a common origin. However, the lack of reliable dielectric relaxation times for this system prevents quantitative calculations. The mobility data suggest that anion solvation remains relatively unchanged up to about 80% acetone. If this is the case, the decrease in K_A in the propanol-rich region can only be attributed to an increase in dielectric constant near an ion pair. Such an effect should parallel increase in acetone content since the higher dipole moment of acetone should result in an effectively higher local dielectric constant. This remains the dominant factor until the propanol concentration is so small that anion solvation by hydrogen bonding disappears. At this point, the K_A values rapidly approach those observed in pure acetone. A comparison of Figs. 8 and 9 shows that this explanation is consistent with both mobility and association behavior.

The K_A values in pure acetone are also higher than coulombic theory [Eqs. (1)–(2)] predicts. Darbari and Petrucci[24] have carried out ultrasonic studies of tetraalkylammonium salts in acetone and conclude that a two-step mechanism [Eq. (5)] is also operative in this solvent. The anion dependence of K_{23} in acetone which results from such an analysis is difficult to explain. A multiple-step mechanism undoubtedly occurs in the solvent mixture as well, but it is difficult to devise an experiment which would allow it to be elucidated in detail.

5. POLYFLUORINATED ALCOHOLS

Since hydrogen-bonding solvation of anions seems to be responsible for the ionic association pattern of tetraalkylammonium salts in alcohols, we looked next at polyfluorinated alcohols which have much stronger hydrogen-bonding ability than hydrocarbon alcohols. Results of these studies show that electrolytes in 2,2,2-trifluoroethanol[25] and 1,1,1,3,3,3-hexafluoro-2-propanol[26] behave in a radically different way than in the analogous hydrocarbon alcohols. It is instructive to compare the properties of ethanol and 2,2,2-trifluoroethanol, as presented in Table III. The two solvents have similar

Table III. Properties of Ethanol and 2,2,2-Trifluoroethanol

	Ethanol	2,2,2-Trifluoroethanol
ρ^{25}	0.78511[a]	1.3826[i]
ε	24.33[a]	26.67[b]
D, debye, cyclohexane[c]	1.65	2.03
n^{25}	1.360[b]	1.291[b]
Kirkwood g factor[j]	3.0	3.0
η, cP	1.084[a]	1.78[d]
pK_A, water[e]	15.9	12.37
ΔH form, acetone complex in CCl_4, kcal/mole[f]	−2.90	−3.72
HCl solubility, moles HCl/mole[g]	0.950	0.063
Boiling point, °C	78.4	73.75[i]
Entropy of vaporization, eu	26.8	28.0[i]

[a] Reference 6.
[b] H. C. Eckstrom, J. E. Berger, and L. R. Dawson, *J. Phys. Chem.* **64**, 1458 (1960).
[c] A. Kivinen, J. Murto, and M. Lehtonen, *Suomen Kemistilehti B* **40**, 336 (1967).
[d] J. Murto and E.-L. Heino, *Suomen Kemistilehti B* **39**, 263 (1966).
[e] P. Ballinger and F. A. Long, *J. Am. Chem. Soc.* **81**, 1050 (1959); *ibid.*, **92**, 795 (1960).
[f] A. Kivinen, J. Murto, and L. Kilpi, *Suomen Kemistilehti B* **40**, 301 (1967).
[g] W. Gerrard and E. D. Macklen, *J. Appl. Chem.* (*London*) **9**, 85 (1959).
[h] L. M. Mukherjee and E. Grunwald, *J. Phys. Chem.* **62**, 1311 (1958).
[i] *Trifluoroethanol*, Booklet DC-1254, Pennsalt Chemicals Corp., Philadelphia, Pennsylvania, 1956.
[j] Reference 25.

dielectric constants and dipole moments. There is comparable intermolecular hydrogen bonding, as indicated by the similar Kirkwood correlation factors and entropies of vaporization. 2,2,2-Trifluoroethanol forms a stronger hydrogen bond with acetone, but it is a poorer solvent for HCl. This indicates that the molecule interacts more effectively with lone-pair electrons but less effectively with positive centers than ethanol. Thus one would predict better solvation for anions but poorer solvation for cations in the fluorinated alcohol. Ionic association in trifluoroethanol differs from that in the hydrocarbon alcohols in three respects: the anion dependence, the cation dependence, and the overall magnitude of the association constant.

The anion dependence of K_A in trifluoroethanol is compared to that in ethanol in Fig. 10. In contrast to the other alcohols, in trifluoroethanol there is little if any anion dependence within experimental uncertainty. In terms of the postulated multiple-step mechanism, Eq. (5), this can be explained if only one kind of ion pair predominates. Since anions are more strongly solvated by trifluoroethanol than by ethanol, this enhanced solvation should stabilize the solvent-separated ion pair relative to the contact ion pair. Consequently, K_2 would be smaller in trifluoroethanol, and the observed association constant would approach K_1, which, as calculated from Eq. (1), is not strongly anion-dependent when the cation radius is large.

The association constants of chlorides in ethanol[6, 27, 28] and trifluoroethanol are shown in Fig. 11. In trifluoroethanol, K_A decreases with increasing cation size in accord with the predictions of Eqs. (1) and (2). The complex behavior in ethanol, then, must be caused by solvent–cation interactions which do not occur in the fluorinated alcohol.

This is consistent with the higher cation-solvating ability of ethanol, as indicated by HCl solubility and nucleophilicity[29] (Table III). The small lithium ion appears to be highly solvated in ethanol, and sodium and potassium

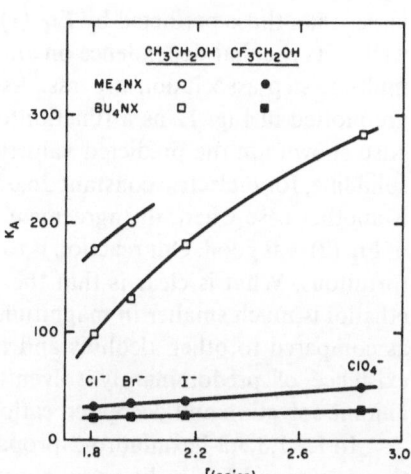

Fig. 10. Anion dependence of K_A for R_4N^+ salts in trifluoroethanol and ethanol at 25°C.

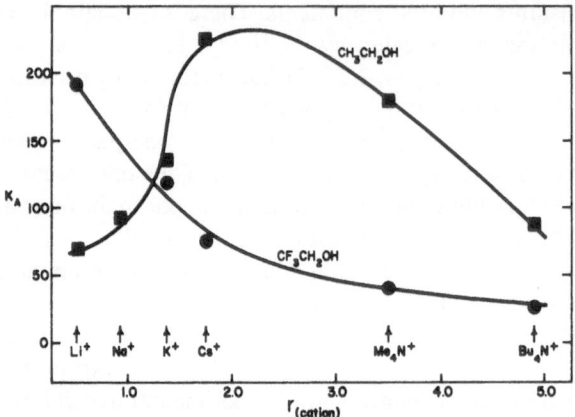

Fig. 11. Cation dependence of K_A for chlorides in trifluoroethanol and ethanol at 25°C.

less highly solvated. By the time ionic radius reaches the size of Cs^+ or Me_4N^+, solvation is no longer important in either alcohol, and K_A decreases with further increase in size. Thus, effective electrostatic radii in ethanol may be in the order

$$Cs^+ < Me_4N^+ < K^+_{(solvated)} < Na^+_{(solvated)} < Bu_4N^+ < Li^+_{(solvated)}$$

Due to the less effective cation solvation in trifluoroethanol, however, the order determined by crystallographic radii, $Li^+ < K^+ < Cs^+ < Me_4N^+ < Bu_4N^+$ is observed. It should be noted that this solvation argument does not explain the anion dependence of K_A discussed above because the crystallographic radii of the halides, $Cl^- = 1.81$ Å, $Br^- = 1.95$ Å, and $I^- = 2.16$ Å, do not change enough for drastically different degrees of solvation to be expected.

In the hydrocarbon alcohols, the association constants are generally larger than those predicted by Eq. (1) or Eq. (2). It is the magnitude of K_A as well as its peculiar dependence on anion size that requires the postulation of a multiple-step association process. Association constants for trifluoroethanol are plotted in Fig. 12 as a function of the sum of the crystallographic radii. Also shown are the predicted values from Eq. (1), dashed line, and Eq. (2), solid line, for dielectric constant 26.67, trifluoroethanol. It is doubtful if there is another case where the agreement between experiment and the prediction of Eq. (2) is as good. Our reaction is to regard this close agreement as somewhat fortuitous. What is clear is that the degree of ionic association in trifluoroethanol is much smaller in magnitude and normal in dependence on ion size as compared to other alcohols and that this behavior is consistent with the existence of predominantly solvent-separated ion pairs due to enhanced anionic solvation and decreased cationic solvation.

In 1,1,1,3,3,3-hexafluoro-2-propanol a similar picture emerges. Although this compound has a dielectric constant of only 16.7 compared to 19.4 for

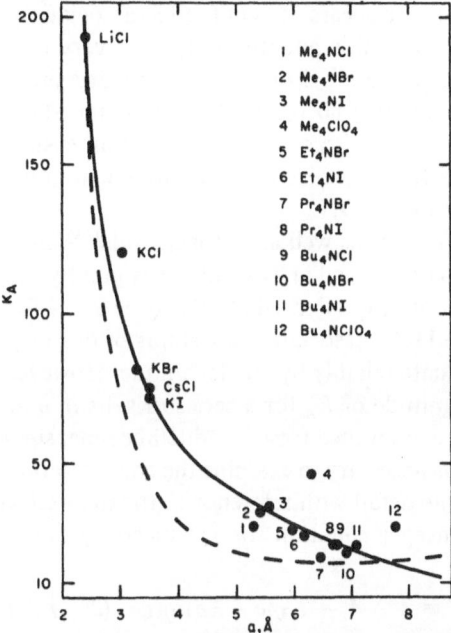

Fig. 12. Comparison of observed association constants with those calculated using the Fuoss (dashed line) and Bjerrum (solid line) equations for electrolytes in trifluoroethanol at 25°C as a function of sum of crystallographic radii.

2-propanol, the tetraalkylammonium salts show a surprisingly small degree of association. In fact, the conductance data can be fit almost as well with the Fuoss–Onsager[30] equation for nonassociated electrolytes as with the Fuoss–Onsager equation for associated electrolytes or the modification of this equation by Justice.[31]

6. EVIDENCE FOR IONIC ASSOCIATION IN OTHER HYDROGEN-BONDING SOLVENTS

In alcohols of dielectric constant less than 35, the pattern of ionic association is affected greatly by the anion–solvent interaction. Where this is moderately strong, as with the hydrocarbon alcohols, a multiple-step mechanism occurs with a degree of association greater than that predicted by theory, and both contact and solvent-separated ion pairs are present. Where the anion–solvent interaction is exceptionally strong, as with the fluorinated alcohols, results are best interpreted in terms of solvent-separated ion pairs only.

It now remains to consider what happens in solvent of higher dielectric constant. Although 1:1 electrolytes are minimally associated in water, the association constants for 2:2 electrolytes can be readily determined. It is the quantity $Z_1 Z_2 / \varepsilon$ that appears in coulombic-association theories, and for 2:2 electrolytes in water this quantity is 1/20 or very nearly the same as for 1:1 electrolytes in propanol. For example, theory predicts an association constant of approximately 80 for $MnSO_4$ in water at 25°C. The measured result is 160.[32,33] The ultrasonic work on this system[22,34] confirms a multiple-step association process which is similar in nature to that observed for 1:1 electrolytes in 2-PrOH.

In aqueous solutions as well as in formamide[35] and ethylene glycol,[36] the degree of ion pairing for 1:1 electrolytes is greatly diminished. For a dielectric constant of 40, Eq. (1) predicts K_A to be 9 ± 4 for interionic contact distances from 3 to 11 Å. Association constants of this magnitude and smaller are difficult to evaluate reliably by any technique. However, a clear indication of the relative magnitude of K_A for a series of salts in a given solvent can be obtained with some assurance from conductance measurements. In order to accomplish this, it is necessary to examine the concentration dependence of the conductance in some detail with reference to the theoretical equations.

The Fuoss–Onsager equation for 1:1 electrolytes can be rearranged to give

$$\Lambda' \equiv \Lambda - \Lambda^\circ + S\sqrt{c} - Ec \log c = (J - F\Lambda^\circ)c \qquad (11)$$

This predicts that a plot of Λ' vs c should yield a straight line of slope $(J - F\Lambda^\circ)$. The term $F\Lambda^\circ$ which corrects for the effect of the solute on the viscosity of the solution can be ignored in this analysis. The terms S, F, and J depend upon solvent properties and Λ°. The quantity J depends, in addition, on the ion-size parameter, $\overset{\circ}{d}$ which, according to theory, should increase as the size of the ion increases. As can be seen in Fig. 13, for the alkali metal halides in water, the

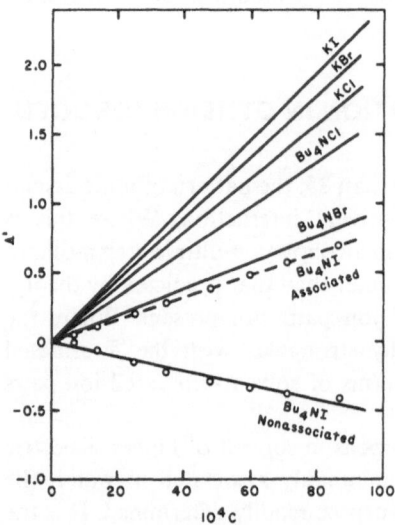

Fig. 13. Plot of Λ' vs concentration showing opposite dependence on anion size for Bu_4N^+ salts as compared to K^+ salts in water at 25°C.

slope of the line and hence the value of $\mathring{a}$ increases in the order I > Br > Cl, in accord with theory.

If an associated electrolyte is analyzed by Eq. (11), and if the degree of association is large, Λ' vs c will not be linear but rather curved with a small or even negative slope. The reason for this can be seen from the following argument. For associated electrolytes, the equation analogous to (11) is

$$\Lambda'' \equiv \Lambda - \Lambda^\circ + S\sqrt{c\gamma} - Ec\gamma \log c\gamma + K_A c\gamma f_\pm^2 \Lambda = (J - F\Lambda^\circ)c \qquad (12)$$

where γ is the fraction of free ions and $f_\pm$ the activity coefficient. Using the Onsager equation $\Lambda = \Lambda^\circ - S\sqrt{c}$, the term $K_A \gamma c \Lambda f_\pm^2$ can be approximated as $K_A \gamma f_\pm^2 (\Lambda^\circ c - Sc^{3/2})$. The $c^{3/2}$ dependence results in curvature of the Λ' plot which increases as K_A becomes large. However, when K_A is small and γ close to unity, Λ' becomes essentially linear in c, and K_A cannot be separated from J. Thus a reliable value of K_A cannot be obtained. In these cases, if an analysis by Eq. (11) is attempted, the J term will always be too small and, in fact, will become smaller the greater the degree of ionic association.

If one analyzes Bu_4NI in MeOH by Eq. (11), the slope of the Λ' vs c plot is too small, and the points show considerable deviation (ca. 0.5 conductance unit) from the least-squares best straight line (Fig. 14). When the same data are analyzed by Eq. (12), a positive slope results, and the points lie on the line to within ± 0.027.

In methanol and ethanol, the tetraalkylammonium salts are more highly associated than sodium and potassium salts. Thus, if any ionic association occurs in water, glycol, or formamide, it would be expected to occur for the tetraalkylammonium salts. An inspection of Λ' plots for halides in water[37] (Fig. 13) and formamide[38] (Fig. 15) shows that potassium halide slopes are in the expected order, $I^- > Br^- > Cl^-$. For the tetraalkylammonium halides, however, very different behavior emerges. Not only are the slopes smaller, but the order has been reversed. The methanol plot (Fig. 14) shows that this is to be expected if a small degree of ionic association is occurring. Although

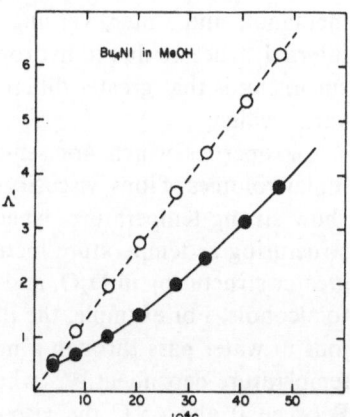

Fig. 14. Plot of Λ' (solid circles) and Λ'' (open circles) vs concentration for Bu_4NI in methanol at 25°C.

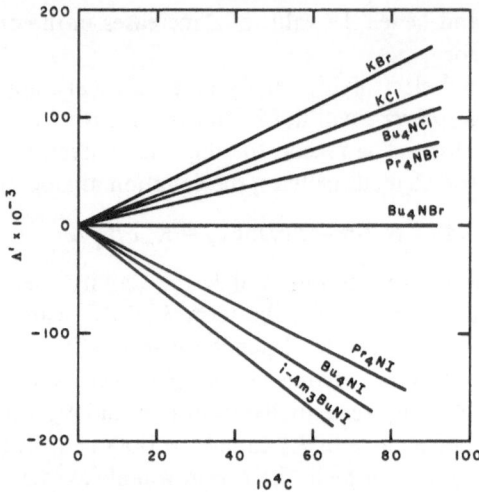

Fig. 15. Plot of Λ' vs concentration for electrolytes in formamide at 25°C.

the amount of association is small, iodide must be more associated than bromide and than chloride. The same trend can be seen in Λ' plots for tetraalkylammonium bromides and iodides in ethylene glycol[36]: the slopes for the iodides are smaller than the slopes for the bromides, and furthermore a slight degree of curvature is seen in the plots.

The peculiar behavior of the Λ' plots for tetraalkylammonium halides in water is thus seen as part of a general pattern in hydrogen-bonding solvents.

The goal of this extensive investigation of ionic association in hydrogen-bonding solvents was to assess the relative influence of ionic solvation and solvent structure. What has emerged is a surprisingly extensive parallelism in behavior encompassing 2:2 electrolytes in water, tetraalkylammonium halides in water, formamide, and ethylene glycol, and 1:1 electrolytes in a series of hydrocarbon alcohols. Solvents as diverse as 2-propanol, formamide, methanol, and water, varying widely in dielectric constant and degree of internal structure due to hydrogen bonding, exhibit similar patterns. What is surprising is that greater differences are not seen, since the solvent structure varies widely.

Properties which are sensitive to solvent structuring, such as partial molar volumes of ions, viscosity B coefficients, and limiting ionic mobilities,[38] show strong temperature dependence consistent with a decrease in solvent structuring as temperature increases, a solvent isotope effect consistent with greater structuring in D_2O, and very different behavior in water as compared to alcohols. For example, the partial molar volumes of tetraalkylammonium ions in water pass through a minimum with concentration and are strongly temperature-dependent,[39] whereas in ethanol they behave normally.[40] Between 0 and 45°C the viscosity B coefficients for the larger tetraalkyl-

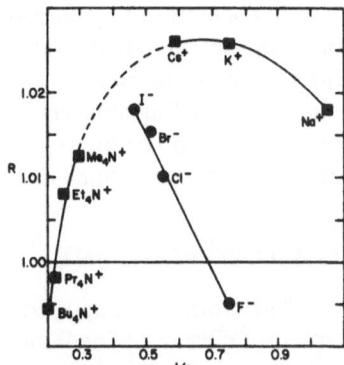

Fig. 16. The limiting conductance–viscosity product in D_2O divided by that in H_2O at the same temperature as a function of ion size.

ammonium ions change by a factor of 2 in water, whereas in methanol they are essentially constant over this temperature range.[41] Limiting ionic mobilities change significantly in going from H_2O to D_2O. The Walden product ratios $(\lambda_0\eta)_{D_2O}/(\lambda_0\eta)_{H_2O}$ change in a way consistent with the structure-making or structure-breaking characteristics of the ion,[42] as shown in Fig. 16.

Ionic-association patterns, on the other hand, show none of these variations. Plots of Λ' for tetraalkylammonium salts in H_2O at 10, 25, and 45°C are almost identical,[37, 38] whereas the degree of structuring in the solutions decreases substantially over this temperature range. Furthermore, Λ' plots in D_2O are identical to those in water, in spite of the higher degree of structure in D_2O. Since association patterns do not seem affected by structure in water, one would predict a similar insensitivity to temperature for association constants in methanol. This is indeed observed.

Thus, it is a combination of electrostatic effects and specific solvation of the ions by acidic and basic sites in the solvent molecules that gives rise to the concentration dependence of electrolyte conductance in dilute solution. This behavior appears to be insensitive to solvent–solvent interactions of the type that lead to structuring of the solvent. Therefore, what one obtains is information about the common features of water and other hydrogen-bonding solvents caused by their solvating characteristics. Those characteristics of water arising from its ability to solvate both cations and anions can thus be separated from structural effects by this comparison with conductance behavior in other such solvents.

7. CONCLUSION

It was indeed a pleasure to have been introduced to research in water structure both as a postdoctoral student and as a close associate of Henry Frank, and certainly it has been a great honor to be part of this symposium on the occasion of his 70th birthday.

In conclusion I would like to refer to a statement made by Henry Frank in the introduction to this talk. The value of studying other solvents in attempting to understand water is implied in Kipling's statement[43]: "What should they know of England who only England know?" With the hope that our discussion of other hydrogen-bonding solvents places water in better perspective, we thus conclude this paper.

ACKNOWLEDGMENT

This work was supported by Contract No. 14-01-001-1281 and 14-30-2615 with the Office of Saline Water, U.S. Department of the Interior.

REFERENCES

1. H. S. Frank and M. W. Evans, *J. Chem. Phys.* **13**, 507 (1945).
2. H. S. Frank and W. Y. Wen, *Disc. Faraday Soc.* **24**, 133 (1957).
3. R. M. Fuoss, *J. Am. Chem. Soc.* **80**, 5059 (1958).
4. N. Bjerrum, *Kgl. Danske Videnskab. Selskab.* **7**, 9 (1926).
5. D. F. Evans, J. Thomas, J. A. Nadas, and M. A. Matesich, *J. Phys. Chem.* **75**, 1714 (1971).
6. D. F. Evans and P. Gardam, *J. Phys. Chem.* **72**, 3281 (1968).
7. R. L. Kay, C. Zawoyski, and D. F. Evans, *J. Phys. Chem.* **69**, 4208 (1965).
8. S. R. C. Hughes and D. H. Price, *J. Chem. Soc.*, 1093 (1967).
9. M. A. Matesich, J. A. Nadas, and D. F. Evans, *J. Phys. Chem.* **74**, 4568 (1970).
10. D. F. Evans and P. Gardam, *J. Phys. Chem.* **73**, 158 (1969).
11. R. P. Taylor and I. D. Kuntz, Jr., *J. Phys. Chem.* **74**, 4573 (1970).
12. W. M. Latimer and W. H. Rodebush, *J. Am. Chem. Soc.* **42**, 1419 (1920); L. Pauling, *The Nature of the Chemical Bond* (Cornell University Press, Ithaca, New York, 1939).
13. G. G. Stokes, *Trans. Camb. Phil. Soc.* **VIII**, 287 (1845).
14. R. Zwanzig, *J. Chem. Phys.* **52**, 3625 (1970).
15. H. S. Frank, *Chemical Physics of Ionic Solutions*, B. E. Conway and R. G. Barradas, eds. (John Wiley and Sons, Inc., New York, 1966), p. 61.
16. D. F. Evans and B. Lamartine, unpublished data.
17. T. Noveske, J. Stuehr, and D. F. Evans, *J. Solution Chem.* **1**, 93 (1972).
18. P. Debye, *Trans. Electrochem. Soc.* **82**, 256 (1942).
19. M. Eigen, *Z. Physik. Chem.* (*Frankfurt*) **1**, 176 (1954).
20. T. Noveske, J. Stuehr, J. Evans, and D. F. Evans, *J. Phys. Chem.* **77**, 912 (1973).
21. M. J. Blandamer, D. E. Clark, N. J. Hidden, and M. C. R. Symons, *Trans. Faraday Soc.* **64**, 2683 (1968).
22. L. Jackopin and E. Yeager, *J. Phys. Chem.* **74**, 3766 (1970).
23. G. Choux and R. L. Benoit, *J. Am. Chem. Soc.* **91**, 6221 (1969).
24. G. S. Darbari and S. Petrucci, *J. Phys. Chem.* **74**, 268 (1970).
25. D. F. Evans, J. A. Nadas, and M. A. Matesich, *J. Phys. Chem.* **75**, 1708 (1971).
26. M. A. Matesich, J. Knoefel, H. Feldman, and D. F. Evans, *J. Phys. Chem.* **77**, 366 (1973).
27. J. R. Graham, G. S. Kell, and A. R. Gordon, *J. Am. Chem. Soc.* **79**, 2352 (1957).
28. J. L. Hawes and R. L. Kay, *J. Phys. Chem.* **69**, 2420 (1965).
29. F. L. Scott, *Chem. Ind.* (*London*), 224 (1954); W. S. Trahanovsky and M. P. Dole, *Tetrahedron Letters*, 2155 (1968).
30. R. M. Fuoss and F. Accascina, *Electrolyte Conductance* (Interscience, New York, 1959).

31. R. Bury, M. C. Justice, and J. C. Justice, *Compt. Rend. Acad. Sci.* (*Paris*) **268**, 670 (1969).
32. G. Atkinson and H. Hallada, *J. Am. Chem. Soc.* **83**, 3759 (1961).
33. T. L. Broadwater, Ph.D. thesis, Case Western Reserve University, September 1968.
34. M. Eigen and K. Tamm, *Z. Elektrochem.* **66**, 107 (1962).
35. J. Thomas and D. F. Evans, *J. Phys. Chem.* **74**, 3812 (1970).
36. R. P. DeSieno, P. W. Greco, and R. C. Mamajek, *J. Phys. Chem.* **75**, 1722 (1971).
37. D. F. Evans and R. L. Kay, *J. Phys. Chem.* **70**, 366 (1966).
38. R. L. Kay and D. F. Evans, *J. Phys. Chem.* **70**, 2325 (1966).
39. W. Y. Wen and S. Saito, *J. Phys. Chem.* **68**, 2639 (1964).
40. W. Y. Wen, Saline Water Conversion Report, Office of Saline Water, U.S. Department of Interior, 1966, p. 13.
41. R. L. Kay, T. Vituccio, G. Zawoyski, and D. F. Evans, *J. Phys. Chem.* **70**, 2336 (1966).
42. R. L. Kay and D. F. Evans, *J. Phys. Chem.* **69**, 4216 (1965).
43. R. Kipling, *Rudyard Kipling Verse* (Doubleday, New York, 1922), inclusive edition, 1885–1918, p. 252.

DISCUSSION

Professor R. L. Kay (*Carnegie-Mellon University*). You made the statement at the end of your talk that the Zwanzig equation overestimated the solvent relaxation effect. It is quite possible that this is the case since it is based on the assumption that the solvent in the ionic cosphere had identical properties to those of the pure solvent, and we know that must be a poor assumption. Consequently, it would appear sensible to add a cosphere term, as Professor Friedman has done for thermodynamic quantities, which will take into account such effects as dielectric saturation and viscosity changes in the cosphere.

Professor Evans (*Carnegie-Mellon University*). That is quite correct. The Zwanzig equation can essentially be used to take into account the effect of the ionic charge and attribute any discrepancy to structure or lack of it in the cosphere.

Professor J. B. Hyne (*University of Calgary, Canada*). Do you think that you can explain the solvent dependence of ion-pair-association constants simply or largely on the basis of dielectric-constant behavior, particularly in mixed solvents, without taking into consideration specific interactions between the ion pair and the component ions with the solvent?

Professor Evans. No, indeed I don't, and I actually had decided at one time that I was never going to do any mixed solvents, but it seemed to me that the possibilities of having sufficiently larger association constants in this case made it worthwhile doing. I think maybe there is something one can do to help, i.e., to study the ultrasonic behavior of these solutions, which essentially would allow one to at least break down in a more fundamental way the association constants. I think clearly one has to go after ion-solvent interactions in an individual way, and that is not going to result from the kind of colligative measurement that association is by itself.

Professor Hyne. It seems to me that this would be the basis of an explanation for the minimum in the association constant when one gets a minimum in the dielectric constant in the isopropanol–acetone mixture.

Professor H. S. Frank (*University of Pittsburgh*). Hasn't C. P. Smyth pointed out that some of his measurements of dielectric relaxation frequencies in nonaqueous systems give reason to believe that the local viscosity that governs the rate of molecular rotation is different from the bulk viscosity? This would then show up in your measurement.

Professor Hyne. But is that local viscosity due to solvent–solute interaction of a specific kind?

Professor Frank. I would say yes because it is in the neighborhood of the particle which is being watched.

Professor H. L. Friedman (*State University of New York, Stony Brook*). In the beginning you discussed the dependence of the pair-association constants upon the size of the tetraalkylammonium ions. Maybe the implication is that this should level off as the radius gets large. That would neglect the lipophilic effect which one finds in the solvation enthalpy where, as the solute gets bigger, its interaction with an oily solvent also gets bigger because there is more molecular interaction across the interface between the solute particle and the medium.

Professor Frank. One might also remember that there is reason to believe that a lipophobic effect can take place in nonaqueous solvents provided there is a high dielectric constant. That is, for certain purposes, oil is antagonistic to any high dielectric medium.

Professor Evans. I think that is right. Our measurements were mostly on solvents of high dielectric constant because we were interested in solvents for electrolyte solutions.

Professor G. Némethy (*Université de Paris-Sud, France*). A high solvent dielectric constant may be conducive towards solute–solute association, but solvent effects on such behavior are not necessarily correlated with the dielectric constant. Dr. Ray, an associate of mine, recently studied micellar aggregation in a variety of pure organic solvents [A. Ray, *Nature* **231**, 313 (1971)]. Micelles are formed in several of them, not just in water. It appears from her study that the ability of solvent to allow micelle formation is not correlated with the dielectric constant, nor several other physical properties, but depends primarily on the ability of the solvent to form intermolecular structures by hydrogen bonding. In other words, in systems such as the one discussed here, the dielectric constant may be an indication of the nature of solvent effects, but a thorough analysis of solvent–solute interactions requires the consideration of the structure of the solvent itself.

The Role of Solvent Structure in Ligand Substitution and Solvent Exchange at Some Divalent Transition-Metal Cations[1]

E. F. Caldin[2] and H. P. Bennetto[3]

Received November 1, 1972

The role of the solvent in reactions involving ions is considered in relation to the structure of liquids. The rate constants and activation parameters for ligand substitutions at divalent transition metal cations in various solvents are compared with those for solvent exchanges. The differences are related to structural properties of the solvents, represented by their heats of evaporation and fluidities, and interpreted with the aid of a model developed from that of Frank and Wen. Water is not a typical solvent.

KEY WORDS: Cation; energies; H. S. Frank, M. Eigen, R. G. Wilkins; kinetics; ligand substitution; mechanism; mixed solvents; nonaqueous; solvation; solvent, solvent struoture, solvent exchange; transition metal.

1. EFFECTS OF SOLVENT STRUCTURE ON RATES OF CHEMICAL REACTIONS

This paper is concerned with a problem in chemical kinetics in solution. In this field the importance of the work of Henry Frank and his collaborators has grown with the realization that rates of reaction are affected by solvent structure and by the influence of solutes upon it. In particular, the model put forward by Frank and Wen for aqueous solutions of ions has played an important part. A brief historical sketch of the development of views on the role of the solvent in ionic reactions may not be out of place.

The importance of the solvent as something more than an incompressible medium in reactions between ions was already recognized in the 1930's. It was handled in terms of the dielectric constant,[1, 2] which is a rough measure

---

[1] This paper was presented at the symposium, "The Physical Chemistry of Aqueous Systems," held at the University of Pittsburgh, Pittsburgh, Pennsylvania, June 12–14, 1972, in honor of the 70th birthday of Professor H. S. Frank.

[2] University Chemical Laboratory, Canterbury, Kent, England.

[3] Department of Chemistry, Queen Elizabeth College, Campden Hill, Kensington, London W.8, England.

of the polarity and solvating power of the solvent molecules but bears no direct relation to the structure of the liquid. The concept of liquid structure was developed in 1933 by Bernal and Fowler in their pioneer theory of water.[5] A molecular picture in terms of electrostriction or orientation of polar solvent molecules around ions[3, 4] was adopted in the 1940's as a broad interpretation of the entropies of reactions involving ions. It has also been shown very elegantly that a detailed model for the orientation of water in the field of an ion could account for the change in heat capacity for the ionization of carboxylic acids in water.[6] The details of the molecular picture were not then clear; the arrangement of solvating water molecules was not known for any ion in solution, and experimental estimates of the number of water molecules concerned varied enormously according to the method employed. The distinction between primary and secondary solvation (corresponding to first and subsequent solvation shells) became recognized in the later 1940's.[7]

The classic paper of Henry Frank and M. W. Evans[8] in 1945 not only gave figures for the entropies of solvation of ions but made use of them to elucidate the effect of ions in water on the structure of the solvent beyond the first coordination shell. For most univalent cations, the entropy of solvation was found to be larger than could be attributed to the formation of this shell ("the first 'saturated' region"). The interpretation was that these ions have a "structure-breaking" effect, attributable to the mismatch between the spatial arrangement of water molecules around the ion and those in the bulk solvent. Neutral molecules were found to produce a decrease in entropy on dissolving in water, an effect attributed (with due qualifications) to the formation of "icebergs." The behavior of water was shown to be abnormal in comparison with nonpolar solvents.

This interpretation was extended in the paper by Henry Frank and W. Y. Wen in 1957 entitled "Structural Aspects of Ion–Solvent Interaction in Aqueous Solutions: A Suggested Picture of Water Structure".[9] For ions in water, the entropy data show a gradation of net structure-altering influence. Small or highly charged ions are net structure formers, while alkali metal ions (except Li^+) and anions (except F^-) are net structure breakers. Experimental results on viscosity, heat capacities, and dielectric-relaxation times were in agreement with these conclusions. In the model that emerged, there are three zones around an ion: A, a region where the solvent is more or less immobilized (the first solvation shell); B, the region of structure-breaking and disorder; and C, normal bulk water. It was noted that such effects need not be peculiar to water.

Nonpolar compounds (or groups) in water appeared from the entropy data to be structure makers, in proportion to their size, without specificity. This was interpreted in terms of the "flickering-cluster" model. Hydrogen-bond formation and breaking in water are cooperative processes; when one bond forms, the resulting polarization will assist others to form, producing a cluster of molecules with a lifetime of perhaps 10^{-10} sec, corresponding to the observed dielectric-relaxation times. A nonpolar solute would prolong the life of such a cluster by protecting it from charge fluctuations that would destroy it. The important point here is that uncharged solute molecules, as well as ions, can affect the structure of the solvent.

With the advent of nuclear magnetic resonance methods in the late 1950's, it has become possible to form a more detailed molecular view of solvation. The number of solvent molecules in the first solvation shell has been determined for many metal ions[10]; it is usually 6. The rate of exchange of solvent molecules between the first coordination shell and the bulk solvent (regions A and C) can be determined for paramagnetic ions[11]; for a number of ions in water, the energy of activation is around 10 kcal-mole^{-1}, while in other solvents it may be as low as 4 or as high as 18 kcal-mole^{-1}. Molecular motions in liquids can also be studied.[12] (The existence of zone A has been disputed,[13] and the extent of zone B depends no doubt upon the cation.) Clearly, all this information is of importance for the chemical kinetics of ions in solution; solvation changes must be taken into account and will involve solvent structure.

Ideas on solvent structure have been applied in reaction kinetics to interpret results on the solvolysis of alkyl halides.[14] In a series of alcohol–water mixtures, these reactions show a monotonic variation of the rate constant and therefore of the activation free energy $\Delta F^{\ddagger}$; but $\Delta H^{\ddagger}$ and $\Delta S^{\ddagger}$ show larger changes, partly compensating, with maxima or minima. These changes can be understood in terms of the structural properties of aqueous alcohols.[15] Addition of a little alcohol to water appears to stiffen the structure (this is shown, for instance, by a rise in the temperature of maximum density), and so does addition of a little water to alcohol; further additions lead to breakdown of the structure. The maxima or minima in $\Delta H^{\ddagger}$ and $\Delta S^{\ddagger}$ are attributable to these variations in solvent structure.

We have sought to extend this type of interpretation to some inorganic systems and to a range of pure solvents, as well as to aqueous alcohols.[16] We also extend Frank and Wen's model to structural solvents other than water and consider the effect of uncharged solutes as well as ions on the structure. A full understanding of the observations would require a well-developed theory of the structure of the solvents commonly used. Theories of liquids have not yet progressed to the point where predictions can be made for such systems, however, and our approach must therefore be largely empirical.

2. THE KINETICS OF LIGAND SUBSTITUTION AND SOLVENT EXCHANGE AT DIVALENT TRANSITION-METAL CATIONS IN AQUEOUS SOLUTION

The ligand-substitution reactions with which we are concerned may be represented, as far as stoichiometry goes, by

$$L + MS_6^{2+} \underset{}{\overset{k_L}{\rightleftharpoons}} MS_5L^{2+} + S \tag{1}$$

where MS_6^{2+} is a metal cation with six solvent molecules S in its first solvation shell[4] and L represents an uncharged ligand such as ammonia or bipyridyl. Solvent exchange is simply the particular case where $L = S$:

$$S + MS_6^{2+} \underset{}{\overset{k_s}{\rightleftarrows}} MS_6^{2+} + S \qquad (2)$$

The energetics of process (2) may well differ from those of (1) because an L molecule, unlike an S molecule, is surrounded by molecules differing from itself and will affect the solvent structure differently. The energetics of neither reaction is fully understood,[17] but we shall be mainly concerned with differences rather than absolute values, and these are more easily interpreted.

Ligand substitution [the forward step in reaction (1)] is found to be first order with respect to ligand and to metal cation: rate $= k_L(M^{2+})(L)$. This rate law is compatible with either dissociation via the 5-coordinate intermediate MS_5^{2+} (3) or a concerted interchange (4), in which the energetics might be controlled either by the contribution of the association or of the dissociation (the I_a or I_d "pathways" of Langford and Gray's treatment[18]).

$$MS_6^{2+} \rightleftarrows S + MS_5 \xrightarrow[\text{slow}]{L} \underset{\text{(transition state)}}{L..MS_5^{2+}} \longrightarrow LMS_5^{2+} \qquad (3)$$

$$L + MS_6^{2+} \longrightarrow \underset{\text{(transition state)}}{L..MS_5^{2+}..S} \longrightarrow LMS_5^{2+} + S... \qquad (4)$$

(Ligand dependence will be expected unless the I_d pathway is followed.) Solvent exchange may be treated similarly. In all cases, the ratio of the observed second-order rate constants for ligand-substitution and solvent exchange is equal to the ratio of the rate constants for the reactions of L and S with a metal cation, either MS_6^{2+} or MS_5^{2+}.[16b]

The model currently accepted for ligand substitution in water, due to Eigen and Wilkins[19, 20] is an I_d interchange in which the ligand is envisaged as first forming an outer-sphere complex:

$$L + MS_6 \underset{\text{(outer-sphere}}{\overset{K_0}{\rightleftarrows}} \underset{\text{complex)}}{L, MS_6} \xrightarrow{k_1} \underset{\substack{\text{(transition} \\ \text{complex)}}}{L..MS_5..S} \longrightarrow LMS_5 + S... \quad (5)$$

The value of k_1 is assumed to be equal to that of k_{ex}, the first-order rate constant for the solvent-exchange (2); that is, the proximity of L is assumed to make no

---

[4] We use the coordination number 6 in Eq. (1) for the sake of illustration; it should, however, be recognized that the coordination number has often been measured only at low temperatures and should not be assumed to be necessarily independent of temperature. In the kinetic discussion, it is not necessary to choose a value of the coordination number, provided that we use the second-order rate constant k_S for solvent exchange rather than the first-order rate constant k_{ex} commonly employed.[16b]

difference to the rate of solvent loss from the cation. Then it is easily shown that, under the usual experimental conditions, k_L is given by

$$k_L = K_0 \, k_{ex} \qquad (6)$$

or, in terms of standard free energies,

$$\Delta G_L^{\ddagger} = \Delta G_0^0 + \Delta G_{ex}^{\ddagger} \qquad (7)$$

An approximate value of K_0 for the outer-sphere complex may be calculated from the sizes and charges of L and MS_6^{2+}; for neutral ligands one assumes a value based on size alone, $K_0 = \frac{4}{3}\pi a^3 N_0$, where a is the least distance of approach of metal cation and ligand (regarded as spheres); with $a = 5$ Å, $k_0 \simeq 0.1$ cm³-mole⁻¹.[21] The kinetics are thus controlled by the energy required to dissociate S from MS_6 ($\Delta G_{ex}^{\ddagger}$) and the electrostatic interaction (if any) between L and MS_6 (ΔG_0^0).

The Eigen–Wilkins model, as represented by Eq. (6), is fairly successful for reactions in water, as the following evidence shows. (a) For the series of transition metal sulfates, $M^{2+} + SO_4^{2-} \rightleftarrows M^{2+}SO_4^{2-}$, there is a parallelism of k_L (determined by ultrasonic absorption methods) with k_{ex} within about a power of 10, and also of $\Delta H_L^{\ddagger}$ with $\Delta H_{ex}^{\ddagger}$. (b) For nickel(II) ion, which has been much studied, with ligands of varying charge, the equation is obeyed within a power of 10 by many ligands, and the values of k_L/K_0 agree quite well with that of k_{ex}.[20, 38] (c) For nickel(II) ion with various ligands of a given charge, however, the model predicts only small changes of k_L, whereas the observed values vary 40-fold for L^{2-}, 5-fold for L^-, and 20-fold for neutral ligands[19] (excluding the diamines for which a special mechanism[22] is proposed). There is evidently considerable ligand specificity, so that Eq. (6) is by no means exact, even in water.

3. LIGAND-SUBSTITUTION AND SOLVENT-EXCHANGE RATES IN NONAQUEOUS SOLVENTS

To test the mechanism further, it is evidently desirable to vary the solvent. For a given neutral ligand reacting with the cations of the first transition series (whose ionic radii do not greatly differ), the value of K_0 calculated as described above is approximately constant (ca. 0.1 liter-mole⁻¹). Consequently, by Eq. (6), the ratio

$$n \equiv k_L / 0.1 \, k_{ex} \simeq 1 \qquad (8)$$

Comparing the rate constants k_L and k_{ex}, therefore, the Eigen–Wilkins model would lead us to expect the value of n to be about 1 in all solvents or, at any rate (if we have taken an incorrect value for a), to be approximately constant. This is not observed, except for water and methanol. Table I shows the results so far accumulated for cations where values of k_L and k_{ex} have been determined in several solvents.[16, 23, 24, 25] In general the values of n diverge

Table I. Comparison of Rate Constants and Activation Parameters for Ligand Substitution and Solvent Exchange

Reaction	Solvent	$10^{-3}k_L$	n	$\Delta H_L^{\ddagger}$	$\Delta\Delta H^{\ddagger a}$	$\Delta S_L^{\ddagger}$	$\Delta\Delta S^{\ddagger}$
Ni²⁺ + bipyridyl	CH₃CN	$4._7(4.1^b)$	17	$6.6(6.5^b)$	−5.1	−20	−16
	CH₃OH	$0.08_3(0.094^b)$	0.8_0	17.0	+1.2	+7.4	0
	DMF	0.54	0.7_0	12.7	+3.3	−3.3	+6
	H₂O	$1.6_5(2.2_6^b)$	0.5_4	$12.6(11.7^b)$	+1.8	−1.7	0
	D₂O	1.0_8	—	14.0	—	—	—
	DMSO	0.069	0.09	12.6	+4.6	−7.7	+6
	DMMP	0.069	0.09	13.9	+6.4	−0.9	+10
	EtGly	0.033	0.08	16.8	—	+4.7	—
Ni²⁺ + terpyridyl	CH₃CN^b	2.2	0.8	8.4	−3.3	−15	−19
	CH₃OH^b	0.046	0.5	16.8	+1.0	+5	−3
	H₂O	$0.49(1.49^b)$	0.16	$13.8(12.3^b)$	+3.0	0	+2
	DMSO	0.026 $(0.025^b, 0.033^c)$	0.035	11.8 $(15.0^b, 11.8^c)$	+3.8	-12^a	+2
	EtGly^b	0.010	0.025	14.0	—	—	—
Ni²⁺ + o-phenanthroline	CH₃CN^b	50	180	4.7	−7.0	−21	−17
	CH₃OH^t	0.35	3.5				
	H₂O^b	4.14	1.4	11.5	+0.7	−4	−2
	DMSO^b	0.35	0.55	8.3	0	−19	−5

Ni²⁺ + pyridine	CH₃CN[b]	0.83	0.3	14.7	+3.0	+4	+8
	H₂O[d]	ca. 4	ca. 0.4	ca. 11	ca. 0	ca. -5	ca. -3
	DMSO[b]	ca. 2	ca. 0.4	ca. 4	ca. -4	ca. -30	ca. -16
Ni²⁺ + pyridine-2-azo-p-dimethylaniline (PADA)	CH₃CN[e]	16.9	6	5.8	-6.2	-20	-16
	H₂O[f]	0.90	0.3	12.4	+1.6	-3.5	-2
Ni²⁺ + NH₃	CH₃OH[e]	1.3	0.4	—	—	—	—
	H₂O[g]	4.7	1.6	10.5	-0.3	-6.6	-5
Co²⁺ + bipyridyl	CH₃OH	1.92	1.1	12.8	-1.0	+0.5	0
	H₂O[d]	63	0.3	ca. 10	ca. -4	ca. 0	ca. -9
	DMSO	3.6	0.02_1	12.6	+3	0	0
Co²⁺ + terpyridyl	CH₃OH	1.0_7	0.5_9	13.1	-0.7	+0.6	-7
	H₂O[d]	25	0.3_2	ca. 9	ca. -1	ca. -7	ca. -12
	DMSO	0.45	0.002_6	11.1	+1.5	-9.0	-9

[a] $\Delta\Delta H^{\ddagger}$ and $\Delta\Delta S^{\ddagger}$ are often dependent on selection of solvent exchange values.[16]

[b] Reference 23.

[c] R. K. Boyd, P. A. Cock, and C. E. Cottrell, Can. J. Chem. 50, 402 (1972).

[d] R. H. Holyer, C. D. Hubbard, S. F. A. Kettle, and R. G. Wilkins, Inorg. Chem. 4, 929 (1965); 5, 622 (1966).

[e] H. P. Bennetto and Z. Sabet-Imani, unpublished results.

[f] E. F. Caldin and co-workers, unpublished work.

[g] Reference 24.

[h] This value is incorrectly given as -7.4 in ref. 16b.

[i] Reference 37.

widely; for the reaction of Ni^{2+} with bipyridyl, for example, they vary about 200-fold from acetonitrile to DMSO; overall, the highest and lowest values differ by a factor of about 10^4, the difference increasing as the temperature is lowered. The ratio does not appear to correlate with either k_L or k_{ex}.

Comparing the activation enthalpies, we see from Eq. (6) that

$$\Delta\Delta H^{\ddagger} \equiv \Delta H_L^{\ddagger} - \Delta H_{ex}^{\ddagger} \simeq 0 \qquad (9)$$

Thus the Eigen–Wilkins model predicts that $\Delta\Delta H^{\ddagger} \simeq 0$ for all solvents. This again is contrary to observation (Table I), except for some ligands in water and methanol. For Ni^{2+} + bipyridyl, $\Delta\Delta H^{\ddagger}$ varies from about -5 to $+5$ kcal-mole^{-1} (in acetonitrile and DMSO, respectively); overall, the highest and lowest values are about $+7$ and -7 kcal-mole^{-1}. These variations are considerably larger than those of $\Delta\Delta G^{\ddagger}$ (which reflects n); they are partly compensated by the corresponding variations of $\Delta\Delta S^{\ddagger}$. Clearly the model needs modification for nonaqueous solvents.

4. COMPARISON OF k_L/k_{ex} AND $\Delta\Delta H^{\ddagger}$ WITH SOLVENT PROPERTIES

We now seek to correlate the experimental values of n or of $\Delta\Delta H^{\ddagger}$ for a given ligand with the physical properties of the solvents, some of which are listed in Table II. We refer primarily to the reaction of Ni^{2+} with bipyridyl[5]; similar trends with other ligands are apparent from the results in Table I.

The value of n shows no correlation with the dielectric constant; it would not be expected to do so, since ε is not relevant to the first coordination shell where the ionic charge is interacting with the local dipole. Solvation parameters such as E_T, which is a measure of polarity derived from spectroscopic shifts in organic systems involving charge separation, likewise show no correlation. A fair correlation is found between $\log n$ and the heat of evaporation ΔH_{evap} of the solvent, which is a measure of the energy required to make a hole of molecular size in the solvent; we may pictorially regard it as a measure of the "stiffness" of the solvent structure. There is also a fair correlation (Fig. 1) between $\log n$ and the fluidity (=density/viscosity), which again is probably related to the "stiffness" and also to the free volume or "open-ness" of the solvent structure. Incidentally, neither k_L nor k_{ex} separately shows these correlations; it is their ratio that does so. It is noteworthy that the variation of n with solvent (several powers of 10) is much greater than the variation in the fluidity, which is about five-fold; a similar small range covers the self-diffusion coefficients, the reorientation times for molecular motion, and the dielectric relaxation times. The correlation of n with fluidity is evidently not

---

[5] The objection might be raised that bipyridyl is a bidentate ligand. Recent results[23] suggest that ring closure is of importance in some cases but that the solvent effect on the rate of the initial substitution generally remains. The ring closure is probably also solvent-dependent.

Table II.[a] Physical Properties of Solvents

Solvent	F.p.	B.p.	V.p.	ε	η	ρ/η	μ	n_D^{20}	Z	E_T	ΔH_{evap} At 25°C	At normal b.p. (1 atm)	ΔS_{evap} At normal b.p. (1 atm)	$10^5 D$
MeCN	−45.7	81.6	92	37.5	0.345	2.27	4.1	1.344_3	71.3	46.0	7.7	7.1	20.06	5.4
MeOH	−97.8	64.5	120	32.6_3	0.54_5	1.43	1.68	1.3288	83.6	55.5	9.0	8.4	24.96	2.2
DMF	−61	153	3.5	36.7	0.796	1.19	3.8	1.4306	68.6	43.8	11.4	10.0	23.46	—
H_2O	0	100	23.7	78.5_4	0.89	1.12	1.85	1.333_4	94.6	63.1	10.5	9.7	26.05	2.27
D_2O	4.1	101.4	—	78.2_5	1.1	1.0_0	1.86	1.328_7	—	—	10.8	9.9	26.51	2.11
DMSO	18.45	189	0.5	46.6	1.99	0.55	4.3	1.4787	71.1	45.0	12.6	10.7	23.16	0.8
EtGly	−12.4	197.4	0.14	37.7	1.8_1	0.6_1	2	1.4317	85.1	56.3	13.8	11.8	25.1	—
DMMP	ca. −50	177	1.3	20.7^c	2.00	0.58	2.78^c	1.4222	—	—	13.3^b	11.3	$25._0$	—

[a] For references see ref. 16a. Reproduced from ref. 16a by permission. Notation: F.p. and b.p. in °C; V.p. at 25°C in mm Hg; ε, dielectric constant at 25°C; η, viscosity at 25°C in cP; ρ/η, fluidity at 25°C in g-ml^{-1}-cP^{-1}; μ, dipole moment at 25°C in Debyes; $^{20}_D n$, refractive index at 20°C for D line; Z, E_T, solvent polarity parameters [E. M. Kosower, *Physical Organic Chemistry* (Wiley, New York, 1968), pp. 301, 305]; ΔH_{evap}, enthalpy of evaporation in kcal-mole^{-1}; ΔS_{evap}, entropy of evaporation in cal-deg^{-1}-mole^{-1}; D, self-diffusion coefficient in cm^2-sec^{-1}.
[b] Estimated
[c] At 30°C.

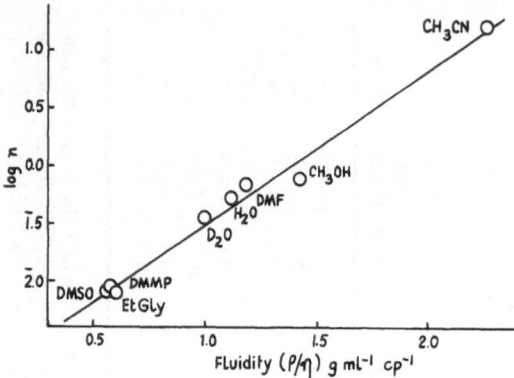

Fig. 1. Plot of log n for the reaction of nickel(II) ion with bipyridyl in various solvents vs fluidity of solvent.[16a]

due to diffusional effects; it is to be interpreted as showing an effect of solvent structure.

Turning to $\Delta\Delta H^{\ddagger}$, we find that there is again a fair correlation with the heat of evaporation of the solvent (Fig. 2); again it is the difference $\Delta\Delta H^{\ddagger} = \Delta H_{L}^{\ddagger} - \Delta H_{ex}^{\ddagger}$, not the separate values, that shows the correlation. The slope of the plot is between 2 and 3, suggesting that several solvent molecules are involved.

These correlations are only approximate (to possibly ± 1 kcal-mole^{-1}), but the effects are sizable. As pointers to an interpretation, we note that: (a) compensation of $\Delta\Delta H^{\ddagger}$ and $\Delta\Delta S^{\ddagger}$ is often attributable to the effects of solvation changes; (b) that $\Delta\Delta H^{\ddagger}$ varies much more than ΔH_{evap} suggests that more than one molecule may be involved; (c) the correlations of $\Delta\Delta H^{\ddagger}$ and n with ΔH_{evap} suggest that the solvent structure is concerned and that it is modified

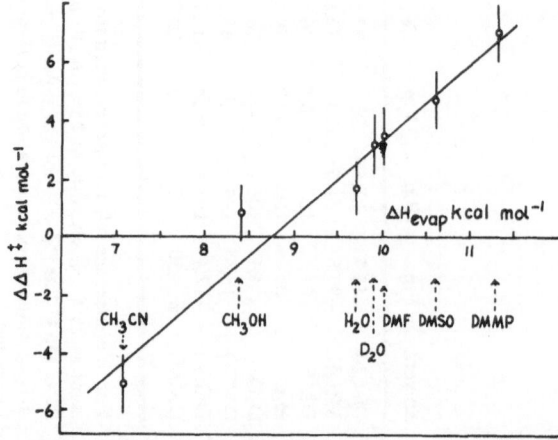

Fig. 2. Plot of $\Delta\Delta H^{\ddagger}$ (enthalpy of activation for bipyridyl substitution relative to that for solvent exchange) in various solvents vs heat of vaporization of solvent.[16b]

locally by ligand molecules; (d) water fits into the scheme along with the other liquids but occupies a middle position: the value of n is about unity, and that of $\Delta\Delta H^{\ddagger}$ is about zero.

5. INTERPRETATION IN TERMS OF SOLVENT STRUCTURE

These correlations can be understood in terms of an adaptation of the Frank and Wen model. Briefly, the hypotheses that we put forward to account for the data are as follows:

(1) Solutions of metal ions in nonaqueous polar solvents can be represented by a model similar to Frank and Wen's model for aqueous solutions.

(2) The kinetics of ligand substitution are controlled partly by the energy required to dissociate the cation–solvent (M–S) bond, as in the Eigen–Wilkins model, and partly by the energetics of the interactions of the leaving S molecule with the bulk solvent into which it passes. The complementary movement of the incoming L or S molecule out of the bulk solvent into the first coordination shell appears to be much less important.

(3) The structural properties of the solvent are modified locally by the presence of ligand molecules; near an L molecule the solvent may be "stiffened" or "loosened."

(4) This effect is related to the intermolecular forces between solvent molecules.

In working out these hypotheses we assume for convenience that ligand substitution follows the interchange pathway [Eq. (4)], but our general conclusions are not dependent on a choice between the dissociative and interchange schemes. We now take points (1) to (4) in turn.

Point 1. Ligand substitution at the metal cations might be represented, if the bulk solvent were not concerned, by a simple three-center model:

$$ L \longrightarrow MS_5..S \longrightarrow \qquad (10) $$

It is clear, however, that more solvent molecules are involved. A model based on that of Frank and Wen, with its three regions, might be represented by

$$
\begin{array}{ccccc}
SSS & S & S & SSS & \\
SSS & & & SSS & \\
SSL.. & \longrightarrow & MS_5..S & \longrightarrow & SSS \qquad (11)\\
SSS & & & SSS & \\
SSS \quad S & & & S \quad SSS &
\end{array}
$$

or, if L approaches on the same side as the leaving S molecule, by

$$
\begin{array}{cc}
SSS. & \longrightarrow \quad L \\
SSS & \qquad \searrow \\
SSS & \qquad\quad MS_5 \qquad (12)\\
SSS & \qquad \nearrow \\
SSS. & \longleftarrow \quad S
\end{array}
$$

Another representation is shown in Fig. 3.

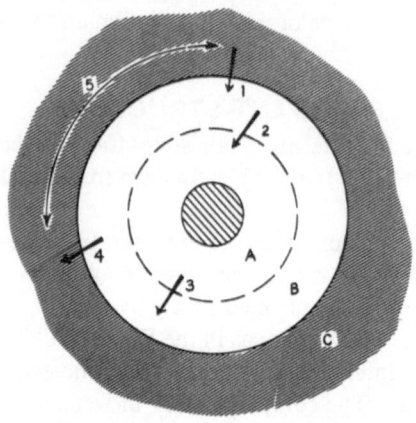

Fig. 3. Model for a solvated ion in a structured solvent. (After Frank and Wen.[9])

Solvent exchange will thus involve five processes, which may be more or less concerted. These (Fig. 3) are: (1) A solvent molecule moves from bulk solvent C to the disordered region B. This is analogous to evaporation and will absorb energy. (2) A solvent molecule from B enters the first coordination shell A of the metal ion, while (3) another solvent molecule leaves it and enters region B, i.e., the bond M–S is broken. (4) A solvent molecule from B joins the bulk solvent—a process analogous to condensation. (5) The sequence of changes may require reorganization of the bulk solvent. For ligand substitution, one may postulate a similar scheme, except that the molecule concerned in processes (1) and (2) is L rather than S.

Point 2. All five processes may contribute to $\Delta G^{\ddagger}$, $\Delta H^{\ddagger}$, and $\Delta S^{\ddagger}$. The contribution of the dissociation step (3) is important. In water it appears to be dominant; in nonaqueous solvents it is modified by contributions from the other processes, especially (4) and (5), as is shown below.

Point 3. Our next assumption is that a ligand molecule L modifies the structural properties of an adjacent solvent region. Suppose first that it stiffens the solvent, i.e., increases the energy required to make a hole. Then, in the presence of L, the heat evolved in the quasi-condensation process (4) is greater, so $\Delta H^{\ddagger}$ is reduced and $\Delta\Delta H^{\ddagger}$ is negative. This fits the results for the reaction of Ni^{2+} with bipyridyl in acetonitrile; it appears that acetonitrile has little structural strength and is stiffened locally by the presence of a bipyridyl molecule. Suppose conversely that L "loosens" the bulk solvent. Then the heat evolved in process (4) is reduced, and $\Delta\Delta H^{\ddagger}$ is positive; i.e., ligand substitution is less favored energetically than solvent exchange. The concomitant entropy changes partly compensate those of $\Delta H^{\ddagger}$, but the effect of $\Delta H^{\ddagger}$ dominates the rate ratio. This fits the more common case, such as that of the reaction of Ni^{2+} with bipyridyl in DMSO; it is reasonable to suppose that the relatively stiff structure of this solvent is weakened locally by a bipyridyl molecule.

Point 4. This interpretation requires us to assume that with a series of solvents the stiffening or loosening effect of the ligand is related to the inter-molecular solvent–solvent forces. If these are weak (as measured by ΔH_{evap}), they are strengthened by the proximity of a ligand molecule; if strong, they are weakened. In the intermediate case of water, they are little changed, pre-sumably because of the strong intermolecular forces between water molecules.

It is not possible either to confirm or refute these ideas by reference to current theories of binary liquids; we do not yet have a clear understanding of the effects of uncharged nonhydrogen-bonding molecules on the structures of liquids. However, it is evident that a solute molecule will produce local changes in the solvent structure by changing the intermolecular forces. It would not be surprising if the effect of the solute were correlated, to a first approximation, with the disparity between the heats of evaporation of the solute and of the solvent from their respective pure liquids; that is, the L–S forces might depend on the difference between the L–L and S–S forces, somewhat as the "inter-change" energy $(E_{11} + E_{22} - 2E_{12})$ is of importance in the theory of regular solutions.[26] The forces here concerned are dipole–dipole and dispersion forces, which might be "relayed" from solute to solvent molecules. Thus, a ligand with relatively high intermolecular attractive forces could increase the attraction between solvent molecules in its neighborhood and so stiffen the solvent structure locally, while a ligand with relatively low intermolecular forces would weaken it.

An extreme example of the influence of solvent structure has appeared in some very recent work on ligand substitution in glycerol (M. W. Grant and E. F. Caldin, in press). The rate constants at 20°C in glycerol for the reactions of Ni(II), Co(II), Cu(II) and Zn(II) with the ligand pyridine-2-azo-p-dimethylaniline (PADA) are all equal within a factor of 4, although in water they range over four powers of 10. For Co(II) and Zn(II) it is possible to determine $\Delta H_L^{\ddagger}$, and this value agrees with that calculated for diffusion control, which for glycerol is exceptionally high (16.4 kcal-mole^{-1} at 25°C). For Zn(II) the volume of activation likewise agrees with that for diffusion control. All this suggests that the solvent structure dominates the energetics; it is the movement of solvent molecules that limits the rate, not the energy required to break the cation–solvent bond.

We should expect that the work required to change solvent structure would become smaller as the temperature approaches the boiling point. It is therefore significant that the extrapolated values of the rate ratio k_L/k_S for different ligands in a given solvent generally approach unity at the boiling point of the solvent. For instance, in water the values for five ligands (pyridine, bipyridyl, terpyridyl, *o*-phenanthroline, and water itself) are respectively 1.36, 1.04, 0.88, 1.75, and 1.00. In acetonitrile, where the variation in rates at 25°C is more than 50-fold, the values are similar: 2.7, 0.84, 0.74, 4.4, and 1.0.

6. CORRELATION OF $\Delta H^{\ddagger}$ AND $\Delta S^{\ddagger}$ FOR LIGAND SUBSTITUTION AND SOLVENT EXCHANGE IN VARIOUS SOLVENTS

Further evidence that in these reactions there are two major influences on the energetics—the dissociation of the M–S bond and the local structure of the solvent into which the leaving S molecule enters—is provided by a general correlation which is found between $\Delta H^{\ddagger}$ (corrected for ligand-field stabilization energy) and $\Delta S^{\ddagger}$. This relation, which covers both types of reaction, is illustrated in Fig. 4.[16] The slope of the best line by least squares is 370 K. Most of the points lie on the line within about 1 kcal-mole^{-1}; exceptions are the point for Be^{2+} (an abnormally small ion) in DMF and several points for solutions in alcohols. The variations in $\Delta H^{\ddagger}$ are from about 1 to 18 kcal-mole^{-1}, and those in $\Delta S^{\ddagger}$ are from about -25 to $+25$ cal-K^{-1}-mole^{-1}; these are much greater than the experimental uncertainties, which are shown in the figure. They are much greater than the variations in $\Delta G^{\ddagger}$, in which they compensate to a considerable extent. Such correlations of ΔH with ΔS, leading to largely compensating contributions to ΔG, are often found associated with solvation effects.[27, 28, 29, 30, 31, 32]

Most of the reactions with which we are concerned have negative values of $\Delta S^{\ddagger}$, so that the points lie in the left-hand half of the plot in Fig. 5 and have values of $\Delta H^{\ddagger}$ between 9 and 4 kcal-mole^{-1}. The lower values of $\Delta H^{\ddagger}$, compensated by low values of $\Delta S^{\ddagger}$, can hardly be attributed to low M–S bond strengths. The "donor strengths" of the solvents, as estimated from their absorption maxima,[33] vary only by about 1 kcal-mole^{-1}. We suppose that the M–S bonds, being all due essentially to the interaction of a lone pair of

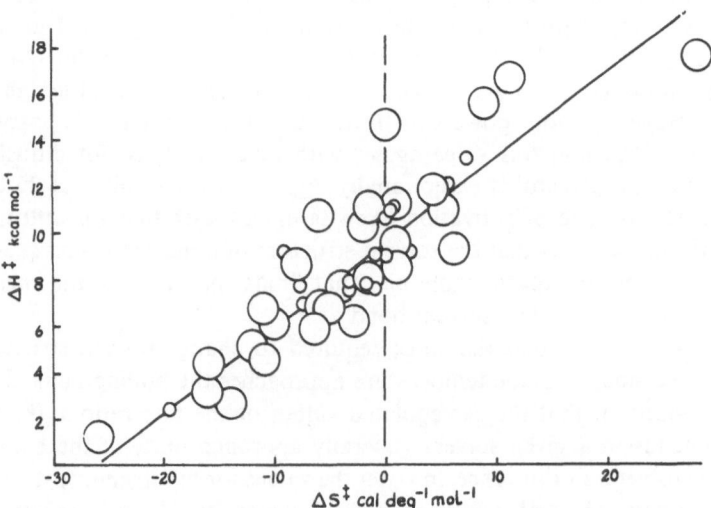

Fig. 4. Plot of enthalpy of activation against entropy of activation for solvent exchange and neutral-ligand substitution at bivalent ions.[16b]

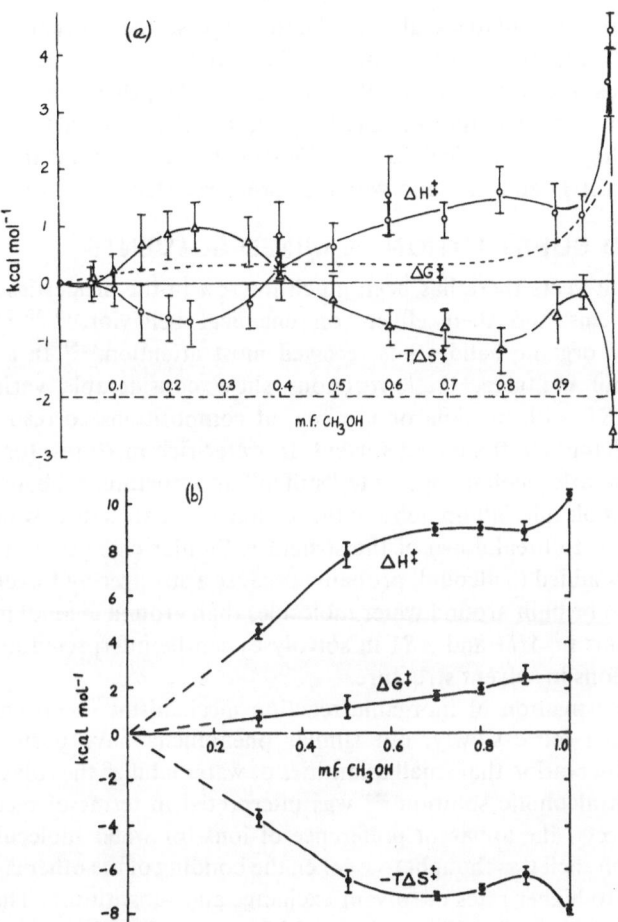

Fig. 5. Kinetic parameters for reaction of nickel(II) ion with bipyridyl in: (a) methanol–water mixtures; (b) methanol–acetonitrile mixtures. Values of $\Delta H^{\ddagger}$ and $\Delta S^{\ddagger}$ in (a) are relative to water, in (b) to acetonitrile.

electrons with ions of about the same size, are all of comparable strength, contributing about 9 kcal-mole^{-1} (corresponding to the observed value of $\Delta H^{\ddagger}$ when $\Delta S^{\ddagger} = 0$); and that the lowest values of $\Delta H^{\ddagger}$ and $\Delta S^{\ddagger}$ are attributable to the negative contribution from the quasi-condensation process (4). This is precisely the process that was found to be important in Sect. 5 above when we compared ligand substitution and solvent exchange by considering the values of $\Delta\Delta H^{\ddagger}$ and $\Delta\Delta S^{\ddagger}$. It is a process that would be influenced by the effect of a ligand molecule on the lqcal solvent structure.

Since the value of ΔH for condensation at 25°C is around 10 to 13 kcal-mole^{-1} (Table II), we can understand the reduction of $\Delta H^{\ddagger}$ from about 9 to as low as 4 kcal-mole^{-1}. Most systems have values of $\Delta H^{\ddagger}$ intermediate

between these values (4 to 9 kcal-mole^{-1}); this suggests that solvent reorganization [process (5)] absorbs some energy. The changes in $\Delta H^{\ddagger}$ and $\Delta S^{\ddagger}$ will be
correlated, as in other solvation phenomena, for the general reason that the
freedom of movement of molecules (which affects $\Delta S^{\ddagger}$) is related to the strength
of the bonding (which affects $\Delta H^{\ddagger}$). We can thus understand the observed
variations of $\Delta H^{\ddagger}$ and $\Delta S^{\ddagger}$ and why they are correlated.

7. LIGAND SUBSTITUTION IN MIXED SOLVENTS

In recent years there has been much interest in the properties of mixed
solvent systems and their effects on chemical behavior.[15, 30, 34, 39] The
solvolysis of organic halides has received most attention.[35] In a series of
water–alcohol mixtures, these reactions show considerable variations in
$\Delta H^{\ddagger}$ and $\Delta S^{\ddagger}$, with maxima or minima, at compositions corresponding to
structural changes in the mixed solvent. In water-rich mixtures, for instance,
addition of a little alcohol appears to "stiffen" the structure, probably because
the alcohol molecules fill up holes in the ice-like water structure, while further
additions lead to breakdown of the structure. Similar changes occur when a
little water is added to alcohol, probably because a stronger hydrogen-bonded
structure can be built around water molecules than around alcohol molecules.
The variations in $\Delta H^{\ddagger}$ and $\Delta S^{\ddagger}$ in solvolyses can be interpreted in terms of
these variations in solvent structure.

The investigation of inorganic reaction mechanisms in mixed solvents
has developed more slowly, but similar phenomena have been observed.
The early observation that small quantities of water labilize the solvation shell
of an ion in alcoholic solution[36] was interpreted in terms of metal-ligand
bonding effects; the apparent preference of ions for water molecules in the
first solvation shell was thought to weaken the bonding of the other component
and so lead to higher rates of solvent-exchange and substitution. The importance of effects of this kind is emphasized by some authors in recent work.[37]
A kinetic investigation of the reactions of nickel(II) ion with bipyridyl and with
thiocyanate ion in methanol–water mixture at 25°C showed variations of
rate constant that could be interpreted on the Eigen–Wilkins model without
modification.[38] Later work, however, indicates that such reactions are
sensitive to solvent structure. The results of further experiments over a range
of temperature on the reaction of nickel(II) ion with bipyridyl in water–
methanol[16c] are shown in Fig. 5a. The free energy of activation $\Delta G^{\ddagger}$ (representing log k_L) varies monotonically over the whole range and shows relatively
little variation. The variations of $\Delta H^{\ddagger}$, however, are considerable (ca. 5
kcal-mole^{-1}) and show two minima; there are similar changes in $\Delta S^{\ddagger}$ (not
shown) which compensate those of $\Delta H^{\ddagger}$ to produce the much smaller variations
of $\Delta G^{\ddagger}$. The positions of the minima correspond reasonably well with the
compositions where the physical effects of solvent structure are most marked
(ca. 0.25 and 0.95 mole fraction). In aqueous t-butanol, the minimum in the
water-rich solvent shifts to the region of 5 % t-butanol where again the physical

effects of solvent structure are known to be most marked.[23] These results are consistent with the interpretation that there is a contribution to $\Delta H^{\ddagger}$ from the passage of a solvent molecule from the disordered region B into the bulk solvent [process (4) above] which becomes progressively "stiffer" as alcohol is added at the water-rich end or as water is added at the alcohol-rich end.

This kind of interpretation in terms of solvent structure is supported by the contrasting results obtained for the same reaction in methanol–acetonitrile mixtures[16e] (Fig. 5b). The structural peculiarities of these mixtures are much less marked (the viscosity, for instance, varies monotonically with composition); there appears to be a smooth change from a hydrogen-bonded liquid to a polar liquid as acetonitrile is added. Correspondingly, there is no minimum in $\Delta H^{\ddagger}$ or $\Delta S^{\ddagger}$ at the acetonitrile-rich end of the composition range, and the minimum at the methanol-rich end is much less marked than with aqueous methanol.

Other effects must also be considered in mixed solvents. Chief among these is the possibility of changes in the first solvation shell as the solvent composition is altered; such changes may be random or selective, and the various solvated species may react at different rates. In a study of the reaction of nickel(II) ion with ammonia in methanol–water mixtures at 25°C, Rorabacher[24, 25] found that the fourfold variation of rate constant with solvent composition corresponds to the variation of the concentration of the ion $[\mathrm{Ni(CH_3OH)(H_2O)_5}]^{2+}$ as determined by UV spectroscopy. More work is needed on the composition of solvation shells and on the bonding of the different solvent molecules and their rates of exchange with the solvent.

8. CONCLUSION

We have observed wide differences between solvents in their behavior in substitution reactions of metal cations and have derived various correlations which hold with an accuracy of the order of 1 kcal-mole^{-1}. Our results point clearly to an influence of solvent structure on rates and energetics of reaction. In the absence of a developed general theory of polar and hydrogen-bonded liquids, we have adopted an empirical approach along with certain hypotheses which seem to us to be worth trying out. A large amount of accurate data will be needed, on the effects of varying the ligand and the cation as well as the solvent. It will be essential to measure rates over a range of temperatures; results at high pressure will also be informative. There is a general need for greater accuracy, which will require more attention to such matters as temperature control and concentration dependence. The degree of agreement between results obtained by different workers using stopped-flow methods, though generally good, gives no ground for complacency; and there are discrepancies in the NMR data, which lead to some uncertainties in our correlations. When reliable results for a wider range of solvents, ligands, and ions become available, a more detailed analysis of the solvent effects may be attempted. It is also clear that more work is required on the theoretical side with regard to the definition of "solvation" and the composition of solvation shells.

This line of investigation, relating reaction kinetics to molecular theories of liquid structure, is still in its early stages. In broaching it, we have made use of Henry Frank's ideas to a considerable extent. The paper by Frank and Wen developed the three-zone model with a region of disordered solvent around the solvated cation, the idea that hydrogen bonding in water is cooperative, and the view that interactions between water molecules are enhanced in the neighborhood of neutral solute molecules. We have attempted to generalize and adapt this model to nonaqueous solvents[39] and to develop it to include solvent exchange. We postulate the following: that there may be a disordered region around cations in any solvent, though its extent may depend greatly on the particular solvent structure; that the interactions in nonaqueous solvents are less cooperative than in water; that the presence of a solute will, in general, affect the local solvent structure and may either enhance or reduce the solvent–solvent interactions; and that since solvent exchange between solvated cations and bulk solvent has been demonstrated, a dynamic model is required.

Like all good models, the model of Frank and Wen proves to be adaptable to new fields. Reaction kinetics must be added to the other fields in which Henry Frank's ideas have been seminal in the study of liquids and solutions. It is a privilege to add the tribute of kineticists to those of others and to express our thanks to Henry Frank.

ACKNOWLEDGMENTS

We thank Professor J. F. Coetzee and his co-workers for helpful discussions and for permission to use their results before publication.

REFERENCES

1. E. A. Moelwyn-Hughes, *Kinetics of Reactions in Solution* (Clarendon Press, Oxford, 1933).
2. E. S. Amis, *Kinetics of Chemical Change in Solution* (Macmillan, New York, 1949).
3. W. F. K. Wynne-Jones and H. Eyring, *J. Chem. Phys.* 3, 492 (1935).
4. R. P. Bell, *J. Chem. Soc.*, 629 (1943).
5. J. D. Bernal and R. H. Fowler, *J. Chem. Phys.*, 1, 515 (1933).
6. D. H. Everett and C. A. Coulson, *Trans. Faraday Soc.* 36, 633 (1940).
7. J. O'M. Bockris, *Quart. Rev. Chem. Soc.* 3, 179 (1949).
8. H. S. Frank and M. W. Evans, *J. Chem. Phys.* 13, 507 (1945).
9. H. S. Frank and W. Y. Wen, *Disc. Faraday Soc.* 24, 133 (1957).
10. S. F. Lincoln, *Coord. Chem. Rev.* 6, 309 (1971).
11. T. R. Stengle and C. H. Langford, *Coord. Chem. Rev.* 2, 349 (1967).
12. a. J. G. Powles, M. Rhodes, and J. H. Strange, *Mol. Phys.* 11, 515 (1966), and later papers; b. H. G. Hertz, *Ber. Bunsenges. Physik. Chem.* 71, 979, 999, 1008, 1032 (1967), and later papers.
13. H. G. Hertz, *Angew. Chem. Intern. Ed.* 9, 134 (1970).
14. a. E. M. Arnett, W. G. Bentrude, J. J. Burke, and P. Duggleby, *J. Am. Chem. Soc.* 87, 1541 (1965); b. E. M. Arnett, P. Duggleby, and J. J. Burke, *J. Am. Chem. Soc.* 85, 1350 (1963); c. E. M. Arnett and D. R. McKelvey, *Rep. Progr. Chem.* 26, 185 (1965); d.

J. B. Hyne, in *Hydrogen-Bonded Solvent Systems*, A. K. Covington and P. Jones, ed. (Taylor and Francis, London, 1968), p. 99; e. I. Lee and J. B. Hyne, *Can. J. Chem.* **46**, 2333 (1968); f. R. E. Robertson, *Progr. Phys. Org. Chem.* **5**, 121 (1967).

15. F. Franks and D. J. G. Ives, *Quart. Rev. Chem. Soc.* **20**, 1 (1966).
16. a. H. P. Bennetto and E. F. Caldin, *J. Chem. Soc.* (A), 2191 (1971); b. *ibid.*, 2198 (1971); c. *ibid.*, 2207 (1971); d. H. P. Bennetto and E. F. Caldin, *Chem. Comm.*, 599 (1969); e. H. P. Bennetto, *J. Chem. Soc.* (A), 2211 (1969); f. H. P. Bennetto, R. S. Bulmer, and E. F. Caldin, in *Hydrogen-Bonded Solvent Systems*, A. K. Covington and P. Jones, eds. (Taylor and Francis, London, 1968), p. 335.
17. C. H. Langford and T. R. Stengle, *Ann. Rev. Phys. Chem.* **19**, 193 (1968).
18. C. H. Langford and H. B. Gray, *Ligand Substitution Processes* (Benjamin, New York, 1966), p. 7.
19. M. Eigen and R. G. Wilkins, in *Mechanisms of Inorganic Reactions*, Advan. Chem. Ser., No. 49 (American Chemical Society, 1965), p. 55.
20. R. G. Wilkins, *Accounts Chem. Res.* **3**, 408 (1970).
21. a. R. M. Fuoss, *J. Am. Chem. Soc.* **80**, 5059 (1958); b. M. Eigen, *Z. Phys. Chem.* (*Frankfurt*) **1**, 176 (1954); c. J. E. Prue, *J. Chem. Soc.*, 7534 (1965); *J. Chem. Ed.* **46**, 12 (1969).
22. D. B. Rorabacher, *Inorg. Chem.* **5**, 1891 (1966).
23. P. K. Chattopadhyay and J. F. Coetzee, *Inorg. Chem.* **12**, 113 (1973).
24. W. J. MacKellar and D. B. Rorabacher, *J. Am. Chem. Soc.* **93**, 4379 (1971).
25. F. R. Shu and D. B. Rorabacher, *Inorg. Chem.* **11**, 1496 (1972).
26. a. E. A. Guggenheim, *Mixtures* (Clarendon Press, Oxford, 1952), Chaps. 3 and 4; b. J. S. Rowlinson, *Quart. Rev. Chem. Soc.* **8**, 186 (1954).
27. a. J. Leffler and E. Grunwald, *Rates and Equilibria of Organic Reactions* (Wiley, New York, 1963); b. C. D. Ritchie and W. Sager, *Progr. Phys. Org. Chem.* **2**, 323 (1964); c. J. Shorter, *Chem. in Britain* **14**, 269 (1967); *Quart. Rev. Chem. Soc.* **24**, 433 (1970).
28. D. W. G. Ives and P. D. Marsden, *J. Chem. Soc.*, 649 (1965).
29. R. Lumry and S. Rajender, *Biopolymers*, **9**, 1125–1227 (1970).
30. E. M. Arnett and D. R. McKelvey, in *Solute–Solvent Interactions*, C. D. Ritchie and J. F. Coetzee, ed. (Marcel Dekker, New York, 1969), Chap. 6.
31. J. H. Norman, P. Winchill, and R. J. Thorn, *Inorg. Chem.* **10**, 2265 (1971).
32. J. Greyson, and H. Snell, *J. Phys. Chem.* **73**, 3208 (1969).
33. V. Gutmann, *Coordination Chemistry in Non-aqueous Solutions* (Springer, Vienna, 1968).
34. a. *Physico-chemical Processes in Mixed Aqueous Solvents*, F. Franks, ed. (Heinemann, London, 1967); b. *Hydrogen-Bonded Solvent Systems*, A. K. Covington and P. Jones, ed. (Taylor and Francis, London, 1968); c. *Solute–Solvent Interactions*, J. F. Coetzee and C. D. Ritchie, eds. (Marcel Dekker, New York, 1969).
35. a. S. Winstein and A. H. Fainberg, *J. Am. Chem. Soc.* **79**, 5937 (1957); b. S. J. Dickson and J. B. Hyne, *Can. J. Chem.* **49**, 2394 (1971); c. J. M. W. Scott and R. E. Robertson, *Can. J. Chem.* **50**, 17 (1972).
36. Z. Luz and S. Meiboom, *J. Chem. Phys.* **40**, 2686 (1964).
37. F. Dickert, H. Hoffmann, and W. Jaenicki, *Ber. Bunsenges. Physik. Chem.* **74**, 500 (1970).
38. R. G. Pearson and P. Ellgen, *Inorg. Chem.* **6**, 1379 (1967).
39. R. L. Kay, G. P. Cunningham, and D. F. Evans, in *Hydrogen-Bonded Solvent Systems*, A. K. Covington and P. Jones, eds. (Taylor and Francis, London, 1968).

DISCUSSION

Professor R. H. Stokes (*University of New England, Armidale, Australia*). In the Fuoss equation for the equilibrium constant for outer-sphere association, a is the distance of closest approach of the ligand to the solvated ion; a^3

will be of the order of the molal volume and will vary by almost an order of magnitude in the series of solvents discussed. Hence, one cannot expect that K_0 will have the same value in all solvents.

Professor Caldin (*University Chemical Laboratory, Canterbury, England*). The molal volumes of the solvents cover a range of about 5. But, if we consider the size of the solvent molecule rather than the density of the bulk liquid, we get a considerably smaller range, say 2 or 3; as we are only concerned at present with approximate generalizations, we ignore this variation.

It would not be possible to explain our results simply by taking these variations into account in calculating K_0; there is no correlation between the observed rate constant and the size of the solvent molecule.

Professor H. L. Friedman (*State University of New York, Stony Brook*). It would be interesting to know how much of your correlation remains if the reference states are changed so that the ligand molecule comes from the gas phase and the solvent molecule ends up in the gas phase; I think the data for such calculations are available or could be estimated.

Professor Caldin. One would expect the correlations to vanish.

Professor J. F. Coetzee (*University of Pittsburgh*). I have two comments. First, since the ligand bypyridine is bidentate, there is a possibility that the rate-determining step is ring closure, which perhaps could be responsible for the large solvent effects that you observe. In the case of dimethyl sulfoxide as solvent, we feel that ring closure may be rate-limiting for bipyridine and terpyridine because pyridine reacts normally, while the rate constants for the multidentate ligands at 25°C are smaller by a factor of 100 or so.

My second comment concerns the basic assumption of the theory, namely, that there should be a disordered region surrounding the solvated metal ion in all solvents. I am glad that Dr. Criss is here, since he has evaluated absolute ionic entropies in a number of nonaqueous solvents. I am wondering whether in a solvent such as acetonitrile, which has no three-dimensional structure but only fairly strong dipole association, the solvated metal ion necessarily would be incompatible with the order existing in the bulk solvent.

Professor Caldin. We do not, of course, rule out an effect of ring closure. Experiments with a variety of ligands should throw light on the matter. The results so far do not invalidate the use of our generalizations as rough approximations with an uncertainty of perhaps ± 1 kcal-mole^{-1} in $\Delta G^{\ddagger}$ or $\Delta H^{\ddagger}$.

Professor C. M. Criss (*University of Miami*). There are two things one has to consider about structural order: the degree of hydrogen bonding and the dipolar–dipole interaction. From our entropy correlations, it seems clear that dimethylformamide, for example, which has a very high dipole moment, is structured, but not nearly as much as a hydrogen-bonded solvent. I should expect acetonitrile to be less structured than dimethylformamide.

Professor R. L. Kay (*Carnegie-Mellon University, Pittsburgh*). Have these reactions been studied in pure water and in solvent mixtures as a function of pressure?

Professor Caldin. We have set up a laser temperature-jump apparatus with a high-pressure cell, in which these reactions can be studied, at pressures up to 4000 atm (E. F. Caldin, M. W. Grant, B. B. Hasinoff, and P. A. Tregloan, in press). Some results on reactions in water have been published [E. F. Caldin, M. W. Grant, and B. B. Hasinoff, *Chem. Comm.* 1351, 1971; *J. Chem. Soc., Faraday Trans. I* 68, 2247 (1972)], and we hope to report later on reactions in nonaqueous solvents. For these we should like to use an anion other than perchlorate.

Professor R. Lumry (*University of Minneapolis*). Do you know of other cases in the literature where changes of solvent and changes of substituent give values of ΔH and ΔS lying on the same isokinetic plot? We have searched high and low in the literature and have never found a pattern in which a solvent change would produce a compensation line with the same slope in the different solvents.

Professor Caldin. This is an interesting point, which I have not examined. Ligand substitution and solvent exchange are unusually simple processes. The reason for the wide scope of our correlations appears to be that the ion–solvent interaction, which is the dominating factor apart from solvent-structural effects, is fairly constant for all divalent transition metal ions after ligand-field effects have been allowed for. In other isokinetic relationships, the substituents may produce greater variation in the energy required to rupture the bond concerned.

Professor H. S. Frank (*University of Pittsburgh*). You make use of the concept of stiffness of the solvent. There is another sense in which the word stiffness can be used, in the comparison of H_2O and D_2O. The breaking of hydrogen bonds is probably more difficult in D_2O, and from that point of view D_2O should be "stiffer." But from another standpoint, H_2O is stiffer than D_2O. The moment of inertia of D_2O is bigger than that of H_2O, and that means that there is a reduction in whatever stiffness is related to in getting to the next librational quantum level. This seems to show up in a number of phenomena in which it appears that whatever a solute does to H_2O it does to D_2O "only more so." I was telling Dr. Millero that we have some measurements of partial molal volumes of salts in heavy water, and in the case of lithium chloride, where the volume is "too big" in H_2O, it is still bigger in D_2O. For the other alkali halides, structure breaking presumably predominates, and the volumes are probably a little small in H_2O. They are still smaller in D_2O. Millero has found also that the compressibility of D_2O is slightly greater than that of H_2O. This fits in with the idea that if the energy levels are closer together and one tries to disturb the system, it disturbs more easily. An extreme example of the converse is the very low heat capacity of diamond, where the crystals are hardly disturbed at all because the vibrational energy levels are so far apart. This concept of stiffness, in the case of very small solvent molecules, must involve the degree to which quantal restrictions affect their motions relative to each other. I don't know if Dr. Stillinger likes this idea or not, but certainly the thermodynamics of transfer of a solute from H_2O to D_2O does seem to bear out the notion that D_2O "pushes easier"; from this point of view, it isn't that the gain in standard entropy comes from D_2O having more structure and the solute breaking more because there is more there to break. Instead, it is that the entropy gain from structure breaking is bigger in D_2O because the entropy itself is bigger. This is an idea that has come to us rather recently, and next year maybe we will have a different one, but for the moment I would say that D_2O seems to "push easier" than H_2O does and from that standpoint seems to be softer.

Professor Caldin. Professor Frank's remarks are of considerable interest to us because we regard the difference between the rates of ligand substitution in H_2O and D_2O as one of the most significant pieces of evidence for a solvent-structural effect on kinetics. It may be of importance that in our discussion we used the term "stiffness" as a colloquial term to mean the energy (ΔE or ΔH) required to make a hole of molecular dimensions in the liquid. This must be distinguished from the work required (ΔG); thus, at the boiling point, ΔG is zero, but ΔH is not. The energy required to make a hole in H_2O appears (from the values of ΔH_{evap}) to be smaller than that for D_2O, but the work might be greater if the entropy change is also smaller.

Nuclear Magnetic Relaxation and Structure of Aqueous Solutions[1]

H. G. Hertz[2]

Received December 21, 1972

By the interpretation of the intermolecular nuclear magnetic relaxation rate of ^{19}F the orientation of the water molecules in the hydration sphere of F^- can be determined. Similarly, the orientation of the water molecules around the methyl group of propionic acid in aqueous solution has been studied. Experiments are described which give information about the nature of association of solute in aqueous solution of a number of carboxylic acids and of ethanol. The local dynamic details of the I^- ion have been investigated. Some new results are briefly discussed regarding the nuclear magnetic relaxation by quadrupole interaction in electrolyte solutions.

KEY WORDS: Hydration of ions; hydrophobic association; ion–ion configurations; intermolecular nuclear magnetic relaxation.

1. INTRODUCTION

The simplest, though not complete, way to describe the structure of a liquid is given by the molecular distribution function. One way of giving this function is $p(r_0)$, the probability density to find a particle at r_0 relative to a reference particle. In a solution there is a number of such distribution functions; e.g., in an aqueous electrolyte solution we have water–water, water–ion, and ion–ion distributions. I wish to present a number of examples of how certain properties of the pair distribution functions in aqueous solution may be determined by the aid of nuclear magnetic relaxation measurements.[1] The

---

[1] This paper was presented at the symposium, "The Physical Chemistry of Aqueous Systems," held at the University of Pittsburgh, Pittsburgh, Pennsylvania, June 12–14, 1972, in honor of the 70th birthday of Professor H. S. Frank.

[2] Institut für Physikalische Chemie und Elektrochemie, Universität Karlsruhe, West Germany.

interrelation between the nuclear magnetic relaxation rate $1/T_1$ and $p(\mathbf{r}_0)$ is as follows:

$$\frac{1}{T_1} \approx A \int_{-\infty}^{+\infty} e^{-i\omega t} \left\{ \int p(\mathbf{r}_0) \frac{Y^{(2)}(\mathbf{r}_0)}{r_0^3} P(\mathbf{r}_0, \mathbf{r}, t) \frac{Y^{(2)}(\mathbf{r})}{r^3} d\mathbf{r}_0 \, d\mathbf{r} \right\} dt \qquad (1)$$

This formula holds for the magnetic dipole–dipole interaction as the mechanism causing nuclear relaxation. The quantity A is a constant. Actually, $1/T_1$ is a linear combination of terms like that given above, with the Fourier transform taken at various frequencies ω. The quantity $\mathbf{r}_0$ is the position of the interacting *nucleus* (relative to the relaxing nucleus) at time $t = 0$, $r_0 = |\mathbf{r}_0|$, and $\mathbf{r}$ is the position of the same nucleus at time t. $Y^{(2)}(\mathbf{r}_0)$ is the spherical harmonic of second order. The argument $\mathbf{r}_0$ means the direction of $\mathbf{r}_0$. $P(\mathbf{r}_0, \mathbf{r}, t)$ is the propagator for the motion of the interaction partner (a nucleus); i.e., it gives the probability density to find the nucleus at $\mathbf{r}$ at time t if we know that it was at $\mathbf{r}_0$ at $t = 0$. $p(\mathbf{r}_0)$ is the pair distribution function for the interaction partner. It is important to note that $p(\mathbf{r}_0)$ concerns a nucleus residing somewhere in the molecule but does not concern the center of mass of the molecule. Very often, the time decay of the quantity in parentheses in Eq. (1) is very much faster than the period of the nuclear Larmor precession. In this case, we measure the 0-frequency behavior, i.e., we get

$$\frac{1}{T_1} = 2A \int_0^{\infty} \left\{ \int p(\mathbf{r}_0) \frac{Y^{(2)}(\mathbf{r}_0)}{r_0^3} P(\mathbf{r}_0, \mathbf{r}, t) \frac{Y^{(2)}(\mathbf{r})}{r^3} d\mathbf{r}_0 \, d\mathbf{r} \right\} dt \qquad (2)$$

Since the propagator $P(\mathbf{r}_0, \mathbf{r}, t)$ and the distribution function $p(\mathbf{r}_0)$ are interrelated, it is obvious that in searching for $p(\mathbf{r}_0)$ one has to have a certain knowledge about $P(\mathbf{r}_0, \mathbf{r}, t)$. It may be shown that, for very tight binding— $p(\mathbf{r}_0)$ is essentially a δ function—integration of Eq. (2) yields the result[2]

$$1/T_1 \sim (1/a^6 \cdot (a^2 / \bar{D}) \qquad (3)$$

where a is the closest distance of approach between the two interacting nuclei and $\bar{D}$ is the mean self-diffusion coefficient. If there is no binding, then one obtains

$$1/T_1 \sim (1/a^3) \cdot (a^2 / \bar{D}) \qquad (4)$$

However, in a number of examples to be described here, the propagator is essentially eliminated because several interacting partners reside on the same molecule. It may be seen from Eqs. (2), (3), and (4) that the measurement of one relaxation rate $1/T_1$ allows the determination of one structural parameter a, i.e., one (intermolecular) closest distance of approach between two interacting nuclei. Now, if a given molecule contains two nonequivalent nuclei and if the interaction with one of these nuclei, respectively, can be switched off by isotopic substitution, then the comparison of the relaxation caused by either of them relative to the other allows a comparison of the closest distance of approach a towards the relaxing nucleus. This in turn gives certain information concerning the relative orientation of the molecules in the liquid.

Equations (1) and (2) and the reasoning just mentioned have been applied in order to give an answer to the following structural questions in aqueous solutions:

(1) The orientation of the water molecule in the first hydration sphere of the F^- ion[2];

(2) The orientation of the water molecule in the hydrophobic hydration sphere[2];

(3) The existence of hydrophobic association and the location of contact of simple carboxylic acids and of ethanol in aqueous solution[3, 4];

(4) The hydration structure of the I^- ion[5, 6];

(5) Ion–ion configuration with local symmetry in simple 1–1 electrolyte solutions.[7]

Item (5) does not involve a direct application of Eq. (1) or Eq. (2).

2. THE ORIENTATION OF THE WATER MOLECULE IN THE FIRST HYDRATION SPHERE OF THE F^- ION[2]

The ^{19}F relaxation rate $1/T_1$ in a 1 m solution of KF in D_2O has been measured. If the D_2O molecules are partly replaced by H_2O, one observes an additional relaxation rate due to the ^{19}F-proton interaction. (The ^{19}F-deuteron interaction is very small.) Application of Eqs. (2) and (3) yields the closest distance of approach between the ionic nucleus and the water protons. In a further experiment the original $D_2^{16}O$ is replaced by 45.8 mole % $D_2^{17}O$ (no protons being present). The ^{17}O nucleus has a magnetic moment which causes magnetic dipole–dipole interaction with ^{19}F. Thus the ^{19}F relaxation rate in this solution gives the closest distance of approach between the F nucleus and the water oxygen. In the symmetrical arrangement of the water molecule shown in Fig. 1a, the normalized ratio of the F-proton and F-^{17}O

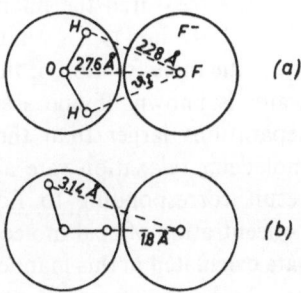

(a)

(b)

Fig. 1. Proposed orientations of the water molecule in the hydration sphere of F^-.

relaxation rates according to Eq. (3) is calculated to be $\kappa = 29.3$, whereas in the hydrogen-bonded configuration in Fig. 1b one calculates $\kappa = 62.8$. The experimental value is found to be $\kappa = 60 \pm 20$. This confirms the hydrogen-bonded configuration as predicted by quantum-chemical calculations.[8, 9]

3. THE ORIENTATION OF THE WATER MOLECULE IN THE HYDROPHOBIC HYDRATION SPHERE[2]

The analogous experiments have been performed with aqueous solutions of propionic acid and with aqueous solutions of methanol. We are asking: What is the hydration configuration around the methyl group? In these experiments the relaxing nucleus is one proton in the methyl group. The starting point was the measurement of the proton relaxation rate in the systems:

$$CD_2HCD_2COOD + D_2O$$
$$CD_2HOD + D_2O$$

The solute concentration was 10 mole %. Partial replacement of D_2O by H_2O gives an additional relaxation rate of the methyl proton due to the methyl proton–water proton magnetic dipole–dipole interaction. Partial replacement of $D_2^{16}O$ by $D_2^{17}O$ gives an additional methyl proton relaxation rate due to the methyl proton–water ^{17}O magnetic dipole–dipole interaction. Comparison of the two increases in the methyl proton relaxation rate yielded the result that the water protons point away from the methyl group. Of course, there are conformations where the methyl proton is close to the polar group of the solute molecule. This is unavoidable. For comparison, we studied the proton relaxation rate in solutions of HCOOD in D_2O, in $D_2O + HDO + H_2O$, and in $D_2^{17}O + D_2^{16}O$. We found that here as well the water protons point away from the carbonyl proton, but to a lesser degree.

4. THE HYDROPHOBIC ASSOCIATION AND LOCATION OF CONTACT IN AQUEOUS SOLUTIONS OF SIMPLE CARBOXYLIC ACIDS AND OF ETHANOL[3,4]

We measured the intermolecular proton relaxation rate of acetic acid, CH_3COOD in D_2O, over the entire composition range $0 < x_{ac} \leqslant 1$, where x_{ac} is the mole fraction of the acid. The self-diffusion coefficient of the acid in water is known. If one assumes a uniform solute distribution $p(r_0)$ for all separations larger than the closest distance of approach, then the intermolecular relaxation rate according to Eq. (2) may be calculated, giving a result corresponding to Eq. (4). The proportionality factor contains the concentration of acid molecules (moles/cm³) in a linear form. The relaxation rate calculated in this manner is shown as the dashed line in Fig. 2a. However, as may be seen from the figure, the observed intermolecular relaxation rate

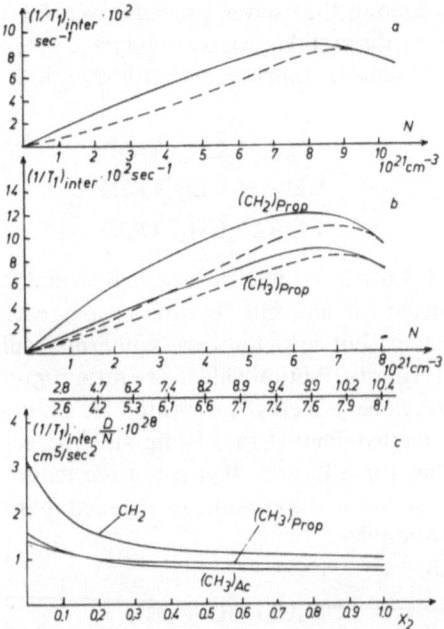

Fig. 2. Intermolecular relaxation rates of protons in (a) $CH_3COOD + D_2O$ and (b) $CD_3CH_2COOD + D_2O$ and $CH_3CD_2COOD + D_2O$ as a function of the acid concentration N, given as acid molecules per cm^3. Dashed curves: calculated intermolecular relaxation rates. (c) $(1/T_1)_{inter} \cdot D/N$ as a function of the mole fraction of acid x_{ac}. At the top of (c) the two concentrations in units of 10^{21} molecules per cm^3 corresponding to the mole fraction are given.[3]

is larger than the predicted one. One concludes that the local acid concentration around a given acid molecule is greater than the mean acid concentration in the solution. The question may be raised: Is this association a hydrophobic association? In order to give an answer to this question we investigated the proton intermolecular relaxation rates in the system

$$CH_3CD_2COOD + D_2O$$

and in the system

$$CD_3CH_2COOD + D_2O$$

The results are shown in Fig. 2b. Again the dashed curves are the calculated relaxation rates if one assumes uniform distribution; here as well one finds a positive deviation of the observed from the predicted relaxation rate, which means association. But it is clearly seen that the positive deviation is greater for the methylene–methylene proton interaction than for the methyl–methyl proton interaction. This indicates that intermolecular contact is more pronounced around the methylene group. Thus the tendency of association of the carboxylic groups among themselves prevails over the tendency of hydrophobic association of the methyl groups.

We have made similar experiments with the three species of butyric acid

$$CH_3CD_2CD_2COOD$$
$$CD_3CH_2CD_2COOD$$
$$CD_3CD_2CH_2COOD$$

dissolved in D_2O. It turned out that the excess intermolecular relaxation rate over the one predicted for uniform distribution is greater here than for the propionic acid system; but now the excess intermolecular relaxation rates relative to the uniform distribution values are rather much alike for all three groups, the β-methylene–β-methylene proton relaxation effect being the strongest and the methyl–methyl rate being still the smallest. This gives a certain evidence that for a longer alkyl chain the tendency of hydrophobic association becomes more important. A detailed paper describing these results is in preparation.[4]

For the systems

$$CH_3CD_2OD + D_2O$$

and

$$CD_3CH_2OD + D_2O$$

we found practically no deviation from the behavior as predicted for uniform solute distribution (see also refs. 10 and 11). Thus, hydrophobic association seems to be absent here, that is, seen from a given methylene group, all other methylene groups have random distribution. The same holds true for the methyl group. In the literature[12] there are measurements of the intermolecular relaxation rate of the alkyl protons of ethanol and propanol in their mixture with CCl_4 which, though not particularly accurate, demonstrate essentially the same behavior as described for the aqueous systems (see also ref. 3).

The next series of experiments involve the systems

$$CH_3CD_3COOD + CD_3CH_2COOD + D_2O$$

and

$$CH_3CD_2OD + CD_3CH_2OD + D_2O$$

In both these systems the methylene proton intermolecular relaxation rate is caused by methylene–methylene and methylene–methyl proton interactions. Since the methylene–methylene contribution is known from previous experiments, the methylene–methyl interaction can be determined. Thus, one gets the closest distance of approach of the methyl and the methylene group towards a reference methylene group separately, which gives information about the orientation of the solute molecule relative to a solute methylene.

We obtained the result that in the first case (propionic acid) there is a slight preferential orientation such that the methylene protons come closer to the reference methyl group, whereas methyl and methylene protons have the same closest distance of approach with respect to the reference methylene group. In the other case (ethanol) the methylene group (or methyl group) points slightly away from the reference methyl group (or methylene group). Further details will be given in a forthcoming paper.[4]

5. HYDRATION STRUCTURE OF THE I⁻ ION[5,6]

Iodide (I⁻) is a typical structure breaking ion in the sense that aqueous solutions of KI have a fluidity which is higher than that of pure water. We wished to answer the following questions: (a) What is the orientation of the water molecules in the first hydration sphere of I⁻? (b) Is the first hydration sphere or the second hydration sphere the region of the faster molecular motion? An experiment as described for the F⁻ ion cannot be performed because the ^{127}I nucleus relaxes by quadrupole interaction, but one can measure the proton relaxation rate in the solution

$$\text{cation} + \text{I}^- + (x_H - 1)\text{D}_2\text{O} + x_H\text{H}_2\text{O}$$

where x_H is the mole fraction of H_2O in its mixture with D_2O. The quantity x_H has to be chosen as small as possible, say, 3%. Then the proton–proton contribution to the relaxation rate is very small, and the contribution due to the proton–iodide magnetic dipole–dipole interaction becomes detectable. This method has been applied previously for an investigation of the hydration structure of Al^{3+}.[13] Equation (2) only gives the product of an interaction factor, which is determined by the structure of the hydration sphere, and the rotational correlation time. Since both these factors are the object of our investigation, i.e., both of them are unknown, Eq. (2) cannot be applied here. We have to use Eq. (1) which allows the determination of the interaction factor and the reorientation time independently. Assume that the time dependence of the propagator in Eq. (1) is essentially given by $\exp(-t/\tau_c)$, where τ_c is the rotational correlation time. As the temperature decreases, τ_c increases. Now it may be shown that $1/T_1$ goes through a maximum when $\tau_c = k\omega^{-1}$, where k is a constant given by the theory and ω is the resonance frequency. The qualitative behavior is depicted in Fig. 3. Thus the experimental determination of the maximum of $1/T_1$ gives τ_c. The absolute value of $1/T_1$ then allows the determination of the interaction factor which implies the structural information. Actually, the interaction factor is proportional to the product $2n_c/a^6$, where n_c is the first coordination number of the ion.

In aqueous solutions it might be difficult to reach a sufficiently low temperature at which $\tau_c = k\omega^{-1}$. For this reason, we made an additional experiment which was contrived in order to support the results to be obtained with the aqueous system. We measured the proton relaxation rate in a solution

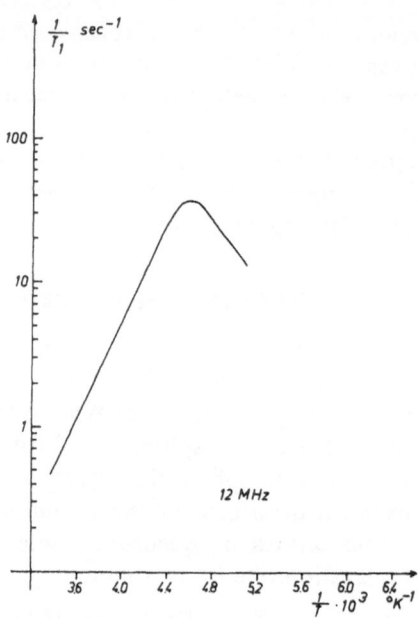

Fig. 3. Typical behavior of $1/T_1$ (sec^{-1}) plotted over the reciprocal temperature $1/T$ ($^{\circ}$K^{-1}). The resonance frequency is known to be 12 MHz; thus, at the maximum $\tau_c = 1/(2\pi \cdot 12 \cdot 10^6)$ sec.

of KI in glycerol. It is known that KI is a "structure-breaking" salt in glycerol as well; i.e., the fluidity of glycerol containing KI is greater than that of pure glycerol.[14, 15, 16] On the other hand, the temperature range where $\tau_c = k\omega^{-1}$ is easily attainable. The corresponding measurements were done with the system

$$\text{K}^+ + \text{I}^- + \text{C}_3\text{D}_5(\text{OD})_3(1 - x_{\text{H}}) + (x_{\text{H}}/2)\,\text{C}_3\text{D}_4\text{H}(\text{OD})_2(\text{OH})$$

where x_{H} is the mole fraction of proton containing glycerol molecules. With our experiments $x_{\text{H}} = 4\%$, but the locations of the protons were not actually known, contrary to the schematic formulation given above. The proton relaxation rate of the above solution was compared with that of pure glycerol $\text{C}_3\text{H}_5(\text{OH})_3$. For both these liquids, the maximum of the relaxation rate was observed. For the solution containing KI, the maximum appeared at a lower temperature. This means that $\tau_c(p\text{I}^-)$, the correlation time for the reorientation of the vector connecting a glycerol proton with the iodide nucleus, is shorter than that of the average proton–proton vector in neat liquid glycerol $\tau_c(pp)$. Using suitable activation energies we found: $\tau_c(pp)/\tau_c(p\text{I}^-) = 2.9$ at $\theta = -18.4^{\circ}$C and 2.4 at $\theta = +6.9^{\circ}$C. Thus, we have demonstrated that the region of faster molecular motion is actually localized at the ion and is not an effect of the second or third solvation layer.

In the aqueous system the corresponding experiments were made in solutions of 6 m ^{7}LiI and ^{6}LiI. ^{7}Li was replaced by ^{6}Li in order to remove the proton–^{7}Li contribution to the relaxation rate. The maxima of the relaxation rates were indeed observable at about $-90°$C, but at room temperature the proton–iodine relaxation contribution turned out to be so high that it would be entirely unreasonable to ascribe it to the magnetic dipole–dipole interaction between the proton and the ^{127}I nucleus. There must be some other relaxation mechanism—perhaps spin rotation interaction. The effect awaits further elucidation, and the question originally envisaged cannot be answered yet.

6. ION–ION CONFIGURATION WITH LOCAL SYMMETRY

The understanding of the nuclear magnetic relaxation by electric quadrupole interaction in aqueous alkali halide solutions has so far presented appreciable difficulties.[18-26] To a first approximation, the ions in the solution represent point charges. The thermal motion of these point charges gives rise to a fluctuating field gradient at the relaxing nucleus, and this in turn should form an effective cause of nuclear magnetic relaxation; but in many cases this relaxation is much smaller than expected. The reason for this discrepancy is now clear: The ion cloud surrounding a given ion screens this ion entirely, and thus an ion which is sufficiently distant from the reference ion behaves as if it had the charge zero.[17] Consequently, only those ions which are in the close neighborhood of the reference ion (the relaxing ion) contribute to the relaxation process, and indeed, for instance, the increase of the relaxation rate of ^{35}Cl with increasing RbCl concentration can be explained in terms of this ionic electrostatic interaction when correlation time effects of water are taken account of properly. Now, if the Rb$^+$ causes a strengthening of the ^{35}Cl relaxation, one should expect that in the same way ^{35}Cl$^-$ ions increase the relaxation of ^{87}Rb. But, contrary to the expected behavior, the ^{87}Rb relaxation rate decreases with increasing RbCl concentration. Part of this effect has to be ascribed to the decrease of water reorientation time, but where is the effect of the point charges, i.e., the Cl$^-$ ions? The explanation of this effect is not clear as yet. One might speculate that the local symmetry of the Cl$^-$ ions around the Rb$^+$ ion is such that their contribution to the field gradient is reduced almost to zero. Another interesting observation is the fact that, similarly, in RbBr solution the effect of the Br$^-$ ions on the ^{87}Rb relaxation almost vanishes. However, if one studies the mixture RbCl$^+$ RbBr, then with 50 % Cl$^-$ there is a slight maximum in the relaxation rate, which would indicate that the local symmetry around Rb$^+$ is disturbed when the two unlike ions, Cl$^-$ and Br$^-$, are approaching. Furthermore, in RbF solution, the F$^-$ ion causes a marked increase of the ^{87}Rb relaxation rate; but if one takes account of the increase of the water correlation time, then he finds that the effect of ^{87}Rb$^+$–F$^-$ interaction is smaller than it should be. Thus, again there seem to be

$Rb^+F^--F^-$ configurations with particular structural properties quenching the quadrupole interaction. These particular configurations should be observable via the ^{19}F intermolecular relaxation rate. ^{19}F relaxes by magnetic dipole dipole interaction. In fact, we were able to observe an excess F^--F^- intermolecular relaxation rate over and above the one to be expected for uniform F^--F^- distribution. As mentioned above, we are as yet unable to give a definite explanation of our observations, but we think it is worthwhile to mention these interesting effects. Further investigation of the details is in progress at our laboratory.

REFERENCES

1. A. Abragam, *The Principles of Nuclear Magnetism* (Oxford University Press, 1961). T. C. Farrar and E. D. Becker, *Pulse and Fourier Transform NMR* (Academic Press, New York and London, 1971).
2. H. G. Hertz and C. Rädle, *Ber. Busenges. Physik. Chem.* (1973), in press.
3. R. Göller, H. G. Hertz, and R. Tutsch, *Pure Appl. Chem.* **32**, 148 (1972).
4. H. G. Hertz and R. Tutsch, to be published.
5. A. Geiger and H. G. Hertz, to be published.
6. K. Hermann and H. G. Hertz, to be published.
7. H. G. Hertz, M. Holz, G. Keller, H. Versmold, and C. J. Yoon, to be published.
8. G. H. F. Diercksen and W. P. Kraemer, *Chem. Phys. Letters* **5**, 570 (1970).
9. P. Russegger, H. Lischka, and P. Schuster, *Theor. Chim. Acta* **24**, 191 (1972).
10. E. v. Goldammer and M. D. Zeidler, *Ber. Bunsenges. Physik. Chem.* **73**, 4 (1969).
11. E. v. Goldammer and H. G. Hertz, *J. Phys. Chem.* **74**, 3734 (1970).
12. M. Grüner and H. G. Hertz, *Advan. Mol. Relaxation Processes* **3**, 75 (1972).
13. H. G. Hertz, R. Tutsch, and H. Versmold, *Ber. Bunsenges. Physik. Chem.* **75**, 1177 (1971).
14. A. S. Golik, A. W. Orishchenko, and A. J. Artemchenko, *Dopovidi Akad. Nauk. Ukr. RSR*, 453 (1954).
15. H. T. Briscoe and W. R. Rinehart, *J. Phys. Chem.* **46**, 387 (1942).
16. G. Engel and H. G. Hertz, *Ber. Bunsenges. Physik. Chem.* **72**, 808 (1968).
17. H. G. Hertz, to be published.
18. H. G. Hertz, *Z. Elektrochem.* **65**, 20 (1961).
19. K. A. Valiev, *Zh. Eksper. i Teor. Fiz.* **37**, 109 (1959) [*Sov. Phys. JETP* **106**, 77 (1960)].
20. K. A. Valiev and B. M. Khabibullin, *Zh. Fiz. Khim.* **35**, 2265 [*Russ. J. Phys. Chem.* **35**, 1118 (1961)].
21. C. Deverell, *Mol. Phys.* **16**, 491 (1969).
22. R. E. Richards and B. A. Yorke, *Mol. Phys.* **6**, 289 (1963).
23. C. Deverell, D. J. Frost, and R. E. Richards, *Mol. Phys.* **9**, 565 (1965).
24. C. Hall, G. L. Haller, and R. E. Richards, *Mol. Phys.* **16**, 377 (1969).
25. C. Hall, R. E. Richards, G. N. Schulz, and R. R. Shang, *Mol. Phys.* **16**, 528 (1969).
26. H. G. Hertz, G. Stalidis, and H. Versmold, *J. Chim. Phys.*, Numéro Spécial, Octobre 1969, p. 177.

DISCUSSION

Professor R. Lumry (*University of Minneapolis*). Do the association results you gave for the acetic and propionic acids also hold for the pure liquids?

Professor Hertz (*Universität, Karlsruhe, West Germany*). No, only in the aqueous solutions. The pure liquid is a little more difficult to handle. The so-called Torrey theory for the intermolecular relaxation rate gives what is to be expected for random distribution of the solute molecules. This theory involves the distance of closest approach. We don't know that too well, and therefore it is fitted to the pure liquid acids. Then, in the aqueous mixture, as the water content increases, the apparent closest distance of approach between the alkyl groups decreases, and this means association.

Professor H. S. Frank (*University of Pittsburgh*). I was particularly interested in your remarks about the iodide ion. There is evidence from a variety of fields (spectroscopic, kinetic, and other fields) indicating that the nearest neighbor waters to the iodide ion are indeed ordered, pointing their protons towards this ion. What I have been waiting for is for the theoretical interpretation of the relaxation mechanism to come to the place where it would be able to give the right answer about the ion.

Professor Hertz. In principle, the experiment with the I^- ion was invented so as to give the correct answer. Unfortunately, however, in aqueous solution the relaxation rate turned out to be much too large. We must conclude that there is another relaxation mechanism which is so far unexplained. As regards the statement that there should be a very distinct orientation order in the first hydration sphere of an ion like iodide, I feel that the argument of Professor Frank is as follows: Since there is a hydration energy, then this hydration energy must involve orientation of water around the ion. My argument is that the hydration energy is only the disappearance of the charging energy in vacuum.

Professor Frank. You and Samoilov agree with each other perfectly, although he had the idea first in connection with the temperature dependence of ionic diffusion coefficients. The other evidence doesn't depend very much on theory at all. You have shown very beautifully in the case of the fluoride ion that the proton is pointing in, and there are several other kinds of evidence also. In the crystal structure of the clathrate hydrate fluorides, the fluoride ion doubles for a water molecule. That means it has to be tetrahedrally bonded, and so this is nice. Also, spectroscopically one looks at the OD stretch in HOD, and if some fluoride ion is placed in it, the OD stretch moves way down in frequency because it has become strongly hydrogen-bonded. There are similar things in the case of the iodide ion. People have interpreted the OD stretch frequency as being different in the case of the iodide present because the O–D is now hydrogen-bonded to I^- instead of to H_2O. Again, the enormous increase in intensity of Raman scattering of the hydrogen bending mode and stretching mode, particularly the bending mode, in the presence of the iodide, cannot be accounted for unless the protons are right up against the ion because this effect is so large. There are a variety of other things that are very hard to explain, and I gather that the situation here is really very complicated. If the OHO angle in the water molecule changed when a proton attached itself in hydrogen bonds to the iodide, then the R^6 in the denominator of the conversion factor would make some difference in some of your numbers. Whatever the water or iodide ion is doing, it should be doing it in every experiment performed, whether relaxation or spectroscopic experiments or any of a variety of others, and until they all give the same answers, we cannot claim that we have really explained it.

Professor Hertz. What we are discussing is essentially the dependence of a potential energy in the first hydration sphere on the orientation of the water molecule, that is, on the angle ϕ. Clearly, the protons probably preferentially point to the iodide. I do not think the protons are pointing the other way around. If the depth of this potential well is large as compared with the depth in pure water, this would make the first hydration sphere largely structured, whereas if the depth is smaller, it would be less structured. One should not forget that the self-diffusion coefficient of the iodide ion is 2.02×10^{-5} cm^2 sec^{-1}, and this is a huge value. Water itself has $D = 2.32 \times 10^{-5}$ cm^2 sec^{-1}, and the mass of iodide is 127! So, this enormously large self-diffusion coefficient also tells us that the forces acting on the I^- ion, i.e., the potential depths, must be small.

Professor H. L. Friedman (*State University of New York, Stony Brook*). With regard to

$$1/T_{1,2}(X) = [1/T_{1,2}(X)]_{c_{Y=0}} + \text{const} \times \int_0^{\infty} \{c_Y \int d^3 r(t) H_{XY}[r(t)] P[r(t) \leftarrow r(o)]$$
$$\times H_{XY}[r(o)] g_{XY}[r(o)] d^3 r(o)\} dt$$

I think if one states that $1/T_{1,2}(X)$ is the relaxation time of a nucleus in X, then $g_{XY}[r(o)]$ is the pair distribution function or correlation function between particles of species X, the one that is relaxing, and Y, the one that is perturbing. The r is again in each case the distance between X and Y. When is such a formula strictly applicable? I think the other terms drop out in the following circumstances. One looks at the plot (Fig. 4) of the relaxation rate of X against concentration of species Y and picks out the limiting slope (dashed line). In such a case, one can show really rigorously that it is only the pair correlation function between X and Y which contributes, and not higher ones g_{XYY}, g_{XYYY}, ... Considering T_1^{-1} or T_2^{-1} for $^{23}Na^+(aq)$, which are the same, one finds in the low-concentration range a limiting slope as shown in Fig. 4. The conclusion of Eisenstadt and myself, in this case, was that the electric

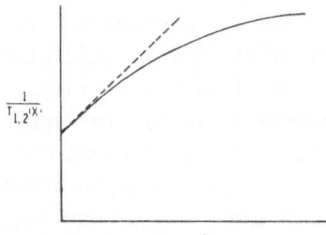

Fig. 4

field gradients of the Y species do not account for the slopes in view of what one expects for the pair correlation functions g_{XY}. That is, if Y is chloride ion the observed slope is practically zero, while Br^-, I^-, ClO_4, and PF_6^- give increasingly large slopes. These data are hard to understand on the basis of the pair correlation function g_{XY}. We agree that there is a three-body or higher-body correlation function which is needed, but it is involved here in an implicit way only. I think the one which is needed is the three-particle correlation function g_{XYW} for a particle X, a particle Y, and water. (Whether the interaction is dipolar or quadrupolar, there is a contribution from the water.) The physical picture is that the interaction of X with a neighboring W is modified when Y approaches closely. I don't know if anybody knows formally how to handle this, and we didn't try, but that is the basis for our conclusion that, at least in this kind of solution, it is quite hard to do anything very conclusive when the quadrupolar mechanism dominates.

Professor Hertz. I think I have made my talk already a bit too long. If I had explained all these things, it would have taken 10 hours. Equation (2) is only written here for dipole–dipole interaction. The difference between dipole–dipole interaction and quadrupolar interaction is essentially a little bit similar to that between coherent and incoherent neutron scattering. In dipole–dipole interactions we have no cross terms. Those terms which are written as triple configurations are not important for dipole–dipole interactions, just as with neutron scattering on protons we have spin-up and spin-down configurations. Here, with quadrupole interaction, the sign of the charge and, consequently, of the interaction at a given relative position, is a constant quantity. Equation (2) presented here is also the self-correlation function for the field gradient of a given particle. One can indeed form, and one has done so, the theory of intermolecular quadrupolar relaxation rate in electrolyte solutions using the self-correlation functions alone; but the results are wrong, and the calculated relaxation rates generally are much too large. The correct theory has to start from the correlation function of the instantaneous total field gradient at the relaxing nucleus:

$$\int p(^{w}\mathbf{R}_1^0, {}^{w}\mathbf{R}_2^0, \ldots {}^{i}\mathbf{R}_1^0, {}^{i}R_2^0, \ldots) V(^{w}\mathbf{R}_1^0, {}^{w}\mathbf{R}_2^0, \ldots {}^{i}\mathbf{R}_1^0, {}^{i}\mathbf{R}_2^0, \ldots)$$

$$\times P(^{w}\mathbf{R}_1^0, {}^{w}\mathbf{R}_2^0, \ldots {}^{i}\mathbf{R}_1^0, {}^{i}\mathbf{R}_2^0, \ldots {}^{w}\mathbf{R}_1, {}^{w}\mathbf{R}_2, \ldots {}^{i}\mathbf{R}_1, {}^{i}\mathbf{R}_2, \ldots t)$$

$$\times V(^{w}\mathbf{R}_1, {}^{w}\mathbf{R}_2, \ldots {}^{i}\mathbf{R}_1, {}^{i}\mathbf{R}_2, \ldots) d^{w}\mathbf{R}_1^0 d^{w}\mathbf{R}_2^0 \ldots d^{i}\mathbf{R}_1 d^{i}\mathbf{R}_2 \ldots$$

Here, $^{w}\mathbf{R}_j$ is the generalized position vector of the jth water molecule and takes into account the orientation of the water molecule, and $^{i}\mathbf{R}_j$ is the position of jth ion. Superscript ^{0} denotes at time 0; without the superscript, at time t. The term $V(\ldots)$ is the electric field gradient in the configuration $^{w}\mathbf{R}_1^0, {}^{w}\mathbf{R}_2^0, \ldots$ $^{i}\mathbf{R}_1^0, {}^{i}\mathbf{R}_2 \ldots$; $p(^{w}\mathbf{R}_1^0, {}^{w}\mathbf{R}_2^0, \ldots {}^{i}\mathbf{R}_1^0, {}^{i}\mathbf{R}_2^0 \ldots)$ is the n-particle distribution function; and $P(^{w}\mathbf{R}_1^0, \ldots {}^{w}\mathbf{R}_1, \ldots, t)$ is the n-particle propagator. The first step in treating this expression is to neglect all higher-particle correlations except two-particle correlations. Thus, one is left with a sum of self-correlation functions and of water–water, water–ion, and ion–ion cross-correlation functions. The latter cross-correlation functions are the first-order corrections of the self-correlation functions. Thus, in Friedman's formula for $1/T_{1,2}(X)$ one has to add not only the one term he has given but also a number of terms which may all be of the same order of magnitude and which may have positive and negative signs. Within this framework I have calculated the quadrupolar relaxation rates. The problem of the collective propagator has been approximated in a suitable way. If there is cubic symmetry around the relaxing ion, then there is complete quenching of the quadrupolar relaxation rate. Of course, this cannot be accounted for within the framework of a two-particle cross-correlation treatment because two particles cannot form a surrounding of such symmetry; however, in the case of cubic symmetry, the computation of the n_c-particle cross-correlation function is relatively easy (n_c = number of particles forming the surroundings of cubic symmetry). All these computations have been done, but they are not yet published for the following reason: We have quite a number of experimental data which have to be explained in terms of the theory. This work is in progress—and part of this work I have mentioned briefly in my talk. If the full job is done and the theory turns out to be satisfactory, it will be published; but there is one thing we have not done yet: this is the interpretation of the experiments with Na^+ which Friedman spoke about. The reason is simply that one thing has to be done after the other, but we shall certainly tackle the $NaClO_4$ problem when we are through with the other systems.

Structural Aspects of Aqueous Tetraalkylammonium Salt Solutions[1]

Wen-Yang Wen[2]

Received November 7, 1972; revised January 10, 1973

Various thermodynamic and spectroscopic investigations on aqueous tetra-alkylammonium salt solutions are reviewed from a structural viewpoint. At present, it is not yet clear what kind of water structure is formed around the alkyl groups of the large cations. However, the importance of "cosphere" overlap and ion–ion interaction (cation–anion, cation–cation, cation–anion-cation, etc.) in determining the solution properties are emerging more clearly. In this regard, model calculations based on the approach of Friedman and his co-workers are expected to be of considerable value.

KEY WORDS: Tetraalkylammonium salts; aqueous solutions; structural aspects; ion–ion interaction.

1. INTRODUCTION

We begin by reviewing various physicochemical methods employed in recent experimental and theoretical investigations of aqueous solutions of tetra-alkylammonium salts. For this purpose we enlarge a similar table presented by Franks[1] and examine a long list as shown in Table I. It covers a wide range of methods but is still of limited scope selected mainly according to the interest of the author. The table is divided into four categories: thermodynamic properties, transport properties, spectroscopic properties, and theory and original concept. In spite of these many different methods of attack by various workers in these fields, at the outset we have to recognize that few agreements have been reached among the investigators on the basic structure of these solutions. This unsatisfactory state of affairs is a reflection on the complexity of the liquid water, in general, and of aqueous solutions of nonpolar solutes, in particular. With a steady improvement in experimental techniques, newer and more refined experimental results are obtained. Some of the newly

---

[1] This paper was presented at the symposium, "The Physical Chemistry of Aqueous Systems," held at the University of Pittsburgh, Pittsburgh, Pennsylvania, June 12–14, 1972, in honor of the 70th birthday of Professor H. S. Frank.
[2] Department of Chemistry, Clark University, Worcester, Massachusetts 01610.

obtained data appear to cast doubt on the older assumptions. As examples, let us consider the following thermodynamic properties.

2. NEW OBSERVATIONS IN THERMODYNAMIC PROPERTIES

First Example. The osmotic coefficients for tetraalkylammonium halide (R_4NX) series at 25°C and at concentrations below 1 m have been found by· the isopiestic comparison method to be in the order given below:

$$Bu_4N^+ > Pr_4N^+ > Et_4N^+ > Me_4N^+ \quad \text{for } R_4NF^{(2)} \text{ and } R_4NCl^{(3)}$$
$$Bu_4N^+ < Pr_4N^+ < Et_4N^+ < Me_4N^+ \quad \text{for } R_4NBr^{(3)} \text{ and } R_4NI^{(3)}$$

Table I. Methods Used in the Investigations of Aqueous Tetraalkylammonium Salt Solutions

I. Thermodynamic properties	Investigators
Osmotic and activity Coefficients	
Freezing point depression	Ebert and Lange (1928)[a]
Isopiestic comparison	Lindenbaum and Boyd (1964)[3]
Emf of concentration cells with transference	Frank and Rupert (1969)[4]; Frank and Ku (1971)[45]
Enthalpy and entropy of dilution	Lindenbaum (1966)[5]; Wood et al. (1967)[6]
Heat capacity	Frank and Wen (1957)[b]; Ackermann et al. (1969)[c]
Partial molal volume	Wen and Saito (1964)[8]; Franks and Smith (1967)[9]; Conway, Verrall, and Desnoyers (1966)[d]
Partial molal expansibility	Franks and Smith (1967)[9]; Millero and Drost-Hansen (1968)[e]
Partial molal compressibility	Conway and Verrall (1966)[f]
Free energy of mixing	Wen, Miyajima, and Otsuka (1971)[g]
Heat of mixing	Wood and Anderson (1967)[h]
Volume of mixing	Wen et al. (1967)[i]
Enthalpy of transfer	Krishnan and Friedman (1969)[j]
Surface tension	Tamaki (1967)[k]
II. Transport properties	
Ionic conductance	Kraus et al. (1951)[l]; Kay et al. (1966)[m]
Viscosity	Bingham (1941)[n]; Kay et al. (1966)[o]
Heat of transport	Agar and Price (1961)[p]
Diffusion coefficient	Hertz, Lindman, and Siepe (1969)[q]
III. Spectroscopic properties	
Dielectric relaxation	Haggis et al. (1952)[r]; Pottel and Lossen (1967)[37]
¹H chemical shift of water	Hertz and Spalthoff (1959)[23]; Davies et al. (1971)[28] Hertz and Pfliegel (1972)[29]
¹H and ²D relaxation of ion and water	Hertz and Zeidler (1964)[20]
¹⁷O relaxation	Fister and Hertz (1967)[s]
⁷⁹Br relaxation	Forslind et al. (1968)[t]

Table I. Continued

IR absorption	Gordon et al. (1964)[u]; Worley and Klotz (1966)[v]; Bunzl (1967)[w]
Ultrasonic absorption	Blandamer et al. (1968)[(38)]; Atkinson et al. (1968)[41]
UV absorption of I⁻ in R₄NCl solution	Blandamer and Fox (1970)[x]
X-ray diffraction	
Solid hydrates	Jeffrey et al. (1959–1963)[(33),y]
Bu₄NF solution	Narten and Lindenbaum (1969)[(34)]
Raman spectra of Bu₄NCl solution	Walrafen (1971)[(36)]
EPR of probes in Bu₄NBr solution	Jolicoeur and Friedman (1971)[z]

IV. Theory and original concept

Concept of "iceberg"	Frank and Evans (1945)[aa]
Model calculation	Friedman, Ramanathan, and Krishnan (1972)[(42)]

[a] L. Ebert and J. Lange, Z. Phys. Chem. **139A**, 584 (1928); J. Lange, ibid., **168A**, 147 (1934).

[b] H. S. Frank and W.-Y. Wen, Disc. Faraday Soc. **24**, 133 (1957).

[c] H. Ruterjans, F. Schreiner, U. Sage, and T. Ackermann, J. Phys. Chem. **73**, 986 (1969).

[d] B. E. Conway, R. E. Verrall, and J. E. Desnoyers, Trans. Faraday Soc. **62**, 2738 (1966).

[e] F. J. Millero and W. Drost-Hansen, J. Phys. Chem. **72**, 1758 (1968).

[f] R. E. Verrall and B. E. Conway, J. Phys. Chem. **70**, 3961 (1966).

[g] W.-Y. Wen, K. Miyajima, and A. Otsuka, J. Phys. Chem. **75**, 2148 (1971).

[h] R. H. Wood and H. L. Anderson, J. Phys. Chem. **71**, 1871 (1967).

[i] W.-Y. Wen and K. Nara, J. Phys. Chem. **71**, 3907 (1967); ibid., **72**, 1137 (1968); W.-Y. Wen, K. Nara, and R. H. Wood, ibid. **72**, 3048 (1968).

[j] C. V. Krishnan and H. L. Friedman, J. Phys. Chem. **73**, 3934 (1969).

[k] K. Tamaki, Bull. Chem. Soc. Japan **40**, 38 (1967).

[l] H. M. Daggett, E. J. Bair, and C. A. Kraus, J. Am. Chem. Soc. **73**, 799 (1951); E. L. Swarts and C. A. Kraus, Proc. Nat. Acad. Sci. (U.S.) **40**, 382 (1954); R. W. Martel and C. A. Kraus, ibid. **41**, 9 (1955).

[m] R. L. Kay and D. F. Evans, J. Phys. Chem. **69**, 4216 (1965); ibid., **70**, 366, 2325 (1966); D. F. Evans, G. P. Cunningham, and R. L. Kay, ibid. **70**, 2974 (1966).

[n] E. C. Bingham, J. Phys. Chem. **45**, 885 (1941).

[o] R. L. Kay, T. Vituccio, C. Zawoyski, and D. Evans, J. Phys. Chem. **70**, 2336 (1966).

[p] J. N. Agar, in Advances in Electrochemistry and Electrochemical Engineering (Interscience, New York, 1963), Vol. 3, Chap. 2. Table II, p. 96.

[q] H. G. Hertz, B. Lindman, and V. Siepe, Ber. Bunsenges, Physik. Chem. **73**, 542 (1969).

[r] G. H. Haggis, J. B. Hasted, and T. J. Buchanan, J. Chem. Phys. **20**, 1452 (1952).

[s] F. Fister and H. G. Hertz, Ber. Bunsenges. Physik. Chem. **71**, 1032 (1967).

[t] B. Lindman, S. Forsen, and E. Forslind, J. Phys. Chem. **72**, 2805 (1968).

[u] H. Yamatera, B. Fitzpatrick, and G. Gordon, J. Mol. Spectry. **14**, 268 (1964).

[v] J. D. Worley and I. M. Klotz, J. Chem. Phys. **45**, 2868 (1966).

[w] K. W. Bunzl, J. Phys. Chem. **71**, 1358 (1967).

[x] M. J. Blandamer and M. F. Fox, Chem. Rev. **70**, 59 (1970).

[y] R. McMullan and G. A. Jeffrey, J. Chem. Phys. **31**, 1231 (1959); ref. 33.

[z] C. Jolicoeur and H. L. Friedman, J. Phys. Chem. **75**, 165 (1971); Ber. Bunsenges. Physik. Chem. **75**, 248 (1971).

[aa] H. S. Frank and M. W. Evans, J. Chem. Phys. **13**, 507 (1945).

However, the emf measurements of concentration cells with transference by Frank and Rupert[4] have shown that the osmotic coefficient curves of Et_4NCl and Bu_4NCl cross each other at about 0.09 m. Below this cross-over concentration, the osmotic coefficient of Et_4NCl is greater than that of Bu_4NCl, suggesting that, at high dilutions, the curves of Pr_4NCl and Me_4NCl also cross so that the order in osmotic coefficient becomes $Bu_4N^+ < Pr_4N^+ < Et_4N^+ < Me_4N^+$, which is the exact opposite of the order of R_4NCl series at concentrations greater than about 0.1 m.

Second Example. Heats of dilution. Previous results of calorimetric measurements at 25°C by Lindenbaum[5] and Wood *et al.*[6] indicated that the order of apparent molal heat content ϕ_L is given by $Bu_4N^+ > Pr_4N^+ > Et_4N^+ > Me_4N^+$ for all halides over the concentration range from a moderately high concentration down to about 0.2 or 0.3 m. This order for ϕ_L is generally expected and assumed to be applicable at concentrations below 0.1 m. Recent measurements of ϕ_L of the bromide series by Wood and Levine[7] have yielded, however, some surprising results, as shown in Fig. 1. The ϕ_L values of Et_4NBr are below those of Me_4NBr, and the curve of Pr_4NBr crosses over that of Me_4NBr at around 0.05 m. At the lowest experimental concentration of slightly above 0.01 m, the order of apparent molal heat content is Bu_4NBr the largest, Me_4NBr next, then Pr_4NBr, and Et_4NBr the smallest. This order of the bromide series does not make sense at all unless, at even lower concentrations, the relative order of ϕ_L is $Me_4N^+ > Et_4N^+ > Pr_4N^+ > Bu_4N^+$. These crossings of curves and reversing of the relative order with respect to ion size, if confirmed, seems to indicate that the "cospheres" or the solvent structural regions surrounding the large cations begin to overlap with each other at concentrations below 0.1 m.

Third Example. Apparent Molal Volumes ϕ_v. It is well known that the apparent molal volumes of the larger tetraalkylammonium salts, when plotted against the square root of concentration, approach the Debye–Hückel

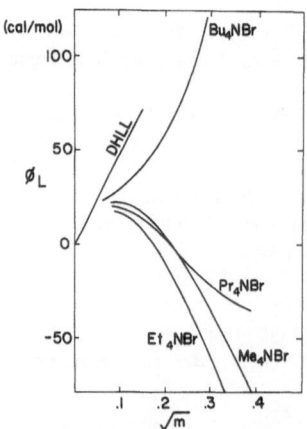

Fig. 1. Apparent molal heat content ϕ_L plotted against the square root of molality $m^{1/2}$ for tetraalkylammonium bromides in water at 25°C.[7]

limiting law from the negative side. The large negative deviation has been attributed to the special water–hydrophobic ion interaction.[8, 9] However, this behavior of the apparent molal volume appears to be not unique for solvent water alone. Gopal and Siddiqi[10] and Gopal and Singh[11] found that the ϕ_v of some larger R_4NX in N-methylacetamide and in formamide also decreases with increasing concentration.

Fourth Example. Osmotic Coefficients in Nonaqueous Solvents. Kreis and Wood[12] have reported that the trends of osmotic and activity coefficients of R_4NX in N-methylacetamide are very similar to that in water. They contended that the water-structure-enhancing properties of the larger R_4N^+ ions cannot be responsible for the trends.

Thus, the hydrogen-bonding organic solvents with high dielectric constants, such as formamide and N-methylacetamide, seem to have something in common with the solvent water when R_4NX salts are dissolved. That something in common may include the spatial arrangement of voids for accommodation of ions, a partial penetration of anion into the space between the alkyl chains of the cation, and the part of free energy brought out by the enthalpy–entropy compensation.

On the other hand, one should also keep in mind that some of our previous observations concerning the uniqueness of solvent water are still valid up to this moment. Large differences still exist when we compare certain physical properties of R_4NX salts in water against that in hydrogen-bonding organic solvents. One of these is the heat of dilution. Though the trends in osmotic coefficients are rather similar, as mentioned earlier, the heats of dilution of Bu_4NBr in N-methylacetamide and in water are quite different. Figure 2 shows the results of Wood and Falcone,[13] where the dashed curves represent ϕ_L of Bu_4NBr and Me_4NBr in water whereas the unbroken curves are ϕ_L of Bu_4NBr, Me_4NBr, and various other salts in N-methylacetamide. The large difference in the case of Bu_4NBr is rather striking.

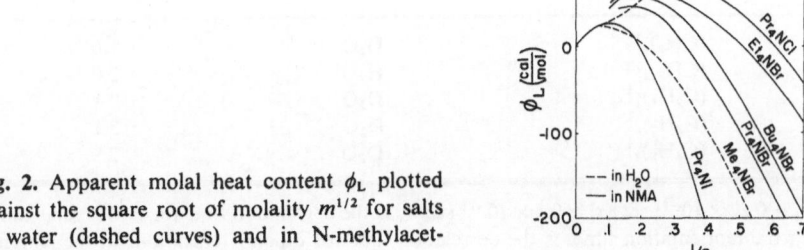

Fig. 2. Apparent molal heat content ϕ_L plotted against the square root of molality $m^{1/2}$ for salts in water (dashed curves) and in N-methylacet-amide (unbroken curves) at 25°C.[13]

In addition to the enthalpy and entropy of dilution, aqueous solution properties which seem to show substantial differences from organic solution properties are the temperature coefficient of the viscosity B coefficient,[14] heat of transfer,[15] heat capacity,[16, 17, 18] and the limiting Walden Product, $\lambda_0^+\eta_0$.[19] These differences have been interpreted qualitatively by the statement that the larger tetraalkylammonium ions (Pr_4N^+, Bu_4N^+, and i-Am_4N^+) enhance the structure of water.

3. NUCLEAR MAGNETIC RELAXATION

As seen in Table II, the well-known work of the spin-lattice relaxation rate ($1/T_1$) of the nuclear quadrupole resonance of D_2O deuterons by Hertz and Zeidler[20] showed that the reorientation time of D_2O molecules around tetraalkylammonium cations increases by a factor of about 1.6 to 3 when compared to that in pure water at the same temperature. At first this result was taken as an evidence for supporting the existence of the "iceberg effect" or the "hydration of the second kind."[20] But soon questions are raised regarding whether the increase in reorientation time is due to the strengthening of the H_2O–H_2O hydrogen-bonding or just due to the increased hindrance in rotational motion caused by the tighter packing of solute particles and water (expected from the smaller partial molal volumes of the salts) with little or no increase in the hydrogen bonding. A factor of 2 in the increase of the correlation time (from 0.27×10^{-11} to 0.53×10^{-11} sec as shown in Table III) corresponds to the decrease of temperature of pure water from 25° to 0°C, which is by no means a small effect as far as the hydrogen bonding of water is concerned. Yet the second possibility—the "wall effect" on the rotational hindrance—cannot be dismissed completely, since even the $(CH_3)_4N^+$ ion, which is not a structure-making ion, was found to increase the reorientational time of water by a factor of 1.6. We have to conclude, therefore, that the argument for the

Table II. Ratio of the Correlation Time (and Reorientation Time) of the Water Molecule in the Hydration Shells to the Corresponding Time in Pure Water at 25°C (Hertz and Zeidler)[20]

Ion	Water molecule	$\tau_c/\tau_c^* \approx \tau_r/\tau_r^{*a}$
$(CH_3)_4N^+$	D_2O	1.6
$(CD_3)_4N^+$	H_2O	2.0
$(C_2H_5)_4N^+$	D_2O	2.1
$(C_3H_7)_4N^+$	D_2O	3.1
$(C_4H_9)_4N^+$	D_2O	2.9

^{a} $\tau_c^* = 0.27 \times 10^{-11}$ sec; $\tau_r^* \approx 0.8 \times 10^{-11}$ sec. τ_c is the correlation time for NMR relaxation; τ_r, the reorientation time, is the correlation time for dipolar relaxation. In the simplest case, $\tau_r = 3\tau_c$.

Table III. Correlation Time τ_c for the Reorientation of the H_2O Molecule in Pure Water (Krynicki[21]; Hertz[22])

t (°C)	$\tau_c \times 10^{11}$ (sec)	t (°C)	$\tau_c \times 10^{11}$ (sec)
0	0.53	60	0.140
10	0.39	70	0.120
25	0.27	80	0.105
40	0.20	90	0.092
50	0.165	100	0.080

hydrogen bonding of water is not yet clear from these results of the nuclear magnetic relaxation rate studies alone.

4. CHEMICAL SHIFTS

As an example of difficulties in interpreting experimental results in structural terms, one can cite the chemical shift of water. The chemical shift of the proton resonance of water in aqueous solutions of interest to us was first reported by Hertz and Spalthoff.[23] The water proton shift in solutions of tetraalkylammonium bromides and nitrates at 25°C were found to be upfield with respect to pure water. It means that in these solutions the diamagnetic screening of the protons by the electron clouds is stronger than that in pure water. Since an upfield shift is observed for pure water with increasing temperature at a rate of about 0.01 ppm/°C, it appears that the H bonds between water molecules are weakening and the water structure is breaking down with the addition of the salts. Kay has confirmed the upfield shift for Bu_4NBr,[24] and, recently, Losurdo and Wirth also reported an upfield shift for the bromide series at both 25 and 65°C.[25] It occurred to us that this upfield shift might be due to the use of the methyl proton of the R_4N^+ ion as the internal reference, and the internal reference may shift with the change of concentration. We therefore measured the chemical shift with respect to the external reference tetramethylsilane and then applied the bulk susceptibility correction by direct NMR measurements of the side bands using the technique of Malinowski et al.[26] The results are shown in Fig. 3.[27] The lines with circles are our results and the dashed lines are those of Hertz and Spalthoff. The internal methyl-proton resonance is found to move slightly downfield with the salt concentration, but the error due to this cause is rather small. In these measurements the chemical shift is seen to increase further upfield with the increase of the cation size—an indication of the structure-breaking effect of the large cations.

Davies, Ormondroyd, and Symons[28] reported, however, that the order of molal salt shift is reversed at temperatures below about 15°C, so that, in the bromide series, Bu_4NBr gives the least upfield shift and Me_4NBr the largest

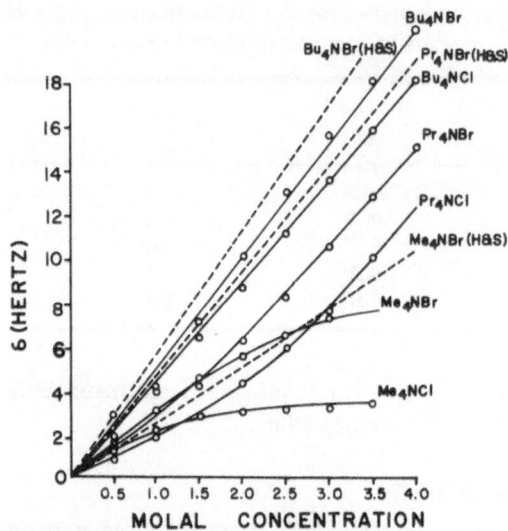

Fig. 3. Water-proton chemical shift δ plotted against the molal concentration of the tetra-alkylammonium halides with internal reference (dashed lines, ref. 23) and with external reference (unbroken lines, ref. 27).

upfield shift. Subtracting the estimated contribution of Br^- ion from the R_4NBr series, they obtained the molal ion shifts for R_4N^+ ions at infinite dilution as a function of temperature. As seen in Fig. 4, the chemical shifts at $0°C$ are found to be downfield with respect to pure water at the same temperature, and the downfield shift increases with the cationic size in going from Me_4N^+ to Bu_4N^+.

Hertz and Pfliegel[29] have determined the hydroxyl-proton chemical shifts for solutions of various tetraalkylammonium halides in methanol,

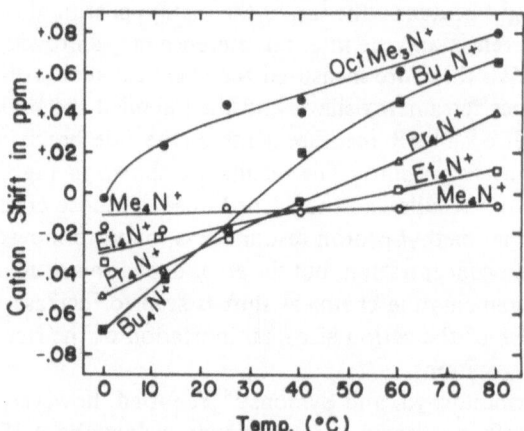

Fig. 4. Water-proton chemical shift per mole of cation plotted against temperature.[28]

ethanol, and propanol and compared them with their results in water at 0, 12, and 25°C. They put forward equations relating the chemical shift to a number of quantities describing the structure of water and alcohol solutions and estimated the shifts caused by polar and nonpolar solute groups.[30] According to their analysis, the proton chemical shift for water surrounding the R_4N^+ ions is found to be slightly downfield relative to pure water at lower temperatures (12 and 0°C), in qualitative agreement with the results of Symons et al.[28] However, their conclusion is that a downfield shift in the solution of nonpolar molecules does not necessarily constitute an increase of tetrahedral configurations for water molecules. A downfield shift may take place when, for a given water molecule, there is a preferred tendency of forming H-bond coupling by its protons rather than by its lone pair of oxygen atom, as shown in Fig. 5. In other words, a downfield shift can appear because a configuration in which a certain number of lone-pair sites of oxygens are facing the hydrophobic surface is more probable than a configuration in which the same number of protons are exposed to the hydrophobic surface.

This conclusion of Hertz and Pfliegel is of particular interest since it has a bearing upon the liquid–vapor interface arrangement of water molecules proposed by Stillinger and Ben-Naim[31] from their theoretical consideration of the quadrupole moment of water. It is also related to the cavity surface effect suggested by Friedman and co-workers for the interpretation of changes of thermodynamic properties due to cosphere overlap.[32] It leads to the possibility of forming some distorted tetrahedral geometry or bent H bonds of water molecules at the nonpolar surface. In clathrates formed by some tetra-n-butylammonium salts and tetra-i-amylammonium salts, water molecules are found to maintain stoutly the local tetrahedral coordination.[33] In view of the above results, the average configuration of water molecules around the alkyl groups in the liquid state may, due to the "surface effect"

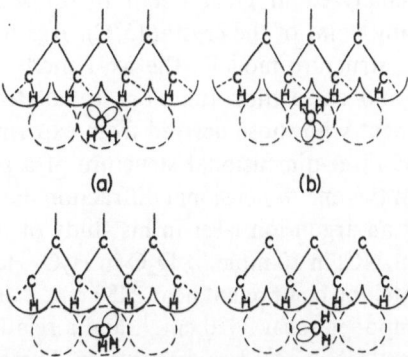

(a)   (b)

Fig. 5. Orientation of water molecules at the hydrocarbon surface. (a) Configuration in which lone-pair site or sites of oxygen atom are facing the hydrophobic surface; (b) configuration in which hydrogen atom or atoms of water are facing the hydrophobic surface.

or "cavity effect," differ appreciably from the near tetrahedral arrangement found in the crystalline clathrates.

5. X-RAY DIFFRACTION AND RAMAN SPECTRAL STUDIES

Narten and Lindenbaum[34] reported the x-ray diffraction study of 1.4 m Bu_4NF solution at 25°C and found that the radial distribution function for the solution is surprisingly similar to that for pure water at the same temperature. However, in the presence of the salt, the near-neighbor distance between water molecules decreases from 2.85 to 2.80 Å which indicates stronger hydrogen bonding in the solution. At the same time, the average number of nearest neighbors decreases from about 4.4 in pure water to about 3.8 in the 1.4 m Bu_4NF solution. In contrast, in 1.2 m NH_4F solution Danford[35] has observed the average near-neighbor distance to be virtually equal to that in pure water. They concluded, therefore, that the degree of water-structure promotion is significantly greater for Bu_4N^+ than for NH_4^+ ion, which is in agreement with thermodynamic measurements. They proceeded to consider two structural models for the solution—the gas-hydrate model and the ice-I model—and concluded that the ice-I model describes the x-ray data of the solution but the gas-hydrate model is incompatible with the experimental radial distribution function of water. They visualized a smooth transition from pure water to the aqueous Bu_4NF solution in terms of the ice-I model. Upon addition of the salt, the "interstitial" water molecules are gradually replaced by the butyl chains of the cations, and this process is accompanied by a contraction of the H-bonded network. This model, though distinguishing itself as one of the quantitative structural models of the aqueous solutions of tetraalkylammonium salts, is an oversimplification.

Recently the diffraction pattern of a solution of trimethylamine in water having the same composition as the solid clathrate hydrate, $(CH_3)_3N \cdot 10.13H_2O$, has been analyzed in great detail by Folzer, Hendricks, and Narten[52] at the melting point of the crystal (5°C). Again they have investigated the two specific structure models: the ice-I model and the clathrate model. The intensity and correlation functions calculated for *both models* are in excellent agreement with those derived from experiment. One of their conclusions is that the three-dimensional structure of a solution cannot be deduced uniquely from the one-dimensional diffraction data for a solution.

Walrafen[36] used an argon-ion laser in his study of the Raman spectra of 1.4 M and 1.8 M Bu_4NCl in 10 mole % D_2O in H_2O. He found an intensification of the OD stretching component near 2483 cm^{-1}, an intensification of the OH stretching component near 3420 cm^{-1}, and a relative lowering of the intensity of the 3525 cm^{-1} compared to pure water. These intensity changes are interpreted by him as a pronounced structuring of the water by the $(C_4H_9)_4N^+$ ion that probably involves nearly linear H bonds. These studies are preliminary in nature, and, though very valuable, it seems that more data are needed before we can draw definite conclusions.

6. DIELECTRIC AND ULTRASONIC RELAXATION

For the dielectric relaxation, the most comprehensive work so far is that of Pottel and Lossen[37] covering the frequency range of 0.5 to 38 GHz. They plotted the conductance-subtracted imaginary part ε'' of the dielectric constant (i.e., permittivity) against the real part of the dielectric constant ε', as shown in Fig. 6. The semicircle of the Cole–Cole plot (i.e., the permittivity locus diagram) for pure water has its center on the ε' axis. For the Pr_4NBr and Bu_4NBr solutions, however, each semicircle has a center lying below the ε' axis. This corresponds to a case of relaxation processes with a continuous distribution of relaxation times. It probably indicates the presence of a series of water structures rather than the simple two states such as represented by the "network" water and "interstitial" water. The principal relaxation time τ_1 of water in solution is found to change linearly with concentration m at least up to 1 m for R_4NBr solutions. The molal changes of the relaxation time of water in solution with respect to the relaxation time τ_0 of pure water, $\Delta\tau_1/(\tau_0\Delta m)$, increase at constant rate with the number of carbons and agree quantitatively with the results of NMR measurements by Hertz.[20] The relative relaxation-time change for a solution with Pr_4N^+ and Bu_4N^+ is 0.054 per mole of CH_2 per kilogram of H_2O at 25°C, which is also common for aliphatic chains with an arbitrary number of CH_2 or CH_3 groups. The fact that the alkyl groups can lengthen the relaxation time of water is attributed to the presence of the hydrophobic hydration.

Blandamer, Symons, and co-workers[38] have investigated the ultrasonic absorption properties of R_4NX solutions. As shown in Fig. 7, the ultrasonic absorption α/f^2, where α is the absorption coefficient and f the frequency, at 70 MHz increases slowly at first but then rises rapidly when the concentration exceeds 0.8 M for Bu_4NBr and Bu_4NCl. In contrast, when Et_4NBr is added, the increase is much more gradual and the absorption much smaller. As mentioned before, the partial molal volumes of Bu_4NBr in formamide de-

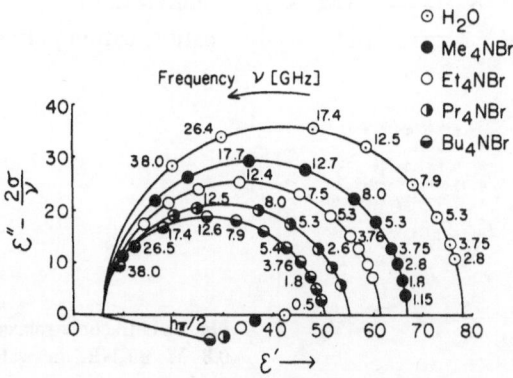

Fig. 6. The permittivity locus diagram for 1 M aqueous tetraalkylammonium bromide solutions at 25°C.[37]

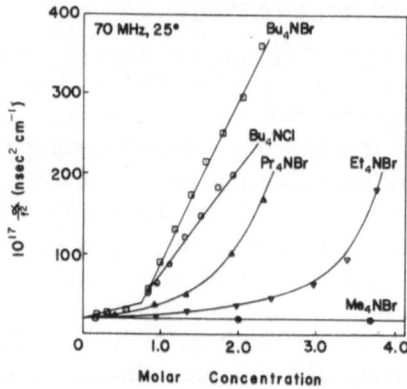

Fig. 7. Ultrasonic absorption α/f^2 plotted against molar concentration of tetraalkylammonium halides at 70 MHz and 25°C.[38]

creases with increase of the salt concentration in a manner similar to that in water. Now the ultrasonic relaxation curves for Bu_4NBr in water and in formamide at different frequencies are also similar, as seen in Fig. 8 which shows the relaxation curves for 0.8 M Bu_4NBr in water, formamide, methyl-formamide, and dimethylformamide. The ultrasonic relaxation occurring in aqueous Bu_4NBr solutions at frequencies around 10^7 to 10^8 Hz is so slow that it must be due to some aggregates formed by the cations. The concentration dependence of the ultrasonic absorption at constant frequency may be understood with the Romanov–Solovyev relations,[39] as suggested by Blandamer and Waddington.[40] They show that for Bu_4NBr solution it is the enthalpy term in the equation which is responsible for the large absorption. The onset of the large ultrasonic absorption for 0.8 M Bu_4NCl and Bu_4NBr solution may be due to the multiplet ion–ion interactions. Atkinson, Garnsey, and Tait[41] have considered the following equilibria in interpreting their ultrasonic absorption results:

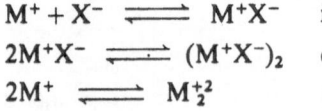

$$M^+ + X^- \rightleftharpoons M^+X^- \qquad \text{ion-pair formation}$$
$$2M^+X^- \rightleftharpoons (M^+X^-)_2 \qquad \text{dimerization}$$
$$2M^+ \rightleftharpoons M_2^{+2} \qquad \text{cation–cation pairing}$$

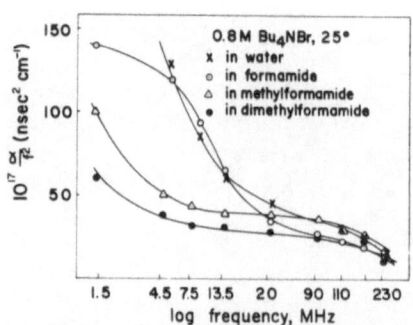

Fig. 8. Ultrasonic relaxation curves for 0.8 M Bu_4NBr in water, formamide, methylformamide, and dimethylformamide at 25°C.[38]

To these we should add the triplet formation equilibria such as

$$2M^+ + X^- \rightleftharpoons M^+X^-M^+$$
$$M^+ + 2X^- \rightleftharpoons X^-M^+X^-$$

7. FRIEDMAN'S COSPHERE-OVERLAP MODEL

The ion–ion pair interaction potential in this model is given by

$$u_{ij}(r) = (e_i e_j)/(Dr) + COR_{ij} + CAV_{ij} + GUR_{ij} \tag{1}$$

where the first term represents the Coulomb interaction of ions of charges e_i and e_j at distance r in a solvent of dielectric constant D. The second term is a repulsive core potential, and the third term is a cavity effect in a dielectric medium. The fourth term, called the Gurney potential, represents the effect of the overlap of the cospheres when the ions come close together and is assumed to be

$$GUR_{ij} = A_{ij} V_w^{-1} V_{mu}(r_i^* + w, r_j^* + w, r)$$

Here A_{ij} is the molar free-energy change of the cosphere solvent returning to its normal state when two cospheres overlap partially, V_w is the molar volume of the solvent water, and V_{mu} is the mutual volume function of a sphere of radius $r_i^* + w$ and a sphere of radius $r_j^* + w$ when the distance between their centers is r. The cosphere thickness w is taken to be the diameter of a water molecule, 2.76 Å. In this model, A_{ij} is the only parameter in Eq. (1) which is adjusted on the basis of comparison with experimental data for electrolyte solutions.

Ramanathan, Krishnan, and Friedman[42] have computed the parameters A_{ij} and found the values of A_{++} to be negative and become increasingly negative with increase of the size of cations ($A_{++} \approx 0$ for Li$^+$, $A_{++} \approx -120$ cal-mole^{-1} for Me$_4$N$^+$, and $A_{++} \approx -200$ cal-mole^{-1} for Pr$_4$N$^+$ or Bu$_4$N$^+$). Their explanation is that a certain increase in free energy is required to make a cavity, and the negative contribution to the Gurney parameter comes from the fact that the total cavity surface is reduced in the overlap process. The trend of A_{+-}, the Gurney free-energy parameter for the $+-$ pair, from F$^-$ to I$^-$ for a given R$_4$N$^+$ ion was found to be qualitatively similar to the trend of free energy of ion-pair formation in H-bonding solvents, as reported by Evans et al.[43] Their interpretation is, therefore, that the cosphere water of halide anions decreases its stability in the order F$^-$ to I$^-$ due to interactions in which the anion is the H-bond acceptor. When contact ion pairs are formed, these cospheres will be partially disrupted, and the cosphere disruption seems the easiest for the R$_4$N$^+$I$^-$ pair and the hardest for the R$_4$N$^+$F$^-$ pair.

Table IV. Gurney Free-Energy Parameters A_{ij} for Some Tetraalkylammonium Halide Solutions at 25°C (Concentration up to 0.2 M)[a]

Cation	A_{++}[b]	A_{+-}[b]			
		F⁻	Cl⁻	Br⁻	I⁻
Me₄N⁺	(−139)		(−124)	(−136)	
Et₄N⁺	−119		−143	−165	(−185)
Bu₄N⁺	−78	(−104)	−178	−213	

[a] Note that A_{--} is assumed to be zero for all these calculations. For Bu₄NBr up to 0.3 M, Friedman, Krishnan, and Jolicoeur obtained $A_{++} = -93$ and $A_{+-} = -210$ cal-mole⁻¹ using the experimental data of Ku[45] (see Table IV of ref. 42).
[b] In cal-mole⁻¹ of water.

W. H. Streng[3] of our laboratory has made some recalculations of the Gurney free-energy parameters A_{ij} on the basis of more reliable experimental data. A "soft core" potential of the form r^{-9} is used for the repulsive core potential COR_{ij}, and the cationic radii used are those given by Robinson and Stokes.[44] The A_{ij} parameters which fit the osmotic-coefficient data of Frank and Rupert[4] for Bu₄NCl and Et₄NCl, of Frank and Ku[45] for Bu₄NBr and Et₄NBr, and of Leifer and Kangvanvongsa[46] for Bu₄NF, Me₄NCl and Me₄NBr at 25°C for the concentration range of 0.01 to 0.2 M are given in Table IV.

As an illustration, the pair correlation functions for aqueous KCl, Me₄NCl, and Bu₄NCl solutions at 0.2 M and 25°C are shown in Figs. 9, 10, and 11, respectively. In these figures unbroken curves are the pair correlation functions g_{+-}, g_{++}, and g_{--} based on the potential function of Eq. (1). The dashed curves are the pair correlation functions g'_{+-} and g'_{++} based on the potential function which includes only the first three terms in Eq. (1) and does not include the Gurney potential GUR_{ij}. The shaded area may, therefore, be considered as the contribution of the cosphere overlap effect to the pair correlation function.

The upper shaded area is the contribution of the cosphere overlap effect to the cation–anion pair correlation function, and the lower shaded area is the contribution of the cosphere overlap effect to the cation–cation pair correlation function. In 0.2 M KCl the relative contribution of the cosphere overlap effect is seen to be much smaller than that of either 0.2 M Me₄NCl or 0.2 M Bu₄NCl, as expected. However, the relative contributions of the overlap effect to g_{++} as shown in Fig. 10 for Me₄N⁺–Me₄N⁺ and in Fig. 11 for Bu₄N⁺–Bu₄N⁺ are unexpected and require further investigations.

[3] W. H. Streng, Postdoctoral Research Associate, Chemistry Department, Clark University, 1972. We are grateful to Professor Harold L. Friedman of the State University of New York at Stony Brook for supplying us with his computer program.

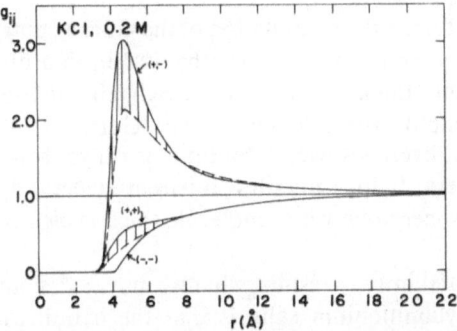

Fig. 9. Pair correlation functions g_{ij} plotted against the distance of separation r for 0.2 M KCl solution at 25°C (unbroken curves). The dashed curves are pair correlation functions based on the potential function which includes only the first three terms in Eq. (1). The upper shaded area is the contribution of the cosphere overlap effect to the g_{+-}, and the lower shaded area is the contribution of the cosphere effect to the g_{++} function.

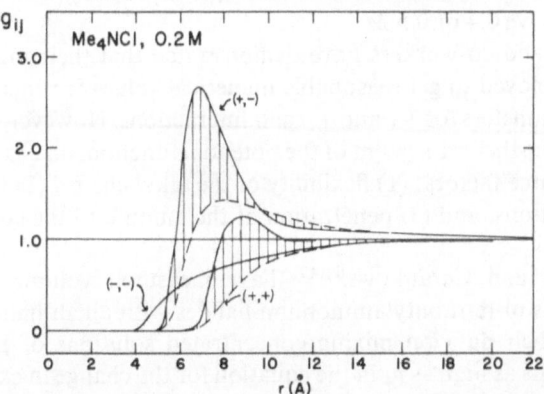

Fig. 10. Pair correlation functions g_{ij} plotted against the distance of separation r for 0.2 M Me$_4$NCl solution at 25°C. (See the caption of Fig. 9.)

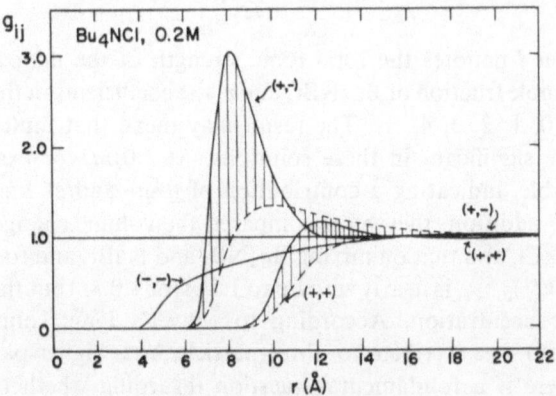

Fig. 11. Pair correlation functions g_{ij} plotted against the distance of separation r for 0.2 M Bu$_4$NCl solution at 25°C. (See the caption of Fig. 9.)

Let us recall briefly the formulation of the Gurney potential in Friedman's model. It should be remembered that the assumption of a cosphere exactly one water molecule thick around every singly charged ion is an artificiality introduced to simplify the calculation. The Gurney free-energy parameters A_{ij} are, therefore, averages over a continuous range from distant encounters to close encounters. If ions manifest different structural effects at different distances, the cosphere one water molecule thick is clearly an oversimplification.

From these preliminary results, what we may say about aqueous solutions of large tetraalkylammonium salts is that the cation–cation interaction is important but is smaller than the cation–anion interaction at a salt concentration of up to 0.2 M. It is also clear that the $+-$ interaction increases its strength with the ionic size. It will be of considerable interest to extend the model calculations to higher concentrations, but there is a problem of nonconvergence for large ions in the hypernetted-chain integral equation when the salt concentration is above 0.4 or 0.5 M.

Friedman and co-workers have demonstrated that their model calculation can be employed to get reasonable numerical values for energy, entropy, and volume parameters for $++$ and $+-$ pair interactions. However, as they have pointed out, in further refinement of the potential function, one has to consider the following three factors: (1) flexibility of the alkyl chains, (2) nonspherical shape of the cations, and (3) penetration of the anion into the space between the alkyl chains.

Kauzmann and Cerankowski[47] have measured volume changes on mixing solutions of tetrabutylammonium halides with alkali halides at 25°C. According to their data on mixing concentrated solutions of Bu$_4$NBr and NaBr, coefficients as high as v_4 in the equation for the change in excess volume on mixing have been detected:

$$\Delta V^{\text{ex}} = y(1-y)I^2 \sum_{i=0} v_i(1-2y)^i \qquad (2)$$

In this equation I denotes the total ionic strength of the mixed electrolyte solution, y the mole fraction of Bu$_4$NBr, and v_i the coefficients in the expansion series with $i = 0, 1, 2, 3, 4, \ldots$. The result may mean that multiple-particle interactions are significant in these solutions. At 1.0 m, for example, v_2 is clearly observable, indicating a contribution of four- and/or higher-particle interactions. In addition, they have compared the volume change on mixing Bu$_4$NCl and NaCl, and that on mixing Bu$_4$NBr and NaBr, and found that the difference, $\Delta(\Delta V^{\text{ex}})_{\text{Br-Cl}}$, is nearly zero up to 1.5 m (or 1.0 M) but then increases rapidly with concentration. According to Mayer's ionic solution theory $\Delta(\Delta V^{\text{ex}})_{\text{Br-Cl}}$ can be ascribed to three-particle and higher-particle interactions.[48] There is a fundamental question regarding whether the cluster expansion techniques can be applied to these high concentrations for tetra-butylammonium halide solutions. If we, nevertheless, persist in using the functional form derived from the theory, then we may say that the onset of three-particle interactions occurs at 1.5 m or 1.0 M. Bu_4N^+ cations are the functional group on certain ion-exchange resins. These results of Kauzmann and Cerankowski show that beyond a certain concentration, this cation can indeed differentiate chloride from bromide and may give some insight into the selectivity of the resins.

8. CONCLUDING REMARKS

Aqueous solutions of symmetrical tetraalkylammonium salts show many interesting properties. (For a recent review, see ref. 49.) In dilute solution, overall evidence is strong that the large cations such as Pr_4N^+ and Bu_4N^+ increase the structure of water. The evidence for this contention comes mainly from the thermodynamic and transport studies. Though not yet conclusive, various spectroscopic investigations are beginning to yield results which are more or less consistent with the results of many thermodynamic and transport studies.

At present, it is not yet clear what kind of water structure is formed around the alkyl groups of the large cations. Is it ice-I-like, gas-hydrate-like, a structure involving some nontetrahedral H bonds, or possibly mixtures of all these? We are still far from being able to give any clear-cut answer to these questions. We are doing a little better concerning the interaction between ions. With increase of the salt concentration from infinite dilution, there is overlap of cospheres followed by ion pairing of cation–anion and cation–cation types. Whether the ion pairing is the solvent-molecule-separated ion pair or contact ion pair remains to be investigated.[50] With a further increase in concentration, triplets ($+\!-\!+$, $-\!+\!-$, etc.) and higher multiplets may be formed.

In order to make further progress and to gain a deeper understanding of the structural aspects of these solutions, in addition to more and better experiments, model calculations such as demonstrated by Friedman, Rasaiah, and co-workers will be needed.[51]

ACKNOWLEDGMENTS

This research has received support from the National Science Foundation and the Petroleum Research Fund. Acknowledgment is made to the donors of the PRF, administered by the American Chemical Society, for partial support of this research.

REFERENCES

1. F. Franks, Effects of solutes on the hydrogen bonding in water, in *Hydrogen-bonded Solvent Systems*, A. K. Covington and P. Jones, eds. (Taylor & Francis Ltd., London, 1968), p. 41.
2. W.-Y. Wen, S. Saito, and C. M. Lee, *J. Phys. Chem.* 70, 1244 (1966).

3. S. Lindenbaum and G. E. Boyd, *J. Phys. Chem.* **68**, 911 (1964).
4. J. P. Rupert, Ph.D. Thesis, University of Pittsburgh, 1969.
5. S. Lindenbaum, *J. Phys. Chem.* **70**, 814 (1966).
6. R. H. Wood, H. L. Anderson, J. D. Beck, J. R. France, W. E. deVry, and L. J. Soltzberg, *J. Phys. Chem.* **71**, 2149 (1967).
7. A. Levine, Ph.D. Thesis, University of Delaware, 1971.
8. W.-Y. Wen and S. Saito, *J. Phys. Chem.* **68**, 2639 (1964).
9. F. Franks and H. T. Smith, *Trans. Faraday Soc.* **63**, 2586 (1967).
10. R. Gopal and M. A. Siddiqi, *J. Phys. Chem.* **73**, 3390 (1969); *Z. Phys. Chem.* **67**, 122 (1969).
11. R. Gopal and K. Singh, *Z. Phys. Chem.* **69**, 81 (1970).
12. R. W. Kreis and R. H. Wood, *J. Phys. Chem.* **75**, 2319 (1971).
13. J. Falcone, Ph.D. Thesis, University of Delaware, 1972.
14. P. P. Rastogi, *Bull. Chem. Soc. Japan* **43**, 2442 (1970).
15. C. V. Krishnan and H. L. Friedman, *J. Phys. Chem.* **75**, 3606 (1971).
16. R. K. Mohanty, T. S. Sarma, S. Subramanian, and J. C. Ahluwalia, *Trans. Faraday Soc.* **67**, 305 (1971).
17. M. J. Mastroianni and C. M. Criss, *J. Chem. Thermodyn.* **4**, 321 (1972).
18. C. de Visser, *J. Chem. Thermodyn.* **5**, 147 (1973).
19. R. L. Kay, G. P. Cunningham, and D. F. Evans, in *Hydrogen-Bonded Solvent Systems*, A. K. Covington and P. Jones, eds. (Taylor & Francis Ltd., London, 1968), p. 249; *J. Phys. Chem.* **73**, 3322 (1969).
20. H. G. Hertz and M. D. Zeidler, *Ber. Bunsenges. Physik. Chem.* **68**, 821 (1964).
21. K. Krynicki, *Physica* **32**, 167 (1966).
22. H. G. Hertz, in *Progress in Nuclear Magnetic Resonance Spectroscopy*, J. W. Emsley, J. Feeney, and L. H. Sutcliffe, eds. (Pergamon Press, 1967), Vol. 3, Chap. 5, p. 192.
23. H. G. Hertz and W. Spalthoff, *Z. Elektrochem.* **63**, 1096 (1959).
24. R. L. Kay, in *Trace Inorganics in Water, Advan. Chem. Ser.*, No. 73, 15 (1968).
25. A. LoSurdo and H. E. Wirth, *J. Phys. Chem.* **76**, 130 (1972).
26. E. R. Malinowski and A. R. Pierpaoli, *J. Mag. Res.* **1**, 509 (1969).
27. W.-Y. Wen, P. T. Inglefield, Y. Lin, and G. Minott, unpublished results, Clark University, 1971.
28. J. Davies, S. Ormondroyd, and M. C. R. Symons, *Chem. Commun.*, 1204 (1971).
29. H. G. Hertz and H. Pfliegel, private communications, Institut für Physikalische Chemie. Universität Karlsruhe, 1972.
30. W.-Y. Wen and H. G. Hertz, *J. Solution Chem.* **1**, 17 (1972).
31. F. H. Stillinger, Jr., and A. Ben-Naim, *J. Chem. Phys.* **47**, 4431 (1967).
32. P. S. Ramanathan and H. L. Friedman, *J. Chem. Phys.* **54**, 1086 (1971).
33. D. Feil and G. A. Jeffrey, *J. Chem. Phys.* **35**, 1863 (1961); M. Bonamico, R. K. McMullan, and G. A. Jeffrey, *ibid.*, **37**, 2219 (1962); *ibid.*, **39**, 3295 (1963).
34. A. H. Narten and S. Lindenbaum, *J. Chem. Phys.* **51**, 1108 (1969).
35. M. D. Danford, Diffraction Pattern and Structure of Aqueous Ammonium Fluoride Solutions, ORNL-4244, 1968.
36. G. E. Walrafen, *J. Chem. Phys.* **55**, 768 (1971).
37. R. Pottel and O. Lossen, *Ber. Bunsenges. Physik. Chem.* **71**, 135 (1967).
38. M. J. Blandamer, M. J. Foster, N. J. Hidden, and M. C. R. Symmons, *Trans Faraday Soc.* **64**, 3247 (1968).
39. V. P. Romanov and V. A. Solovyev, *Akust. Zh.* **11**, 84 (1965) [Sov. Phys. Acoust. **11**, 68 (1965)].
40. M. J. Blandamer and D. Waddington, *J. Chem. Phys.* **52**, 6247 (1970).
41. G. Atkinson, R. Garnsey, and M. J. Tait in *Hydrogen-Bonded Solvent Systems*, A. K. Covington and P. Jones, eds. (Taylor & Francis Ltd., London, 1968), p. 161.

42. P. S. Ramanathan, C. V. Krishnan, and H. L. Friedman, *J. Solution Chem.* **1**, 237 (1972).
43. D. F. Evans and P. Gardam, *J. Phys. Chem.* **73**, 158 (1969).
44. R. A. Robinson and R. H. Stokes, *Electrolyte Solutions* (Butterworths, London, 1965), 2nd ed., p. 125.
45. J. C. Ku, Ph.D. Thesis, University of Pittsburgh, 1971.
46. T. Kangvanvongsa, Ph.D. Thesis, Michigan Technological University, 1971.
47. L. D. Cerankowski, Ph.D. Thesis, Princeton University, 1969.
48. J. Mayer, *J. Chem. Phys.* **18**, 1426 (1950).
49. W.-Y. Wen, in *Water and Aqueous Solutions*, R. A. Horne, ed. (Wiley, New York, 1972), Chap. 15, p. 613.
50. D. W. Larsen, *J. Phys. Chem.* **74**, 3880 (1970); *ibid.* **75**, 509 (1971).
51. J. C. Rasaiah and H. L. Friedman, *J. Chem. Phys.* **48**, 2742 (1968); J. C. Rasaiah, *ibid.*, **52**, 704 (1970); P. S. Ramanathan and H. L. Friedman, *ibid.*, **54**, 1086 (1971).
52. C. Folzer, R. W. Hendricks, and A. H. Narten, *J. Chem. Phys.* **54**, 799 (1971).

DISCUSSION

Dr. F. H. Stillinger (*Bell Laboratories, New Jersey*). I would like to ask some information on one of your slides. Was 3.8 the average number of oxygen nuclei nearest neighbors for each oxygen nucleus in a tetraalkylammonium chloride solution?

Professor Wen (*Clark University, Worcester, Massachusetts*). Yes.

Dr. Stillinger. Did you get that from Narten and Lindenbaum? How can one claim that these tetraalkylammonium ions form structure while at the same time permit the existence of only the 3.8 nearest neighbors, which, to my mind, must violate, disrupt, and destroy any ice-like structure—any tetrahedral structure? One cannot have it both ways unless, of course, the experimental error is sufficiently wide to incorporate 4.0. Could you clarify this situation?

Professor Wen. Since tetrabutylammonium ion is a large cation with a radius of about 5 Å (the distance from the center of the nitrogen atom to the end of a butyl chain), even in moderate concentrations, the ion can occupy a considerable fraction of the solution volume and consequently diminish the number of water molecules involved per unit volume of the solution. Thus, the average number of nearest-neighbor water molecules can decrease to a value well below that of pure water (4.4) in spite of the fact that the hydrogen bonding among them has increased its strength.

Professor H. S. Frank (*University of Pittsburgh*). Actually that is at 1.4 *m*, but the volume fraction of the salt is much more considerable, and this means that there are a large number of water molecules that are kind of stuck up against something else that does not have any structure. This would be partly responsible, don't you think? You are not talking about an oxygen–carbon coordinate; this is an oxygen–oxygen coordinate.

Professor S. Lindenbaum (*University of Kansas, Lawrence*). The number 3.8 comes from a least-squares fit to the radial distribution function. It should not be regarded as differing significantly from 4.0. The real evidence for water structure promotion is in the estimate of the O–O nearest neighbor distance, which is significantly smaller than that of pure water.

Professor J. H. Gibbs (*Brown University*). There is another possible answer, in addition to the one given, concerning this point. There is an accident that seems not to be very well known: The oxygen–oxygen hydrogen-bonding distance, 2.8 Å, seems to be identical to the equilibrium nearest-neighbor oxygen–oxygen distance obtained if two molecules approach "back-end-to." A Van der Waals radius for oxygen in this situation is 1.4 Å, exactly half of the 2.8 Å. Further evidence of this kind of thing is observed in ice-VIII where eight nearest-neighbor oxygens are at a distance of about 2.8 Å. One certainly does not have eight water molecules hydrogen-bonding to a central one.

Professor Frank. No, but there is a lot of repulsion there, and that is why one has to have a 20,000 atm pressure. How firm is the estimate of the Van der Waal radius that you use?

Professor Gibbs. Offhand, I cannot tell you. But this is not likely to be an important point because the first peak of the radial distribution function is broad enough to represent two (or more) different types of near-neighbor situations even if these do not have very similar average oxygen–oxygen spacings.

Professor R. H. Stokes (*University of New England, Armidale, Australia*). Probably many of the experiments on the tetraalkylammonium salts qualitatively give the suggestion of cation–cation pairs, which you suggested was a part of the electrostatic energy. It might be an inference, but it should not, I think, necessarily be a part of the classical electrostatics. To clarify the situation here, one of the ways the energy of a system like this could be determined is to integrate the square of the electric field density, divided by the dielectric constant, over the whole of the space containing all these charges, which is a classical way of finding the electrostatic energy. It appears to me that if one brings two positively charged spheres of low dielectric constant together, the electric field will be annulled over a large part of the configuration space and, therefore, possibly reduce the integral of E^2/ε in the region where ε is small, giving in effect an attractive force.

Professor H. L. Friedman (*State University of New York, Stony Brook*). I think it is the other way around. For two spheres of low dielectric constant, with positive charges at their centers, in a medium of higher dielectric constant, there is a repulsion in addition to that of the simple Coulomb law.

Professor Frank. Actually Dr. Wen did not tell us about one of the beautiful thermodynamic evidences of an attraction between cation and cation. This is his study of the free-energy change on diluting equimolal mixtures of tetrapropylammonium bromide and KBr [W.-Y. Wen, K. Miyajima, and A. Otsuka, *J. Phys. Chem.* **75**, 2148 (1971)]. When treated according to the Friedman formalism, Wen showed that two tetrapropylammonium ions can be separated only by doing a considerable amount of work. The mixing curves are quite striking in this regard. The mixing curves for volume were done first and are better known [W.-Y. Wen, K. Nara, and R. H. Wood, *J. Phys. Chem.* **72**, 3048 (1968)], but the free energies give the first evidence of an actual attractive force.

Dr. F. Franks (*Unilever Research Laboratories, England*). As I understand it, in the postulated ion pair the anion can penetrate the cosphere and move even closer to the cation. When one uses electrolytes to precipitate proteins, this is due to neutralization of the charge. Some work done at our laboratories indicate that, as far as charge neutralization is concerned, the tetraalkylammonium ions behave just like sodium chloride in that they reduce the electrophoretic mobility of the protein in the same way. But, whereas sodium chloride eventually precipitates globular proteins, the alkylammonium halides do not. Instead, clear gels are obtained. I wonder how this agrees with your comments. It seems that the charge interaction is the same.

Professor Wen. The gel formation indicates that there is a linking up of water structure around the tetraalkylammonium ions with the water structure around globular proteins. As long as anions are partly hydrated, they can be accommodated very close to the cation, particularly when the salt concentration is high.

Professor K. S. Pitzer (*University of California, Berkeley*). In looking at the hypersonic absorptions, I was reminded that for hydrocarbons or other organic compounds with large alkyl groups, one obtains absorptions that have to do with conformational changes within the alkyl group. There may well be different alkyl group geometries, all of which are present, and under pressure oscillations their proportions change; and one knows that gives an absorption in other cases, which may be why there are complications in propyl and butyl which just do not appear in the methyl and ethyl.

Professor D. F. Evans (*Case Western Reserve University, Ohio*). That is part of the problem we had to deal with in studying ultrasonic absorption in the propanol system. One of the pieces of evidence that was cited for conformational changes within the alkyl groups was that absorption was found in tetraethylmethane, which, as a matter of fact, showed up in the correct frequency range. This would suggest that indeed there was some contribution from methyl rotation. On the other hand, rotation is detected for the tetraalkyltins only in some cases. Depending, therefore, on which compound is chosen for the model, there may or may not be rotation. For tetrabutyl- and tetrapropylammonium ions in methanol, the absorption which corresponds to the ionic interactions is very much further out in the frequency range (something like 10^7 Hz) so that rotation should be easy to detect if it was present. However, the curve shows no indication of any sort of contribution from rotation. It is unlikely that the rotation is shifted substantially by the solvents. Thus, it would appear that, at least in the frequency range studied, in the propanol system there was no contribution from hydrocarbon rotation.

Professor Wen. Are you saying the frequency of absorption should be quite different?

Professor Evans. There is no indication of any absorption in that region in methanol or, for example, in acetonitrile.

Dr. Franks. Would such absorption be expected?

Professor Wen (communicated). Blandamer et al.[38] postulated that the ultrasonic absorption properties over the range 1.5 to 230 MHz may be the result of some form of rotational isomerism within the tetraalkylammonium ion. This postulate has been refuted by Atkinson et al.[41] and, more recently, Blandamer and Waddington[40] attributed the marked changes in the absorption properties at 0.8 M as a consequence of changes in water structure.

Professor H. G. Hertz (*University of Karlsruhe, West Germany*). You showed two pictures of the tetrabutylammonium ion, and in both cases the four side chains were stretched into the water. Why are they not folded together into more of a sphere? Is there any argument for the geometry you have suggested? Has anyone else demonstrated this?

Professor Wen. I think this is the most probable arrangement in terms of stabilities. To tape two arms together, one has to bend them and distort them, and there is energy to be paid for compensation.

Professor Frank. Another answer is that this is what they do when they get cold, and this is observable in the crystal. In the crystal structure, they are all sticking out.

Professor Hertz. In the crystal, yes, but is it also necessarily true in solution?

Professor Friedman. I thought the solvation thermodynamics of the tetraalkylammonium ions in water really are easiest to understand if one assumes the chains are outstretched rather than curled up. It is pretty hard to put it together the other way.

Professor Hertz. I would rather be curled up if I were a hydrophobic particle!

Professor J. B. Hyne (*University of Calgary, Canada*). Is there any definitive experimental evidence for this interalkyl chain penetration by the anion?

Professor Wen. There is no definitive evidence, but there is a possibility of such penetration. A suggestion comes from the studies in N-methylacetamide or formamide. In these organic solvents, salts show very similar properties to those shown in water, at least as far as apparent molal volumes and activity and osmotic coefficients are concerned. It seems to indicate that the radius of a large tetraalkylammonium ion in solution is not a simple thing because of its peculiar shape. If one makes a model calculation, one can see that the anion has to approach nitrogen rather closely to fit the experimental data.

Professor Frank. This may be true in the case of volumes and activity coefficients, but it is not true for the heats. Water is definitely a unique solvent when heats are considered.

Professor Hyne. I find it hard to understand or to accept this penetration of the alkyl chains by the anion. If one puts a negative ion, a highly charged entity, into such a low dielectric area, one is heading for the lowest energy situation. But if so, I would think spectroscopic techniques such as NMR would reflect a measurable change in the shift of the protons in these alkyl chains if some high-charge-density ions are in among them.

Professor Wen. In our laboratory we have measured the methyl proton shifts of tetrabutylammonium ion in water relative to the external reference tetramethylsilane using coaxial cells and corrected for the differences in bulk diamagnetic susceptibility.[27] The methyl protons are found to shift downfield with increasing salt concentration. (A downfield shift of 2.1 Hz is, for example, observed in going from 0.56 to 3.62 m.) This downfield shift may be caused by the electric field of the approaching anions. It must, however, be emphasized that this interpretation is by no means unique.

Professor Hertz. Didn't Larsen [D. W. Larsen, *J. Am. Chem. Soc.* **91**, 2920 (1969); *J. Phys. Chem.* **74**, 3880 (1970); *ibid.*, **75**, 509 (1971)], who has studied the nitrogen relaxation via the protons in aqueous solution, conclude that there is no approach of the anions to the cations?

Professor Wen. I think it was only applicable to tetramethyl- and tetraethylammonium salts, and we do not postulate any anion penetration into these ions.

Dr. Franks. Can I suggest also that the ^{79}Br relaxation results of Forslind *et al.* suggest that the bromine ion never penetrates near to the alkylammonium ion?

Professor Frank. Didn't Forslind find that there were two states to the bromide ion [B. Lindman, S. Forsen, and E. Forslind, *J. Phys. Chem.* **72**, 2805 (1968)]?

Professor Friedman. It is very hard to get an unambiguous answer from NMR relaxation when the mechanism of the relaxation is quadrupolar because there are so many mysterious things going on.

Professor Hertz. I think that at least a part of this mysteriousness can now be removed. The large dielectric hole which is represented by the tetraalkylammonium ions causes a rearrangement of the water molecules. Now, in electrolyte solutions of small ions—and in the limit of infinite dilution—there are strong water–water correlations which reduce the electric field gradient at the relaxing nucleus. But the rearrangement of the water molecules close to the dielectric hole destroys these original water–water correlations and may, moreover, cause other water–water correlations which increase the electric field gradient at the Br nucleus. This causes the strong quadrupolar relaxation rate in these solutions.

Dr. G. E. Walrafen (*Bell Laboratories, New Jersey*). It is clear from Raman spectra that ions such as Cl⁻ and Br⁻ are hydrated. Hence, one might suppose that Cl⁻ and Br⁻ would approach a large tetraalkylammonium ion without losing all anionic water of hydration.

Professor Wen. Yes. Actually, in my model the anion is shown as a sphere; it does not indicate that there is no hydration. I just indicated that the entity, the anion, can approach somewhat inside the space between alkyl chains.

Professor Friedman. I have a comment about another matter. It concerns the sound absorption as a function of solute concentration. The graph always starts out almost horizontal at $C = 0$ and picks up somewhere between $C = 0.5$ and 1 M. This suggests strongly that what is relaxing is not intrinsic to one solute particle plus its surroundings. It is either slow relaxation of the water between two or more adjacent solute particles or slow formation of solute–solute pairs or large clusters. A misunderstanding of this motivated our study of paramagnetic hydrophobic solutes in water. We thought we could find a faster probe to measure the relaxation of cospheres on hydrophobic solutes, but, roughly speaking, the same thing happened. We only find slow motion when we have clusters rather than single hydrophobic particles in water.

Toward a Model for Liquid Water[1]

Julian H. Gibbs,[2] Claude Cohen,[2] Paul D. Fleming III,[2] and Harold Porosoff[2,3]

Received December 15, 1972; revised January 15, 1973

A new model is proposed for liquid water. It is obtained by consideration of the two transitions (melting and boiling) which define the liquid phase. These transitions are discussed with the aid of two analogies to well-known phenomena in polymer physical chemistry. In analogy to the helix-coil transition in polypeptides and polynucleotides, the melting of ice is viewed as a process consisting essentially of the destruction of the orderly interconnected small rings of hydrogen bonds characteristic of the crystal. The fact that the breakup of interconnected small rings is cooperative, even when unaccompanied by the breaking of bonds which are not parts of rings, is clearly seen by inspection of the theory for the putatively analogous helix-coil transition. The condensation of water vapor is viewed in analogy to gelation in reversibly polymerizing systems, an analogy which interprets its cooperativity. Taken together, these interpretations of the phase transitions indicate that the liquid can be viewed as an infinitely and randomly branched "gel" of (rapidly interchanging) hydrogen bonds in which closures of rings (primarily large rings) are present at random but in which there is no significant preference for an ordered array of small rings. These concepts also lead naturally to an interpretation of the triple point and sublimation. The random gel model is seen to be consistent with most of the known properties of liquid water. In particular, the radial distribution function, infrared and Raman spectra, dielectric properties, density maximum, and properties of the supercooled region are discussed briefly here.

KEY WORDS: Water; statistical mechanics; phase transitions; melting; condensation; gelation; helix-coil transition; triple point; rings; hydrogen bond; liquid; radial distribution function; glassy state; dielectric constant; infrared and Raman spectra.

[1] This paper was presented at the symposium, "The Physical Chemistry of Aqueous Systems," held at the University of Pittsburgh, Pittsburgh, Pennsylvania, June 12–14, 1972, in honor of the 70th birthday of Professor H. S. Frank.
[2] Department of Chemistry, Brown University, Providence, Rhode Island 02912.
[3] Present address: American Cyanamid Company, 1937 West Main Street, Stamford, Connecticut 06904.

"It must be jelly, 'cause jam don't shake like that."

anon.

1. INTRODUCTION

Theories of liquids fall into two classes: (1) effects, based solely on an assumed intermolecular potential and the principles of statistical mechanics, to predict the very existence of the liquid state as an entity separated from gaseous and crystalline states by discontinuities (first-order phase transitions) and (2) "model" theories in which at least some aspects of supramolecular liquid "structure" are assumed, on the basis of which conclusions are drawn and compared with experiment. Unquestionably, a convincing theory of type (1) is what one ultimately wants. Equally unquestionably, efforts to develop such a theory should be focused on substances simpler than water.

On the other hand, the need for a theory in the case of liquid water is very pressing, particularly for the discussion of many biophysical problems. Therefore, many type (2) theories have been presented for liquid water in the last half century or so.[1]

In the area of theories of type (2) it is essential not to forget that agreement of calculations based on any model with experimental results does not prove the validity of the model; such agreement may be fortuitous, particularly if the number of arbitrary parameters is large. Furthermore, disagreement of the theoretical conclusions with experimental results can disprove the model only if the calculations on the model are performed without further assumptions or approximations, a circumstance seldom encountered. Therefore, it is clearly important to explore, however qualitatively, every conclusion which may be drawn from a model and to effect comparison of theoretical conclusions with every relevant experimentally determined property rather than merely to achieve precise comparison of such an approximate calculation with only one or a few experimentally observed properties.

Among the experimental observations which model theories of liquid water have tended to ignore are the two phase transitions (melting and boiling) which define the liquid, the very properties which the more rigorous type (1) theories have been at such great pains to explain.[4] There has also been a tendency, though not a universal one, to forget the supercooled region of the liquid.

---

[4] It is important not to confuse the ability of model theories to predict the locations of these phase transitions with the ability to predict their very existence (i.e., the existence of the liquid). Once one assumes the existence of the liquid by postulating a specific model for it, one can, of course, calculate the partition function corresponding to that model and thereby obtain a curve displaying the temperature dependence of the (appropriate) free energy (at specified volume or pressure). The points where this curve crosses those for the gas and crystal (also calculated from specific models) are the calculated boiling and melting points, respectively. Unless, however, one has obtained these three free-energy curves as different solutions of the *same* general model, one cannot claim to have predicted or interpreted the phase transitions.

The present considerations represent an attempt to carve a path intermediate between the two types of theories described above. They approach the problem of the two phase transitions via analogies to two phenomena which have been widely investigated in polymer physical chemistry. Specifically these analogies relate a major feature of the helix-coil transition in polypeptides and polynucleotides to a putative corresponding feature in the melting of ice and relate the essential physics of reversible gelation in a three-dimensional polymerizing system to that of condensation of water vapor.

We shall consider these two phase transitions in Secs. 2 and 3, respectively, and comment on the nature of the triple-point phenomenon in Sec. 4. In Sec. 5 we show that the picture of liquid water that emerges from these considerations of the bounding phase transitions is in qualitative agreement with most of the known properties of the liquid itself, e.g., its radial distribution function, dielectric properties, infrared and Raman spectra, etc.

2. MELTING AND THE ANALOGY TO THE HELIX-COIL TRANSITION

Here we approach the problem of the nature of liquid water by asking how it may differ from ice. Recognizing that melting is describable as a mathematical discontinuity, we resist any temptation to view the liquid as some sort of extrapolation of the crystal (however "disordered").

A cooperative process such as a phase transition can be described crudely, though essentially correctly, as one which, though difficult to start, tends to proceed to completion under conditions adequate to start it. Thus, ice melts not at all below 0°C (at 1 atm) but melts completely upon only an infinitesimal increase of temperature at this pressure. However, a closer look at the situation reveals a basic puzzle. The cooperativity is clearly in this case somehow to be ascribed to a process of breaking hydrogen bonds (H bonds). This cooperativity causes the process to proceed catastrophically at 0°C only up to a point— a point which must be far short of the point where all H bonds are broken since the evidence for intact H bonds in liquid water is indisputable. Further catastrophic (i.e., highly cooperative) H-bond breaking must wait until 100°C is reached (at 1 atm). Looking at the problem of the existence of liquid water from the vantage point of the ice crystal, we can say that the problem is to explain why hydrogen-bond breaking occurs in two separate cooperative processes.

The helix-coil transition displayed by polypeptides (and polynucleotides) is a cooperative process involving the breakage of hydrogen bonds which is understood. The one-dimensional character of this phenomenon has permitted exact evaluation of the partition function for a realistic model for the possible states of the whole cooperative polymer molecule. Agreement of the various, virtually equivalent, treatments[2, 3, 4] of this problem with experimental results has been excellent. It may, therefore, be instructive to take a closer

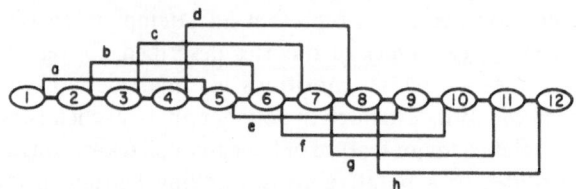

Fig. 1. Schematic diagram of the alpha helix. When, for example, the three hydrogen bonds *d, e, and f* bridging the residue 7 are broken, rotation is possible around the chemical bonds in this residue, allowing it to assume nonhelical postures.

look at the physical interpretation implicit in the mathematical theory of this transition (although this "transition" is imperfectly sharp) to see if any clue can be obtained concerning our problem of the two-stage breaking of H bonds in the three-dimensional case at hand.

In the case of the polypeptide helix-coil transition, the basis of the co-operativity is easily described.[2, 3, 4] Beginning at one end of the α-helical chain, let us label the sequence of hydrogen bonds, *a, b, c, d* and the sequence of amino acids 1, 2, 3, 4, etc. (Fig. 1). Then, in this α-helical structure we see that H bond *e* closes a ring involving what would otherwise be rotatable C–C bonds of amino acids, 6, 7, and 8; H bond *f* closes 7, 8, 9, etc. There is, thus, a network of interconnected rings. Focusing attention on amino acid 7, we see that the rotations around its C–C bonds are restricted by its involvement in three rings, those closed by H bonds *d, e,* and *f*. All three of the latter must be broken if rotation around the C–C bonds of amino acid 7 are to occur. Terms in the partition function involving, for example, the breakage of *e* alone among the set *d, e,* and *f* will tend to carry little weight, for they represent an expenditure of energy (small Boltzmann factor) in return for which no rotational entropy (no degeneracy factor) has been gained. Terms involving the breakage of both *d* and *e*, but only *d* and *e*, among the set *c, d, e,* and *f* will be even more suppressed in the partition function, for in these cases two H-bond energy units have been paid with no entropic return. Terms involving isolated breakage of an adjacent trio, e.g., breakage of *d, e,* and *f* with *c* and *g* intact, will be somewhat less suppressed, in the appropriate temperature range, for one amino acid rotational entropy unit (that of amino acid 7) has been gained for the expenditure of the three units of H-bond energy. Once this has been accomplished, however, a roughly equal amount of extra entropy can be gained by breakage of only one more H bond, either *c* or *g*, for the breakage of four adjacent H bonds releases two amino acids from the ring constraints. Furthermore, the breakage of five adjacent H bonds releases three amino acids from ring restraints, six releases four . . ., and *j* releases *j* − 2. Thus, whereas we have difficult "nucleation" of so-called random-coil regions, we have easy growth of these regions. At temperatures where any H bonds at all are broken, relatively long strings of broken H bonds will tend to be formed; the magnitudes (governed by degeneracies and Boltzmann factors) of the various terms in the partition function reflect this situation.[2, 3, 4]

What we can learn from this situation that may be applicable in our three-dimensional problem is clearly that the breakup of orderly interconnected small rings of interaction carries a cooperative entropic component.[5] In the melting of ice we may also expect cooperative entropic effects associated with the breaking of orderly interconnected small rings. It is thus tempting to postulate that the breaking of these may be the principal process involved in melting. This suggestion has the virtue of providing part of an answer to the puzzle alluded to above—the incompleteness of melting as a process of breaking H bonds (i.e., the fact that melting is not sublimation). It portrays melting as a cooperative process of breaking ordered small rings which, though it may go to completion as a proper cooperative process should, does not require the breaking of all the H bonds; highly branched structure (containing ring closures at random[6]) persists and constitutes the liquid phase.[7]

To make the analogy with the helix-coil situation more tractable, we outline here another way of representing it. Although we do not pursue here the quantitative results already obtained by the more complete methods,[2, 3, 4] we do demonstrate the existence of the "transition," this being our current principal concern. Before proceeding with this, however, we note that two questions are posed by the postulate we are now pursuing: (1) Would hydrogen bonds which are not parts of small rings indeed persist above a temperature high enough (0°C at 1 atm) to break orderly interconnected small rings? (2) If they do persist at higher temperatures, would their ultimate breakup also be

---

[5] Consideration of the polynucleotide helix-coil transition yields somewhat similar conclusions. Using the familiar analogy between the DNA double helix and a (twisted) ladder, we see that each pair of adjacent (base pair) "rungs" of the ladder forms, in conjunction with the two included (ribosephosphate) side rails, a small closed ring, when both adjacent potential base pairs are paired through the Watson–Crick H bonds. The breakup of these H bonds on heating involves a cooperativity which is partly entropic in origin. Thus, in addition to an energetic cooperativity associated with the breaking of "stacking" interactions, there is an entropic cooperativity associated with the fact that, as originally pointed out by Stockmayer and Jacobson, [H. Jacobson and W. H. Stockmayer, *J. Chem. Phys.* **18**, 1607 (1950)], the rotational entropy of a ring increases much faster than linearly with the size of the ring. When a base pair originally at the interface of helical and randomly looped regions rotates out of the helical posture, it increases the size of the loop on its "random-loop" side by joining the latter. The larger the latter was originally, the more entropy it gains by the accretion of the new base pair. Thus, we have another cooperative effect associated with the breakup of H bonds in rings.

[6] The random coil "phase" of polypeptides (and polynucleotides) certainly involves configurations possessing ring closures as well as pure chain configurations. It is only the ordered array of small rings characteristic of the α-helix that is lost on the transition from helix to random coil.

[7] The fact that the liquid state is characterized in any instant by essentially one cluster, or "network," will be seen below via the gelation analogy to condensation.

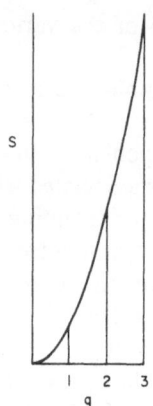

Fig. 2. The entropy S per site is built up quadratically as the bonds holding that site fixed are broken. In ref. 2, the entropy per site is zero until $q = 3$ where it takes its maximum value (step function).

cooperative (vaporization)? These questions, taken in the inverse order, are the subjects of the next two sections. The reader whose interest lies primarily in our qualitative considerations may wish to pass directly to them.

The main difference between the formulation of helix "melting" to be presented now and the familiar ones[2, 3, 4] outlined above is that, whereas in the previous treatments no entropy is assumed to be gained until all three of the bonds capable of fixing an amino acid (site) are broken, in the treatment considered below the entropy per site is assumed to be built up progressively (quadratically) as these bonds are broken (see Fig. 2). The contribution, to the configurational partition function, of an amino acid for which q of these bonds ($0 \leqslant q \leqslant 3$) are broken is then

$$\exp -\beta u(q) = Z^{q^2} \exp -\beta q\varepsilon = \exp -\beta q(\varepsilon - \beta^{-1} \ln Z^q) \qquad (1)$$

where $\ln Z^{q^2}$ is the entropy associated with an amino acid when q of the bonds restricting it are broken, $\beta^{-1} = kT$, and ε is one-third of the energy associated with a broken H bond. The configurational partition function for a helix with N amino acid sites is now written as

$$Q_N = \sum_{\{n\}} g(\{n\}) \exp \left(-\beta \sum_{q=0}^{3} n_q u(q) \right) \qquad (2)$$

where n_q is the number of sites with q broken bonds and $g(\{n\})$ is the number of linear arrangements of bonds for a given set $\{n\}$ with the restriction $\sum_q^3 n_q = N$.

In the spirit of Ising-model calculations we assign a number $\sigma_{\alpha, i}$ to each pair i of fourth-neighbor amino acids which enclose site α and specify that $\sigma_{\alpha, i} = 0$ for an intact H bond and $\sigma_{\alpha, i} = -1$ for a broken H bond at α, i. The number of broken H bonds in any configuration is then given by

$$R = -\tfrac{1}{3} \sum_{\alpha, i} \sigma_{\alpha, i} \qquad (3)$$

where the sum over i goes from 1 to 3, and the sum over α from 1 to N. Using Eq. (1) we can write

$$\sum_{q=0}^{3} u(q)n_q = \varepsilon \sum_{q=0}^{3} qn_q - \beta^{-1}\ln Z \sum_{q=0}^{3} q^2 n_q \tag{4}$$

By defining the new variable $s_{\alpha, i} = 2\sigma_{\alpha, i} + 1$, so that $s_{\alpha, i} = 1$ if there is an intact H bond at α, i and $s_{\alpha, i} = -1$ if there is not, we find

$$\sum_{q=0}^{3} qn_q = -\sum_{\alpha, i}\sigma_{\alpha, i} = \tfrac{3}{2}N - \tfrac{1}{2}\sum_{\alpha, i}s_{\alpha, i} \tag{5}$$

and

$$\sum_{q=0}^{3} q^2 n_q = -\sum_{\alpha, i}\sigma_{\alpha, i} + 2\sum_{\alpha, i<j}\sigma_{\alpha, i}\sigma_{\alpha, j}$$

$$= 3N - \tfrac{3}{2}\sum_{\alpha, i}s_{\alpha, i} + \tfrac{1}{2}\sum_{\alpha, i<j}s_{\alpha, i}s_{\alpha, j} \tag{6}$$

Equation (6) is obtained by enumeration. Using Eqs. (4–6), we can rewrite the configurational partition function Eq. (2) as

$$Q_N = \exp-[\tfrac{3}{2}\beta N\varepsilon - 3N\ln Z]\sum_{(s)}\exp\left\{(\tfrac{1}{2}\beta\varepsilon - \tfrac{3}{2}\ln Z)\sum_{\alpha, i}s_{\alpha, i} + \frac{\ln Z}{2}\sum_{\alpha, i<j}s_{\alpha, i}s_{\alpha, j}\right\} \tag{7}$$

Equation (7) has the well-known form of the partition function for the Ising model.[5] In the one-dimensional case, the "transition" which it displays is not perfectly sharp (except in the limit $\ln Z \to \infty$ and $\beta\varepsilon \to \infty$).

On generalizing Eq. (7), obtained for the one-dimensional problem (helix-coil "transition"), to the corresponding three-dimensional problem (melting of ice), we would expect a sharp transition temperature, in accordance with familiar experience with this equation in its application to other Ising problems. Such a generalization is not, however, straightforward. The co-operativity that we have suggested above for the ice-melting process is one between hydrogen bonds which when intact are responsible for small ring closures and when broken provide the system with a configurational entropy gain. The form of this proposed cooperativity in the three-dimensional case remains an unresolved question. If we assume, for example, that it is principally the six-membered ring pattern in ice that is destroyed upon melting, then we may assume that the contribution to the partition function due to the entropic cooperativity retains the essential features of the aforementioned one-dimensional problem. We can then assign a contribution to the partition function analogous to Eq. (1) to each molecule for which q of the six-membered rings in which it might be restrained are broken. A formulation similar to the one presented above would then lead to a three-dimensional Ising model which would exhibit a transition from a state of many six-membered rings to one containing only very few. It is important to note that in this transition the (fractional) change in the total number of H bonds is actually rather small. The key point is the distinction between the breaking of ring-closing H bonds, with which a large entropy increase is associated, and the breaking of others, with which a smaller entropy increase is associated if the pressure is high enough (see Sec. 4). We identify this ring-breaking transition with melting.[8]

3. LIQUID–VAPOR EQUILIBRIUM AND THE ANALOGY OF GELATION

The next question that we are faced with is whether the breakup of a randomly branched structure can also be cooperative. This question is best considered from the viewpoint of its inverse, that of the formation of such structure. Looked at this way, the question immediately elicits an affirmative answer, for it is well known that a branching system can display a discontinuity. Nuclear-fission bombs and explosive gaseous chemical reactions are particularly well-known examples of branching systems which display such discontinuities (in time). In the case of branching processes in space, a particularly good example of a discontinuity is the gel point displayed in the "three-dimensional" polymerization of polyfunctional monomers.

The close analogy between the Flory–Stockmayer theory[6,7] of gelation and the Mayer[8] cluster theory of vapor–liquid equilibria was pointed out by Stockmayer long ago—and apparently overlooked by all except Scatchard *et al.*[9] and Gordon *et al.*[9] who discussed (respectively) the vapor pressures of H_2O_2–H_2O mixtures and the liquid miscibility of H_2O–benzene mixtures. For an H-bonded condensing system, the analogy to gelation should be particularly close because the discrete character of the propensity of (for example) a water molecule to form H bonds (up to four of them) exactly parallels the integer-value property of the number of chemical bonds emanating from any polyfunctional unit in chemical polymerization.

We show here that the Flory–Stockmayer (FS) statistical theory of gelation can be transcribed into a statistical-mechanical theory of condensation for systems with large directional forces such as H-bonding systems. The independent variable in the FS theory is the extent of reaction α. In our case the analogous quantity is the fraction of OH groups (or lone pairs of electrons) in the whole system which are participating in H bonds. In the case

---

[8] It is interesting to note that Perram and Levine [*Mol. Phys.* **21**, 701 (1971)], using a tetrahedral lattice model for water and a cooperativity of bond formation of mathematical form similar to that in Eq. (1) but with no distinction between bonds involved in rings and bonds which are not, obtained a transition from a state of many bonds to one containing very few. They associated their transition with a liquid–vapor transition, but it could equally be associated with a solid–vapor transition.

of reversible condensation, we need in effect to relate this quantity to a pair of thermodynamic variables such as temperature and volume (or pressure).

We consider a system of N elementary units (e.g., water molecules) which can bond together to form polymeric complexes (including unbonded monomers) and let m_n be the number of n-mer complexes $(1 \leqslant n \leqslant N)$. The m_n are related to the total number of molecules by

$$N = \sum_n nm_n \tag{8}$$

and to the number of complexes by

$$M = \sum_n m_n \tag{9}$$

The major assumption of the FS theory of gelation is that ring formation can be ignored. We shall see that this is a poor approximation for discussion of properties of the liquid (the gel in FS theory). It does not, however, prevent the theory from displaying the essential features of the sol–gel (vapor–liquid) transition. Since our purpose is to show that the formation (or breakup) of branched structures is cooperative by itself, this approximation serves our needs. If there are no rings, each n-mer contains $n-1$ hydrogen bonds, and the potential energy attributable to hydrogen bonds is then

$$V_{HB}(\{m_n\}) = -\varepsilon \sum_n (n - 1)m_n = -\varepsilon(N - M) = -\varepsilon(zN\alpha/2) \tag{10}$$

where ε is the binding energy of an H bond and z is the "functionality" of the monomer units (4 for water).

For a given value of N, the number of ways that m_1 monomers, m_2 dimers, . . ., and m_n n-mers can be formed is[7]

$$\Omega_N(\{m_n\}) = N! \prod_n (w_n/n!)^{m_n}(1/m_n!) \tag{11}$$

where w_n is the number of ways in which n units may form an n-mer without ring formation. The quantity w_n is given by[7, 9, 10]

$$w_n = z^n (zn - n)!/(zn - 2n + 2)! \tag{12}$$

[9] Strictly speaking, this Stockmayer formula for w_n requires modification to account for the indistinguishability of the two OH groups and of the two lone pairs of electrons in each of the water molecules which serve as the units in our case. Appropriate numerical modification has been provided by H. Porosoff.[10] The modification does not affect the qualitative results of this section.

[10] This Stockmayer formula for w_n specifies ways of connecting n monomer units to make an n-mer. Multiplicity associated with the possibility that each such topologically distinct species may possess a variety of different conformations is not included in w_n. Since our use below of the DiMarzio–Gibbs[12] modification of the Flory–Huggins[12] procedure for counting configurations of nonintersecting polymers on a lattice also ignores this possibility, we are assuming here that each topologically distinct n-mer possesses only one conformation around which it may, however, vibrate internally (see footnote 18). Electrostatic calculations by Porosoff[10] and Ben-Naim and Stillinger (see footnote 15) suggest that the potential governing internal rotation around a hydrogen bond possesses one broad minimum at the symmetrical-eclipsed orientation and one broad maximum.

If we were to consider only the hydrogen-bonding contribution to the potential energy, then, using (10), (11), and (12), we would obtain a partition function of the form

$$Q/N! = (v/\lambda^3)^N \sum_{\{m_n\}} (V/v)^M \left[\Omega_N(\{m_n\})/N!\right] \exp\left[-\beta V_{HB}(\{m_n\})\right] \qquad (13)$$

where V is the volume of the system and $\lambda = (\beta h^2/2m\pi)^{1/2}$ is the thermal wavelength of the molecules. (The inclusion of λ assures that the kinetic energy is always $\frac{3}{2} kT$ per molecule.) The quantity v is a measure of the volume of the attractive part of the potential energy. Using Eq. (13), we would find, not unsurprisingly, that the system undergoes a collapse. The collapse occurs when the extent of H bonding is precisely the same as the critical extent of reaction which characterizes the gel point of the FS theory. The source of this collapse is easily seen. We have totally neglected the effects of the hard-core repulsion of two molecules which approach one another too closely. It is well known that a similar collapse occurs if the repulsive term in the van der Waals equation is omitted. Indeed the vapor–liquid phase transition (with its accompanying critical point) is properly viewed as a consequence of the interplay of attractive and repulsive forces. Therefore, we must incorporate the effects of repulsions into our analysis.

Lattice models[11, 12][11] have proven to be extremely useful in the treatment of repulsions in condensing systems. In particular, the use of a lattice model allows us to utilize the Flory–Huggins procedure[12] in calculating the number of ways of placing polymers on a lattice without allowing them to intersect. For each distribution of Eq. (11), the number of ways of placing that distribution on the lattice is[12]

$$P(N_0, \{m_n\}) = [N_0!/N_0^N(N_0 - N)!] \{[N_0 z(z-1)]^M/z^{m_1}(z-1)^{m_1+m_2}\}$$

$$= [1/(1 - N/N_0)^{N_0-N}] \prod_n [A_n N_0 z(z-1)e^{-n}]^{m_n} \qquad (14)$$

[11] Perhaps the best way of introducing lattice models of fluids is through the cell-hole theory (CHT) [see T. Hill, *Statistical Mechanics* (McGraw-Hill Book Co., Inc., New York, 1956) Chaps. 7 and 8]. In CHT the partition function has the form

$$Q/N! = (1/\lambda^{3N}) \sum_{\text{config.}} e^{-\beta N_{AA}u(a)} \prod_{i=1}^{N} v_i$$

where N_{AA} is the number of nearest-neighbor pairs and the sum is over all possible configurations of N molecules on a lattice of N_0 sites with multiple occupation excluded. The term $u(a)$ is the value of the intermolecular potential evaluated at the nearest-neighbor distance, and v_i is the "free volume" of particle i in the configuration.

In the usual treatment of CHT a combinatorial factor $\Gamma(N_0, \{n_j\})$ is introduced, and the sum is taken over all possible distributions of molecules $\{n_j\}$ with j nearest-neighbor vacant cells. In terms of our polymer variables, with neglect of rings and with identification of N_{AA} as the number of hydrogen-bonded neighbors, we write $N_{AA} = N - M$ and $u(a) = -\varepsilon$. Thus, setting $v_i = v_c$ (the volume of the unit cell) we can write Q as

$$Q/N! = (v_c/\lambda^3)^N \sum_{\{m_n\}}' \Gamma(N_0, \{m_n\}) \prod_n [\exp \beta\varepsilon(n-1)]^{m_n}$$

where the primed sum is taken over all sets of m_n which satisfy equation (8). We then obtain equation (15) by setting $\Gamma = \Omega P/N!$

where the second line is obtained with Stirling's approximation and with $A_1 = [z(z-1)]^{-1}$, $A_2 = (z-1)^{-1}$, and $A_1 = 1$ for $i \geqslant 3$.

We can thus write the partition function, including effects of repulsions, as

$$Q/N! = (v_c/\lambda^3)^N (1 - N/N_0)^{N-N_0} \sum_{\{m_n\}} \prod_n (1/m_n!) \, [1/n!]$$

$$\times \, A_n N_0 z(z-1) w_n e^{-n+\beta\varepsilon(n-1)}]^{m_n} \tag{15}$$

where v_c is the volume of the unit cell. The sum in Eq. (15) may be approximated by its maximum term, for which m_n is given by

$$\bar{m}_n = N_0 [A_n z(z-1) w_n/n!] e^{-\gamma n + \beta\varepsilon(n-1)} \tag{16}$$

where γ is a Lagrange multiplier determined by the restriction that Eq. (8) be satisfied.

Stockmayer was able to evaluate sums of the general form of $\sum_n \bar{m}_n$ and $\sum_n n\bar{m}_n$ in terms of the variable α. In our case γ can be expressed in terms of α, and then the second sum essentially gives the density ($\rho = N/N_0$) as a function of α and T.[12] This equation can be inverted to give α (and hence γ) as a function of ρ and T. We obtain an equation of state by differentiating the free energy A [i.e., $-kT\ln(Q/N!)$] with respect to the volume. This straightforward but tedious calculation gives

$$\beta P v_c = -\ln(1-\rho) - \rho + \rho_M \tag{17}$$

where $\rho_M = (1/N_0) \sum_n \bar{m}_n$ is given by a complicated function of ρ and T.

Equation (17) can also be obtained from a grand canonical ensemble via the method of steepest descent.

Equation (17) possesses a coexistence (i.e., phase transition) curve and a critical point; the coexisting phases are a high-density phase (liquid) and a low-density phase (vapor) which correspond, respectively, to high (gel) and low (sol) number of hydrogen-bonded pairs. This is the major result of this section.

Several further comments are in order. Equation (17) cannot be very accurate. It does not yield the correct behavior either at moderately low densities or in the weak coupling (small $\beta\varepsilon$) limit. This defect can be remedied by replacing[13] $e^{\beta\varepsilon}$ by $e^{\beta\varepsilon} - 1$. A more serious defect is that, as a result of the total neglect of rings, the equation of state (17) cannot represent the high-density (i.e., liquid) region.

---

[12] We have used $A_n = 1$ for all n in our calculations. The correct values of A_1 and A_2 should contribute importantly only at densities much smaller than the critical density.

[13] Recently, we have found that this replacement arises naturally out of careful consideration of the role of kinetic (as well as potential) energy in the definition of a cluster bond.

These considerations have nevertheless served our purpose of demonstrating that the breakup of a branched structure is both cooperative and qualitatively consistent with the properties of the liquid–vapor transition. It should be particularly noted that, according to this analysis of condensation via gelation theory, the liquid, at any instant, is indeed essentially one branched structure.

4. FURTHER COMMENTS ON THE PHASE TRANSITIONS

The analogies noted to polymer phenomena suggest that our gel picture for liquid water is capable of accounting for the existence of the two phase transitions (melting and boiling). The arguments presented in Secs. 2 and 3 suggest that the essence of melting lies in the disruption of small, orderly interconnected rings of hydrogen bonds and that the essence of boiling lies in the breakup of an extensively branched random network. It has been shown that each of these processes, taken by itself, is cooperative. It remains to be argued that these processes do indeed occur separately under appropriate conditions.

Now it is only when the total volume into which partially depolymerized fragments can escape is large (low-pressure conditions) that the breakup of random branching yields as much entropy as does the opening of small rings, for it is only in this case (sublimation) that the increase in translational entropy associated with the breakage of non-ring-closing bonds is comparable to or greater than the increase in internal rotational (and librational) entropy associated with the openings of rings. Under higher pressures the first cooperative process encountered on heating ice (identified with melting) will not include the breaking of bonds which are not parts of rings since the translational entropy which might be gained on breaking such bonds would be less than that gained on ring opening (whereas the energy changes associated with these two types of breakage should be nearly the same). The triple point occurs at an intermediate pressure at which the entropy gained on opening small rings is comparable to that gained on breaking non-ring-closing bonds.

The principal process which occurs as the liquid is heated or cooled between 0 and 100°C (at 1 atm) must be the alteration of the number and nature of closed rings. Looking at the liquid from either the vantage point of fusion or that of condensation leads to this conclusion.

From the viewpoint of condensation we can see that, even if we can ignore ring formation in the interpretation of the essence of the vapor–liquid transition, we cannot avoid it in the interpretation of what occurs on further cooling in the pure-liquid range, for not many more interconnections (H bonds) are possible in the gel (liquid) without ring closure. Therefore, unless we insist on completing our model of the liquid in a fashion which will yield a heat capacity for the liquid that is much too small, we see that we cannot ignore ring closure. If, on the other hand, we do not insist on ignoring ring closure during condensation, we come to the same conclusion, for it is still true that all the sol (vapor) will not have been converted to gel (liquid) until the capacity to form H bonds which do not close rings has been saturated. When all the vapor has been condensed, the only types of potential H bonds which remain to be formed on further cooling in the pure liquid are ring-closing ones, the formation of which must therefore be primarily responsible for the large heat capacity of the liquid.

From the viewpoint of fusion, a similar conclusion can be reached. Even if we suppose that fusion involves essentially no net breakage of non-ring-closing bonds (as suggested by our discussion of fusion above and demanded by our liquid–gel analogy drawn from consideration of condensation), we are faced with the conclusion that, unless the liquid at the melting point contains ring closures, the maximum average H-bonded coordination number cannot exceed 2,[14] so that at least half of the H bonds would have to have been broken on melting. The supposition that such a large fraction of the H bonds of ice are broken on melting yields a heat of fusion significantly larger than that observed, unless the H-bond energy is chosen as unreasonably less than the "spectroscopic value" of about 2.6 kcal-mole^{-1}. Allowance for loops in the randomly branched network removes this difficulty and should certainly permit good fits to the known heats of fusion and vaporization with a reasonable choice for the hydrogen-bond energy.

5. PROPERTIES OF LIQUID WATER IN VIEW OF PROPOSED MODEL

5.1. Radial Distribution Function

The radial distribution function indicates that an average oxygen atom in liquid water has between four and five nearest-neighboring oxygen atoms between 2.8 and 2.9 Å away, next-nearest neighbors predominantly between 4.3 and 5.2 Å away, and some small excess density of neighbors between 6.2 and 7.5 Å away. Now, an infinitely long linear chain structure composed of water molecules H-bonded together with H-bond lengths of 2.85 Å and angles between adjacent H bonds of $\cos^{-1}(-\frac{1}{3})$, i.e., the tetrahedral angle, provides each water molecule with two nearest neighbors 2.85 Å away and two next-nearest neighbors 4.7 Å away. If, as calculations of the electrostatic interaction between H-bonded water molecules suggest,[10] the symmetrical eclipsed configuration[15] with respect to rotation around H bonds is significantly

---

[14] This is true for branched chains lacking ring closures as well as for linear chains.

[15] This configuration is defined as follows. Consider the partial plane which is bounded on one edge by a line drawn through the O–H---O axis of the hydrogen bond in question and within which lies the off-axis OH bond of the water molecule which is the proton donor in this H bond. Consider also the partial plane which is bounded by the same hydrogen-bond axis and within which lies the bisector of the two off-axis OH bonds of the water molecule which is the electron pair donor to the H bond. In the symmetrical-eclipsed configuration the dihedral angle between these two partial planes is 180°. See A. Ben-Naim and F. Stillinger, in *Water and Aqueous Solutions*, R. A. Horne, ed. (Wiley–Interscience, New York, 1972), p. 295.

preferred, there will also be an excess population density centered about 6.8 Å away.

The additional two to three nearest neighbors at distance between 2.8 and 2.9 Å cannot be fully accounted for as H-bonded nearest neighbors even with allowance for a reasonably large number of closed rings to increase the average H-bonded coordination number, for the total number of nearest neighbors in the 2.8–2.9 Å range (ca. 4.5) exceeds even that in ice. However, the mean separation of two non-H-bonded near-neighboring oxygen atoms is also expected to fall between 2.8 and 2.9 Å since the van der Waals radius of an oxygen atom is ca. 1.4–1.5 Å. (Indeed, in the structure of ice-VIII there are eight nearest neighbors at a distance of 2.86 Å, only four of which are H-bonded to the central molecule.) The fact that this coincidence has been ignored has apparently been responsible for the often-made assumption that virtually all water molecules must be in ice-like environments. The random network or "gel" model to which we have been led in the preceding discussion is consistent with the radial distribution function, and there is no need for further structural assumptions.

5.2. Infrared and Raman Spectra

Possibly the most discussed feature in the infrared and Raman spectra of liquid water, studied as a function of temperature, is the appearance of isosbestic points in several groups of absorption bands associated with proton motions.[13] We accept the point of view that these points provide strong evidence that the OH groups in liquid water exist in effectively two discrete states in mutual equilibrium and that these states are related to the presence of broken and intact hydrogen bonds.

Several facts have emerged from the spectral investigations of isosbestic points in water:[13, 14]

(a) The fraction of water molecules not involved in hydrogen bonding (i.e., monomeric water) is probably less than 1 % and can be ignored.

(b) The fraction of molecules having no more than one OH group involved in a hydrogen bond increases with increasing temperature.

(c) At low temperature the spectrum of liquid water bears some resemblance to that of ice.

These facts are not inconsistent with the model proposed here. In consideration of fact (c), one must recognize that infrared and Raman spectra are not particularly sensitive to the finer details of structure. This observation is supported by the very close similarity of the spectra of linear and branched hydrocarbons and those of polyethylenes with varying degrees of crosslinking.

5.3. Static Dielectric Constant and Dielectric Relaxation Time

The dielectric constant of water is unusual not only in the largeness of its static value but also in the narrowness of its dispersion, which behaves like that of a single mode of motion. With regard to the former, we limit our remarks here to the qualitative observation that the random gel model provides the local correlations between angular coordinates of molecules which, according to the theory of Kirkwood,[15] are requisite to a large dielectric constant. If most of the correlations that are lost on melting are those associated with small rings, the dielectric constant would not be expected to change dramatically on melting, since dipole-moment vectors in rings tend to cancel out.

With Haggis et al.[16] and in conformity with the accepted mechanism for dielectric relaxation in ice,[17] we attribute the single dielectric relaxation time in water to a mode of motion in which doubly hydrogen-bonded molecules (Bjerrum "DL fault" pair in the case of ice) reorient via rotation around one hydrogen bond with requisite breakage of the other. In the gel model the equilibrium population of doubly hydrogen-bonded molecules cannot be negligible. The energy of activation for such a mode of motion should be approximately the energy of one hydrogen bond, i.e., ca. 3 kcal-mole^{-1}. Curvature in the experimentally observed Arrhenius plot somewhat obscures simple analysis, but the value of 3 kcal-mole^{-1} is cited by Haggis et al. as yielding rough correspondence with the relaxation time and its temperature dependence.

We must now ask why molecules which are singly, triply, or quadruply hydrogen-bonded to the rest of the gel network do not give rise to observed relaxations. The spinning, around its single H bond, of a singly H-bonded molecule should be associated with infrared frequencies. Indeed, the value of the dielectric constant (ca. 5) on the high-frequency side of the relaxation at 10^{-11} sec (which we have attributed to doubly H-bonded molecules) is still significantly larger than the square of the refractive index measured at optical frequencies, indicating the presence of another dispersion somewhere in the infrared region.

The triply and quadruply H-bonded molecules fail to yield observable dispersions for another reason. In these cases the frequencies of relaxation would certainly fall in the range of dielectric measurements, but the intensities associated with these relaxations would be small. In the case of a triply bonded molecule, relaxation by rotation around one H bond requires the breakage of the other two and a consequent activation energy of at least 6 kcal-mole^{-1}, which yields a relaxation time about 100 times larger than that required for the relaxation of a doubly bonded molecule. Long before an appreciable fraction of this longer time has expired, the three branches defining a particular molecule as triply H-bonded will have been snipped away by the relaxation of doubly H-bonded molecules: A triply H-bonded unit does not remain triply bonded long enough to relax as a triply bonded unit. The same argument applies *a fortiori* to the relaxation of quadruply H-bonded molecules, which would be slower than that of doubly H-bonded molecules by a factor at least 10^4.[16]

5.4. The Density Maximum at 4°C

The conventional interpretation of the shrinkage of molar volume between 0° and 4°C at 1 atm attributes this to a decrease in the number of tetrahedrally coordinated water molecules, to each of which an abnormally large molecular volume is attributed. The interpretation of this abnormally large molecular volume of 4-coordinated units need not be sought in the peculiarities of any particular tetrahedral lattice as a whole; it can be expressed in the idea that, if four molecules are located in four very special positions around a central one it may be difficult for as many others (not H-bonded to the central one) to get as close to the central one as they could if all were free to accommodate themselves to the best geometry for a coordination number higher than 4. Thus if would seem to make little difference for the interpretation of the volume–temperature behavior whether these tetrahedrally H-bonded water molecules are presumed to exist together in ice-like structures or presumed to occur at random in a gel-like structure of the type we are proposing here.

One should note that the process which gives rise to this contractile component on heating is not restricted to a narrow range of temperatures (such as 0–4°C) but rather exists over a wide range extending from low temperatures in the supercooled liquid to *ca.* 30°C. That this is true can be seen by subtracting from the total curve of volume vs temperature the monotonically increasing curve which one may expect to arise from increasing amplitudes of vibration of molecules in any fixed configuration, i.e., the thermal expansion one sees in a glassy or crystalline state. When the latter is subtracted, the remainder of the V–T behavior, attributable to configurational changes in the liquid, exhibits a (broad) minimum not at 4°C but rather at about 30°C.

5.5. The Supercooled Region and the Glassy State

The speed of the relaxation mechanism in liquid water, discussed above, facilitates nucleation of ice crystals and, in practice, prevents extensive supercooling. Nevertheless the properties of the metastable supercooled

---

[16] It might be thought that this argument leads to the conclusion that the dielectric relaxation time in ice ought to be only 10^4 times longer than that in water, as compared with the factor 10^6 actually observed. However, the effect of the rotation in ice is to convert a molecule which originally participated in four H bonds into one which participates in two H bonds and two high-energy faults (Bjerrum D and L faults), which may then be expected to travel apart via rotation of the molecules forming the other halves of the faults. This situation has been discussed by others [N. Bjerrum[17] and L. Onsager and M. Dupuis, *Electrolytes*, B. Pesce, ed. (Pergamon Press, London, 1952)] and need not be pursued further here.

liquid, even in the range where they are not directly observable in practice, need to be confronted by any theory.

In the cases of liquids which can be extensively supercooled, a transformation to a glassy phase is normally observed. If extensive supercooling could be achieved in the case of liquid water, a transition to glassy water would presumably be observed. Support for this point of view is provided by the successful production of amorphous water by a vapor-deposition technique.[18]

In and below the temperature region of the glass transition, not even internal metastable equilibrium can be attained in a liquid in reasonable times. Nevertheless, even the hypothetical thermodynamic properties, estimable by extrapolation below the glass transition of equilibrium data obtained above it, are of first importance.

Such extrapolations[19] show clearly that the ultimate fate of the metastable equilibrium liquid on extensive cooling is *not* continuous conversion to the crystal. They show that the liquid, sufficiently supercooled, will have lost all its excess (as compared to the crystal) entropy at a temperature (well above absolute zero) where its energy is still much larger than that of the crystal. That this temperature represents some sort of ground state (which may be highly degenerate *per mole* and still yield a negligible entropy) for the amorphous phase was recognized by Gibbs,[20] who used this recognition as the basis of a theory for equilibrium properties of polymers. His postulate of an empirical correlation between this reference temperature, where the configurational entropy is essentially zero, and the glass temperature, where the relaxation time becomes longer than the duration of an experiment (a reasonably well-defined temperature in spite of variations in the durations of experiments because the relaxation time is a rapidly varying function of temperature in this range), was developed by Gibbs and DiMarzio.[20] This latter correlation was provided with an interpretation by Adam and Gibbs[21] who succeeded in showing a way in which configurational entropy, in a temperature region in which it is in short supply, can be the predominant factor in the determination of relaxation times.

The success of this body of theory in accounting for the influence of molecular weight, composition (in the case of copolymers), concentration (in the case of "plasticized" polymers), crosslinking, etc., on the glass transition and, indeed, in accounting for the fact that the glass transition (a kinetic property) always falls below the melting point (a thermodynamic property), as well as in accounting for the quantitative nature of the variation in relaxation time on the approach to the glass transition (the so-called WLF equation), indicates that its implications concerning the nature of glass-forming liquids other than polymers should also not be ignored. The theory ascribes the dearth of configurational entropy and consequent glassification to the absence of spherical molecular symmetry in glass-forming liquids. In cases where this asymmetry is particularly marked, as in linear or branched polymers, the glass-transition temperature is found, both theoretically and experimentally, to be high (*ca.* 300 ± 100°K for many noncrosslinked polymers).[17] In the case of a liquid composed of small molecules with a propensity to form a hydrogen-bonded network, it should be moderately high, as is the case for glycerol and (putatively) water.

6. SUMMARY AND CONCLUSIONS

The considerations of Secs. 2–4 have provided an interpretation of the phase transitions which distinguish liquid water from ice and water vapor and have thereby provided the essence of a physical model for liquid water. The possibility that the model thus obtained can account for the "peculiar" properties of water within the liquid range has been discussed briefly in Sec. 5, where it is seen that, if difficulties are eventually to emerge, they are not so obvious as to be recognizable at this time.

In this article the problems posed by the two phase transitions have been neither formulated in totally satisfactory ways nor analyzed completely. The effects of intracluster vibrations, for example, are yet to be considered.[18] It would be most surprising, however, if the two-transition feature were to be lost upon more careful treatment of the basic concepts introduced here. Inasmuch as the problem of accounting for two transitions (let alone one!) is of long standing, these concepts merit further examination, both with regard to the phase transitions and with regard to the nature and properties of liquid water.

The model for the liquid which our considerations of the phase transitions strongly suggest is that of a random (constantly rearranging) gel. The term "gel" is intended to convey the concept that, at any instant (time interval $<10^{-11}$ sec for water), the liquid consists essentially of a cluster (hydrogen-bonded in this case) of macroscopic size, in analogy to this use of the term in polymer physical chemistry (where the bonds are chemical and more persistent).

This model is consistent with that indicated by the computer-simulated molecular dynamics effected by Rahman and Stillinger.[22] On inspecting pictorial representations of their results, one is unable to identify an appreciable number of small ring closures unless one adopts absurdly liberal definitions of "hydrogen bond" and "ring." This conforms with the proposition posed here to the effect that it is primarily the ordered small rings of the

[17] If the polymer chains are stiff, in an appropriately defined sense,[20] or if they are cross-linked, the configurational entropy is especially small and the glass temperature especially high.

[18] An approach to this, somewhat analogous to that of G. Némethy and H. Scheraga [*J. Chem. Phys.* **36**, 3382 (1962); **41**, 680 (1964)], has been given by H. Porosoff.[10] He found that the thermodynamic properties in the liquid phase are particularly sensitive to the intramolecular vibrations. For the purposes of this paper, one should view the parameter ε introduced in Sec. 3 as a *free* energy associated with a hydrogen bond.

crystal that are lost on fusion. Both the Rahman–Stillinger computations and our considerations of concepts which will account for two phase transitions thus indicate that clusters[19] in the liquid are more disordered than has sometimes been supposed. They both also suggest, on the other hand, that these clusters are far larger[20] than has often been supposed.

REFERENCES

1. See D. Eisenberg and W. Kauzmann, *The Structure and Properties of Water* (Oxford University Press, New York, 1969).
2. J. H. Gibbs and E. A. DiMarzio, *J. Chem. Phys*. **30**, 271 (1959).
3. B. H. Zimm and J. K. Bragg, *J. Chem. Phys.* **31**, 526 (1959).
4. S. Lifson and G. Allegra, *Biopolymers* **2**, 65 (1964).
5. H. E. Stanley, *Introduction to Phase Transitions and Critical Phenomena* (Oxford University Press, New York, 1971), Chap. 8.
6. P. J. Flory, *J. Am. Chem. Soc.* **63**, 3083, 3091, 3096 (1941).
7. W. H. Stockmayer, *J. Chem. Phys.* **11**, 45 (1943).
8. J. E. Mayer and M. G. Mayer, *Statistical Mechanics* (Wiley, New York, 1940), Chaps. 13 and 14.
9. G. Scatchard, J. M. Kavanagh, and L. B. Tichnor, *J. Am. Chem. Soc.* **74**, 3715 (1952); M. Gordon, C. S. Hope, L. D. Loan, and R.-J. Roe, *Proc. Roy. Soc. (London)* **258A**, 215 (1960).
10. H. Porosoff, Ph.D. Thesis, Brown University, 1970 (unpublished).
11. T. D. Lee and C. N. Yang, *Phys. Rev.* **87**, 410 (1952).
12. P. J. Flory, *J. Chem. Phys.* **10**, 51 (1942); M. L. Huggins, *J. Phys. Chem.* **46**, 151 (1942); *Ann. N.Y. Acad. Sci.* **43**, 1 (1942); *J. Am. Chem. Soc.* **64**, 1712 (1942). We have employed in our calculations the DiMarzio–Gibbs [*J. Chem. Phys.* **28**, 807 (1958)] modification of the Flory–Huggins procedure.
13. G. E. Walrafen, *J. Chem. Phys.* **48**, 244 (1968).
14. W. C. McCabe, *J. Phys. Chem.* **74**, 4360 (1970).
15. J. G. Kirkwood, *J. Chem. Phys.* **7**, 911 (1939).
16. G. H. Haggis, J. B. Hasted, and T. J. Buchanan, *J. Chem. Phys.* **20**, 1452 (1952). See also J. B. Hasted, in *Water, A Comprehensive Treatise*, F. Franks, ed. (Plenum Press, New York, 1972), Vol. I, Chap. 7.
17. N. Bjerrum, *Science* **115**, 385 (1952).
18. J. A. Ghormley, *J. Chem. Phys.* **25**, 599 (1956).
19. W. Kauzmann, *Chem. Revs.* **43**, 219 (1948).
20. J. H. Gibbs, *J. Chem. Phys.* **25**, 185 (1956); J. H. Gibbs and E. A. DiMarzio, *J. Chem. Phys.* **28**, 373 (1958); J. H. Gibbs, in *Modern Aspects of the Vitreous State*, J. D. MacKenzie, ed. (Butterworth Scientific Publications Ltd., London, 1960), Chap. 7.
21. G. Adam and J. H. Gibbs, *J. Chem. Phys.* **43**, 139 (1965).
22. A. Rahman and F. Stillinger, *J. Chem. Phys.* **55**, 3336 (1971); **57**, 1281 (1972).

---

[19] In this paper we have ignored the component of energetic cooperativity arising from the putative nonpairwise additivity of hydrogen-bond formation [H. S. Frank and W.-Y. Wen, *Dis. Faraday Soc.* **24**, 133 (1957)]. Such an effect should operate in the direction of sharpening the profiles of the two phase transitions, the existence of which as true transitions (discontinuities) we have been at pains to interpret here. Furthermore, in principle, at least, such an effect can be incorporated into the gelation analysis.

[20] This feature is an implicit assumption of the model of J. A. Pople [*Proc. Roy. Soc.* **A205**, 163 (1951)]. It can also be obtained from lattice-gas (Ising) models for the condensation of simpler liquids[5,11] and of water (P. D. Fleming and J. H. Gibbs, to be published).

DISCUSSION

Professor G. S. Kell (*National Research Council, Ottawa*). Perhaps you could clarify a point for me. There are four bond directions at each vertex, giving six pairs of bonds, and in hexagonal ice this means that 12 six-member rings pass through the vertex. When the broken rings are counted, it is found that breaking a single bond breaks many rings—in this example one broken bond breaks 6 six-member rings. Rather than talk about rings, it is going to be simpler and more economical to talk about broken bonds as people have all along. Yet, Pauling showed 30 years ago that only one quarter of the bonds disappear from ice on melting. Perhaps at the end of your paper you backed down a little from the extreme position taken at the starting point. Still it appears that at high temperatures, at least, your liquid must show essentially 2-coordination. Are you not, by such emphasis on topological considerations, in danger of getting in the position where it is said that liquid water is essentially the same thing as solid alcohol, which contains linear, hydrogen-bonded chains?

Professor Gibbs (*Brown University, Providence, Rhode Island*). I need to make several points clearer to answer this many-faceted question:

Firstly, the value 2 for the average H-bonded coordination number is correct, even at high temperatures, only in the absence of ring closures. With allowances for ring closures it is found to be larger.

Secondly, even an average coordination number as small as 2 doesn't imply purely linear chains. They may be highly branched, as only ring closures can increase it to values above 2.

Thirdly, I discussed the hypothetical case of condensation without ring closures simply to show that the essence of this transition is not dependent on ring closing. In this way, I could demonstrate simply how it happens that, under appropriate conditions, we observe two disparate phase transitions— one (condensation) associated *essentially* with the formation of a randomly branched network or "gel" and the other (freezing) associated *essentially* with the formation of an ordered array of small rings.

Now, my fourth observation is that we know we shouldn't carry this viewpoint to the extreme of failing to recognize that ring closures at random must accompany the infinite branching process which nonetheless appears to be the essence of the discontinuous, or phase transition, character of condensation. On purely intuitive grounds, one expects a highly branched structure to display numerous "accidental" ring closures. It would have to pay a heavy entropy price to avoid many of them and simply cannot avoid them all. Indeed, a pure Cayley tree (no ring-closing bonds) cannot be contained in space. Recent developments in the treatment of gelation address these difficulties with techniques of "percolation" theory. Although these go beyond our immediate purposes here, we still should note that they are related to the difficulties that we have found in our approach to condensation via simple gelation theory. Mathematically, these difficulties are manifestations of problems associated with an assumed analytic continuation of the sums of infinite series beyond their circle of convergence (that is, beyond the zero-density gel point). Physically, these difficulties are almost certainly attributable to our neglect of ring closures.

My fifth point is a reminder that, even if ring closures could be totally ignored in an approximate discussion of condensation, they could not be ignored in even a qualitative discussion of what ensues on further cooling, for, once gelation (condensation) has occurred, all water molecules are (at any instant) part of the same cluster (the "gel"), and only *intragel* H bonds remain to be formed. Cooling in the liquid range will introduce many ring closures and markedly increase the average H-bonded coordination number, although, for the reason I have given in discussing the radial distribution function, there is no apparent need to assume that it becomes as large as 4.

For my sixth and last point, I want to confront the part of Kell's question which dealt with melting and note that his formulation of this part of the question contains the key to the answer. The very fact that breaking one hydrogen bond in the ice structure destroys many six-membered rings ensures that only rather few hydrogen bonds need to be broken in order to destroy all the hexagonal rings. If this were not true, we would indeed be stuck (in our theoretical concepts) with an average H-bonded coordination number for the liquid which would be too small to be consistent with Pauling's interpretation of the observed heat of fusion.

Professor G. Némethy (*Université de Paris-Sud*). It seems to me that by neglecting all interactions between nonbonded molecules, you are throwing out completely any analogy with liquid argon.

Professor Gibbs. We do not totally ignore them. I did not talk about this in order to keep the presentation simple. When a Flory–Huggins lattice is used, not only can the repulsions be accounted for, but, in a very rough way, a nondirectional kind of attraction can also be taken into account. Simply, every time a lattice site is left empty, it costs a certain amount of "hole" energy. This is a very crude way of introducing another interaction in addition to the hydrogen bonds.

Professor Némethy. Still, if you do not account for directional interaction between the molecules, you will not get actual condensation, either in the case of argon or of water, especially of the non-hydrogen-bonded molecules.

Professor Gibbs. You can if you want to. To pursue this kind of treatment in cases where hydrogen bonding does not exist, you obviously cannot neglect the weaker interactions because you have to have something to serve as the basis for "gelation" (or condensation), but you do not have to assume that these interactions are directional.

We obtained condensation just from reversible gelation theory, which *per se* doesn't require directional bonding.

What gelation theory does require (in addition to saturability of bonding) is the approximation contained in the two-state concept for a "bond," i.e., an intact state and a broken state as opposed to a continuous potential. It is in this sense (and in the neglect of rings) that it differs from its otherwise close cousin, the Mayer cluster integral expansion, and represents a *quantitatively* less reasonable way of describing argon condensation. It is also in this sense that it bears a resemblance to an Ising model (of which the lattice gas is one version) and appears much more tractable than Mayer cluster theory for a case like water.

Now, in our treatment the specification that H bonding is directional actually enters only when the lattice is introduced. Without the lattice we have no repulsions in the theory, but we get condensation. In fact, we get too much condensation in this case! Without the lattice repulsions the liquid–vapor equilibrium curve never terminates at a critical point.

It is interesting to note that the lattice device is often used not only just to account for repulsions (and directionality of bonding), as here, but also (in place of cluster or gelation theory) as the means of enumerating attractive interactions (by assigning them to nearest-neighboring site occupiers). When this is done, one has the so-called "lattice-gas" version of the Ising model, which, in spite of its directional character, has most often been applied to argon!

Professor Némethy. You said that the liquid–vapor equilibrium corresponds to the condition that the average functionality is 2, leading to gelation. Is this a sufficient condition?

Professor Gibbs. Only if you make Stockmayer's assumption that, as the clusters are growing in the gas phase, ring closures can be ignored. This corresponds, in the Mayer cluster theory, to ignoring all irreducible cluster integrals β except β_1. This is an enormous simplification, but it still has the transition in it.

Professor Némethy. If you allow for the absence of ring closures by introducing only β_1, don't you need a critical concentration?

Professor Gibbs. There is a particular vapor density associated with each temperature on the liquid–vapor equilibrium curve and a "critical" one at the critical point, just as with the experimentally observed phenomena.

Professor H. S. Frank (*University of Pittsburgh*). If you are giving a new proposal for water structure, you should not get away without saying how the maximum of density is accounted for, and I am going to do just that for you. I am going to propose something that I got some years ago from Morrison up at Ottawa, who said that he had been toying with the idea that the shrinkage on warming of water is analogous to the shrinkage on warming of rubber. The analogy seems sound statistically, and the only thing that kept him from believing it was the fact that the hydrogen bond in water breaks too often, and this, as a matter of fact, is one of your difficulties. You have been giving comparisons with polymers in which there are honest bonds that live a long time compared to the lifetime for reorientation of the water molecule, which Professor Hertz has been telling us is 10^{-11} sec. We cannot use this analogy, therefore, unless we can count on a water molecule turning around and coming back to the same place. If this can be achieved, then you can have your rubber stretching.

Professor Gibbs. This raises a point I should clarify. As long as a bond persists for enough vibrations, then, as far as the thermodynamics are concerned, you can treat the system in the way I have described. What we do is say that we have all different kinds of clusters, all different kinds of topologies and shapes, etc., that we put on the lattice. Now, as the system flows in time it makes no difference for the thermodynamics whether a cluster of shape A originally in one location and a cluster of shape B originally in another location interchange places by diffusion or whether A and B simply change into B and A, respectively, by breaking bonds and forming new connections. In the theory for equilibrium properties, all that matters is that I sum over all possible states, that is, all kinds of shapes in all kinds of locations. But you are right in your observation that it is necessary for the validity of such a model that bonds persist for a time long compared to the periods of important intracluster vibrations.

One other thing your question brings to mind is that, although we have had in mind an interpretation of the density maximum, this phenomenon is the one thing among all the things we have looked at which has not yet fallen out of our model in a rather natural way. I should not be surprised if we should come to like your suggestion on this point better than the more-or-less conventional one that we have invoked.

Professor P. A. Giguère (*Université Laval, Canada*). Let me point out that any explanation for the density maximum of ordinary water at 4°C must also account for the important fact that in heavy water the temperature range of increasing density is nearly double, *viz*, from 3.8 to 11.5°C.

May I also make one last remark regarding hydrogen peroxide. The melting point of that crystal is very close (−0.4°C) to that of ice in spite of a nearly double molecular mass. Now, the packing forces are quite the same in both crystals, namely, four hydrogen bonds per molecule. However, in ice these forces are much better balanced (tetrahedral about each oxygen atom) than in hydrogen peroxide (one unengaged orbital on each oxygen atom). This makes for a less stable structure.

A View of Electrolyte Solutions[1]

J. C. Rasaiah[2]

Received January 29, 1973

The uncertainties in the route to infinite dilution for 2–2 electrolytes are discussed in relation to the practical difficulties of determining the standard emf's of simple reversible cells containing ZnSO₄ in H₂O and D₂O solutions. These difficulties are due to uncertainties in the theory of highly charged ions in aqueous solution. Recent developments in theories of electrolytes, especially those for which numerical results are available, are critically evaluated for their accuracy and adaptability to changes in the solute potential. Simple refinements to the model (i.e., the solute potential) are described, and the changes are interpreted, in terms of the molecular interactions between sets or pairs of ions in the pure solvent. Recent work on the effect of solvent granularity and other molecular properties of the solvent (e.g., dipole moment) on the solute potential is reviewed.

KEY WORDS: Electrolytes; aqueous solutions; heavy water; electromotive force; extrapolation; electrolyte theory; models; thermodynamics, cospheres; solvent granularity.

1. INTRODUCTION AND THE EMF's OF CELLS

My interest in electrolyte solutions was stimulated in a provocative way when I began graduate work under Professor Henry S. Frank. One of the problems which interested him was the effect of solvent structure on the thermodynamic properties of electrolytes. Since water and heavy water have nearly the same dielectric constant ($\varepsilon \approx 78$), it seemed worthwhile to attempt a comparison of the thermodynamic properties of electrolytes in these two solvents. Robinson and Stokes[1] had already mapped out the osmotic coefficients of electrolytes in water; we sought to do the same in heavy water. Our chief difficulty then was that an isopiestic standard for electrolytes in

---

[1] This paper was presented at the symposium, "The Physical Chemistry of Aqueous Systems," held at the University of Pittsburgh, Pittsburgh, Pennsylvania, June 12–14, 1972, in honor of the 70th birthday of Professor H. S. Frank.

[2] Department of Chemistry, University of Maine, Orono, Maine 04473.

heavy water was not available. This prevented our group,[3] at least for a while, from measuring the osmotic coefficients of electrolytes in D_2O solutions by the same methods that had been so widely and successfully used for electrolytes in aqueous media. It was necessary to start at the beginning and establish a suitable standard, a problem which Professor Frank suggested as a worthwhile topic for a Ph.D. dissertation.[2] I accepted, unaware of the troubles ahead—one of which was so serious that it led years later to a deeper study of electrolyte theory—from which I have not recovered.

The principal difficulty which we ran into arose from the lack of an accurate method for the extrapolation to infinite dilution of emf measurements on cells containing 2–2 valent electrolytes in aqueous solution. Such an extrapolation is usual in the determination of activity coefficients from cell measurements, but 2–2 electrolytes are particularly difficult to handle because of their anomalous properties at high dilution. It concerned us because of our choice of the following cell as the means of establishing an isopiestic standard in heavy water:

$$\text{Zn/Hg} \quad | \quad \text{ZnSO}_4(m), \text{PbSO}_4(s) \quad | \quad \text{Pb/Hg}$$
$$\text{(2-phase)} \quad | \quad \text{(Solvent: H}_2\text{O or D}_2\text{O)} \quad | \quad \text{(2-phase)}$$

We were guided in this by the existence of earlier investigations of the same cell, in which water was the solvent, by Cowperthwaite and LaMer[3] and also by Bray.[4] It was known that, when the usual precautions had been taken, the electrodes were reversible and reproducible. The solubility range of $ZnSO_4$ in D_2O was also wide enough to enable a useful vapor pressure standard for electrolytes to be established. Another possibility would have been to study an appropriate concentration cell with transference, but the transport numbers of a suitable electrolyte in D_2O solution were not known to the extent (precision and range of concentration) that the establishment of an isopiestic standard would have required. The zinc sulfate cell seemed an ideal choice, except for the problem of extrapolation.

The manner in which this difficulty was overcome is illustrated in Figs. 1–3. We made use of two established features of electrolyte solutions, one empirical and the other theoretical. The activity coefficient $\gamma_\pm$ of an electrolyte at a molality m can be represented over different concentration ranges in the following ways:

(1) At low concentrations, not too close to infinite dilution

$$\ln \gamma_\pm = a - bm^{1/3} \tag{1}$$

where a and b are constants which differ from one electrolyte solution to another[5, 6] and m is the molality.

---

[3] Particularly Robert E. Kerwin who carried out the isopiestic measurements in D_2O solvent (Ph.D. Thesis University of Pittsburgh, 1964).

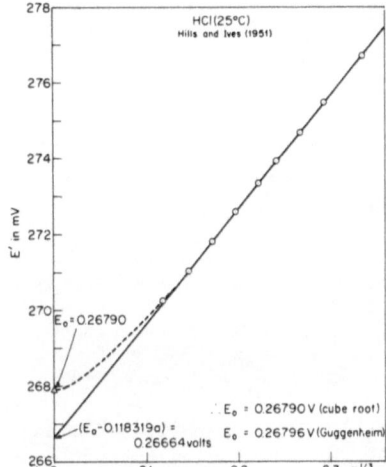

Fig. 1. Extrapolation of the emf of the cell $H_2(Pt)|HCl(m)$, $Hg_2Cl_2(s)|Hg(l)$ to infinite dilution by the cube-root method. E_0 (Guggenheim) is the standard emf obtained by conventional methods.[8] The data are from Hills and Ives [*J. Chem. Soc.*, 318 (1951)]. The figure is from ref. 2.

(2) As $m \to 0$, the Debye–Hückel theory[7] predicts that

$$\ln \gamma_{\pm} = -Am^{1/2} \tag{2}$$

where A is known from the same theory.

Equation (1) is an empirical relation whose generality was understood but overshadowed by the arrival of the Debye–Hückel theory.[7] It may be confirmed directly,[2] avoiding the uncertainties in $\gamma_{\pm}$ associated with a particular extrapolation to infinite dilution, by plotting the function E' against $m^{1/3}$, where E' is related to the electromotive force E of a simple reversible cell, containing the electrolyte of interest, by

$$E' \equiv E + vk \log m \tag{3}$$

$$= E_0 - vk \log \gamma_{\pm} \tag{4}$$

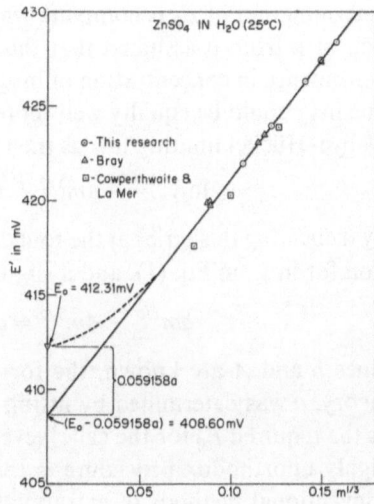

Fig. 2. Extrapolation of the $ZnSO_4$ cell to infinite dilution by the cube-root method. The measurements of Cowperthwaite and LaMer,[3] at low concentrations, do not agree with the independent measurements by Bray[4] and Rasaiah.[2] Figure from ref. 2.

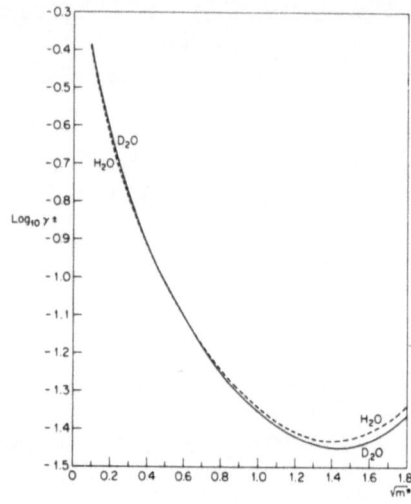

Fig. 3. The activity coefficients of $ZnSO_4$ in H_2O and D_2O solutions at 25°C. The quantity m^* is the aqua-molality of the electrolyte. From ref. 2.

Here vk is a constant determined by the temperature and the cell reaction and E_0 is the standard emf of the cell. When Eq. (1) holds, we can combine it with Eq. (3) to obtain

$$E' = (E_0 - vka) + vkbm^{1/3} \qquad (5)$$

which allows direct confirmation of Eq. (1) through emf measurements and also establishes b and $(E_0 - vka)$. Equation (5) is obeyed to within 5–10 μV by aqueous solutions of NaCl, KCl, HCl, $CaCl_2$, and $ZnSO_4$.[2] The cube-root range for 1–1 electrolytes is from about $3 \times 10^{-3}\ m$ to 0.25 m. The range, for 2–2 electrolytes, is shifted to lower concentrations, with the possibility that the lower limit may lie below the point at which reproducible emf measurements are usually possible. When viewed in terms of Eq. (5), the determination of E_0 is complete when a is calculated from b. Our determination of a from b assumed that the transition from cube-root to square-root dependence in concentration of $\ln \gamma_{\pm}$ was smooth and that along the cube-root line $\ln \gamma_{\pm}$ could be equally well represented by a power series in $m^{1/2}$, with the Debye–Hückel limiting law as the leading term:

$$\ln \gamma_{\pm} = -Am^{1/2} + Bm + Cm^{3/2} + Dm^2 + \ldots \qquad (6)$$

By truncating this series at the fourth term, equating the result with the expression for $\ln \gamma_{\pm}$ in Eq. (1), and a slight rearrangement, we arrived at

$$bm^{1/3} - Am^{1/2} = a - Bm - Cm^{3/2} - Dm^2 \qquad (7)$$

Since b and A are known, the former from experiment and the latter from theory, a was determined by fitting Eq. (7) to the cube-root range. This gave us the required E_0 for the cell. Nevertheless, the only real justification for this highly unorthodox procedure is that, in a crisis, it appears to work. When conventional methods of extrapolation are applicable, as, for example, with

1-1 electrolytes, the unorthodox method gives results in close agreement with the "established" values of E_0 (See Fig. 1; the established value is due to Guggenheim and Prue.[8]) Moreover, when emf measurements for $ZnSO_4$ in aqueous solution are treated by the same method,[2] the osmotic coefficients derived by Gibbs–Duhem integration of $\gamma_\pm$ agree well with Robinson and Jones' independent determination of ϕ.[9] Accordingly, in the critical situation that we faced, the same method of extrapolation was used with the measurements on the $ZnSO_4/D_2O$ cell described earlier, and the activity coefficients of this electrolyte in heavy water solutions were determined at concentrations ranging from 2.5×10^{-3} m to 2.5 m (see Fig. 3). However, until the theory of 2–2 electrolytes is well understood, these results, which depend on an assumed path to infinite dilution, must, as the following discussion shows, be regarded as tentative.

2. MODERN ELECTROLYTE THEORY

With a Ph.D. in my pocket and a strong sense of purpose, I felt it was time to learn more theory. The opportunity arrived when I moved to Stony Brook to work with Professor Harold L. Friedman. We were fortunate at that time in being able to make use of various approximate methods that had been worked out in the statistical-mechanical theory of gases, for example, the Percus–Yevick (PY) and hypernetted-chain theories (HNC).[10a] It was known that the PY theory was good for gases; we found that the HNC theory was better for electrolyte solutions. But before continuing further, I will endeavor to explain, in qualitative terms, how a theory for gases can be applied to solutions.

First, consider gases. It is common knowledge that the pressure of a gas has a density expansion and that the virial coefficients can be written in terms of the pair-potential and higher component potentials.[10b, 11] A useful property for further study is the radial distribution function $g(r)$. This is related to the probability that two gas molecules are at a distance r apart and is completely determined, at a given density ρ of the gas, by the pair-potential $u(r)$, by the higher component potentials, and by the temperature T. The thermodynamic properties of the gas are available through several alternative routes from the radial distribution function.[12] We depict this schematically as follows:

$$u(r) \xrightarrow[\text{Model mechanics}]{\text{statistical}} g(r) \begin{array}{l} \xrightarrow{v} \\ \xrightarrow{c} \\ \xrightarrow{E} \end{array} \begin{array}{l} \text{Thermodynamic} \\ \text{properties, e.g.,} \\ \text{pressure, energy} \end{array}$$

The letters v, c, and E are abbreviations for the virial, compressibility, and energy equations through which the thermodynamic properties may be obtained from $g(r)$. Each of these equations yields the same result only for the exact $g(r)$. The passage from $u(r)$ to $g(r)$ is chiefly what the statistical mechanics of fluids is about. Since the exact solution to this problem is difficult, various shortcuts have been tried. One of these is the PY approximation, and another is the HNC approximation.[10a] The Monte Carlo (MC) and Molecular Dynamics (MD) methods are alternative routes to the thermodynamic properties.[13] They are direct simulation methods which are in principle exact, though they require the expenditure of considerably more computing time than the other methods mentioned here. But what is of prime importance is that they provide, in the best situations, the standards against which various theoretical approximations for a model system, specified by a particular $u(r)$, can be judged.

The relationship between the theories of gases and solutions was deduced by McMillan and Mayer in 1945.[14] They proved an important theorem which states roughly that precisely the same statistical methods used for gases can be applied to solutions provided (a) the solution is under an *additional* pressure P, called the osmotic pressure, which maintains it in osmotic equilibrium with the pure solvent under a pressure P_0, and (b) the potential energy functions for solute molecules, used in the determination of $g(r)$ and its attendant thermodynamic properties, are averaged over the positions and coordinates of the solvent molecules. The first stipulation leads naturally to the description of the thermodynamic state of a solution in terms of the variables $(T, \mu_s, c_1, c_2, \ldots, c_\sigma)$ rather than $(T, P', m_1, m_2, \ldots, m_\sigma)$. Here c_i and m_i are the concentration (moles per liter of solution) and molality (moles per kilogram of solvent) of species i which are σ in number, P' is the external pressure on the solution, and μ_s is the chemical potential of the pure solvent at a temperature T and pressure P_0. Osmotic equilibrium between pure solvent and solution ensures that μ_s is also the chemical potential of the solvent in solution at a temperature T and pressure $P_0 + P$. The transformation from one set of variables to the other, in the thermodynamic description of real as well as model electrolytes, has been worked out by Poirier[15] and Friedman.[16] The osmotic coefficient in the McMillan–Mayer system (variables $T, \mu_s, c_1, \ldots, c_\sigma$) is defined by

$$\phi \equiv P/ckT \tag{8}$$

where k is the Boltzmann constant and c is the total ionic concentration

$$c = \sum_{i=1}^{\sigma} c_i \tag{9}$$

An important practical consequence of the McMillan–Mayer theorem is that all of the approximate theories developed for gases, for example, the PY and HNC approximations, can also be applied to solutions provided the answer to an additional statistical problem is at hand. This is the one referred to under condition (b), the correct formulation of the potential energy U_N for a set of ions in an infinite sea of solvent molecules. Assuming pairwise additivity, we first consider the pair potential $u_{ij}(r)$ for an isolated pair of ions (i,j) in the pure solvent. Since two ions can interact directly as a result of their respective charges e_i and e_j, as well as indirectly through the mediating influence of the solvent molecules, we could depict the more elementary pairwise interactions between two ions in the solvent sea as follows: The diagram

represents the direct interaction (coulomb potential $e_i e_j/r$ plus repulsion) and the diagrams below represent a few of the indirect interactions (e.g., ion–dipole–ion, ion–dipole–dipole–ion, plus other combinations as shown, plus repulsions between ion–solvent and solvent–solvent pairs):

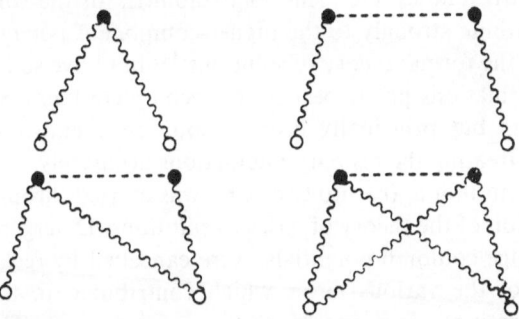

In the above diagrams, the open circles denote ions and the solid circles, solvent. The potential $u_{ij}(r)$ is the free-energy change in bringing together two isolated ions in the solvent from infinite separation to a distance r apart, while

$$\partial[\beta u_{ij}(r)]/\partial \beta \text{ (where } \beta = 1/kT) \text{ and } -\partial u_{ij}/\partial T$$

are the corresponding changes in energy and entropy, respectively, for this process. If we idealize the solvent molecule by a hard core plus dipole, we arrive at the simplest model in which the *total interaction* potential contains a coulomb term modified by the presence of a dielectric constant different from unity in the denominator. There are, of course, additional terms which must be added to this to get the complete $u_{ij}(r)$, and even though progress in this direction has been made, the results are relatively too recent[4] to have been exploited fully in the way that simpler models have been studied. Moreover, the representation of solvent molecules as hard core plus dipole may be too simple. Nevertheless, we know that the correct asymptotic form of $u_{ij}(r)$ for large r is the coulomb potential with the bulk dielectric constant of the pure solvent in the denominator. The potential $u_{ij}(r)$ now separates in two parts

$$u_{ij}(r) = u_{ij}^*(r) + e_i e_j/\varepsilon r \tag{10}$$

where $u_{ij}^*(r)$ is of shorter range than the coulomb term and includes
  (1) the repulsive part of the direct interaction between two ions;

---

[4] G. Stell, to be published.

(2) the effect of the granularity of the solvent;

(3) dielectric effects not included in the coulomb term;

(4) other effects associated with the detailed molecular structure of the solvent, e.g., overlap of cospheres;

(5) van der Waal's and chemical interactions.

This list is not exhaustive, neither are the categories mutually exclusive, but the result of it all is that the solute potential $u_{ij}(r)$ depends on temperature T, on the density of the solvent ρ_s, on the pressure P_0, and on the molecular characteristics of the ions and solvent molecules, for example, the dipole moments, polarizabilities, etc. The polarizabilities of the solvent molecules and ions contribute strongly to the higher-component potentials, which can be included in the formal theory of solutions[11] but have so far been ignored in detailed calculations partly because of incomplete knowledge about their functional form but principally because solution chemists have been preoccupied with treating the primary interactions accurately.

The separation of $u_{ij}(r)$ into two parts was carried through by Mayer[17] in his application of the theory of gases to solutions. Divergent integrals, due to the long-range coulomb potentials, were cancelled by reclassification and resummation of the various terms which contributed to the density (i.e., solute concentration) expansion for the osmotic pressure. The result of this virtuoso feat was a new expansion in which the Debye–Hückel limiting law was the leading term and the higher virial coefficients were functions of concentration. The expression for the Helmholtz free energy per unit volume to terms up to and including the second virial coefficient is the following:

$$-(F^{ex}/kT)_{DHLL+B_2} = \kappa^3/12\pi + \sum_{i=1}^{\sigma} \sum_{j=1}^{\sigma} c_i c_j B_{ij}(\kappa) \tag{11}$$

The second virial coefficients for each pair of ions are simple integrals determined by the short-range potential $u_{ij}^*(r)$, the Debye length κ^{-1}, and the Bjerrum distance $e_i e_j/\varepsilon kT$.[18] This approximation is referred to as the DHLL + B_2 approximation in the next section where numerical results are reviewed.

In contrast to the concentration expansion, the expansion in the Bjerrum parameter $e^2/\varepsilon kT$ developed by Stell and Lebowitz[19] (SL) has the following leading terms for the free energy:

$$F^{ex}/kT = F^{ex,*}/kT + (e^2/2\varepsilon kT) \sum_{i=1}^{\sigma} \sum_{j=1}^{\sigma} c_i c_j z_i z_j \int h_{ij}^*/r \, d\mathbf{r} - \kappa_1^3/12\pi + \dots \tag{12}$$

Here $h_{ij}^* = g_{ij}^* - 1$, z_i is the valence of ions i, e is the electronic charge, and the asterisk characterizes the properties of the uncharged system acting according to the short-range potential $u_{ij}^*(r)$. The quantity κ_1^{-1} is a modified Debye length defined by

$$\kappa_1^2 = (4\pi e^2/\varepsilon kT)\left(\sum_{i=1}^{\sigma} c_i z_i^2 + \sum_{i=1}^{\sigma} \sum_{j=1}^{\sigma} c_i c_j z_i z_j \int h_{ij}^* d\mathbf{r}\right) \qquad (13)$$

in which the first term is identically κ^2. When the short-range potentials are individually the same for all possible pairs of ions (e.g., rigid ions of equal size in a dielectric continuum), electrical neutrality of the solutions ensures that the second term in Eq. (13) is zero when the distinction between κ_1 and κ vanishes. The modification of the limiting law due to the asymmetry of the short-range potentials stands in contrast to the modification of the limiting law which exists for systems possessing an asymmetry in the ionic charges.[11] It is known that the first two terms in the Stell–Lebowitz expansion form an upper bound for the free energy,[20] so unlike the Debye–Hückel limiting law which can be approached from above or below as $c \to 0$ at fixed T (e.g., $ZnSO_4$ and HCl in H_2O solutions), the limiting law as $T \to \infty$ at fixed c must always be approached from one side (e.g., plasmas).

For the most part, detailed and accurate numerical estimates of the higher-order terms which contribute to the thermodynamic properties of electrolytes have been obtained for models in which granularity of the solvent has been ignored. The repulsive part of the direct interaction is often represented as arising from hard cores[21-26] or is assumed to vary inversely as some integral power of the distance.[27, 28, 45] Additional refinements to the short-range potential $u_{ij}^*(r)$ which include a dielectric repulsion term[27, 28] varying as r^{-4} and a term arising from the overlap of structurally altered zones around the ions[27-30] have also been considered.[31] For all of these specialized models, the calculation of the second virial coefficient B_2 is relatively easy,[23b] but the higher virial coefficients become progressively more difficult to evaluate because of their greater complexity. The convergence of the Mayer expansion for aqueous electrolytes is also slow, so that accurate estimates of the thermodynamic properties of a model electrolyte at concentrations in the region of 1 M are difficult to obtain.

One way to overcome this difficulty is to abandon the term-by-term evaluation of the virial coefficients and to consider instead approximations to the radial distribution functions $g_{ij}(r)$ which correspond to the summation of a certain class of terms which contribute to all of the virial coefficients. Hopefully, the terms considered in these partial summations can be chosen to be the more dominant ones. The HNC and PY approximations attempt to do just this. Their success or failure for electrolytes cannot be predetermined but must be discovered either from the self-consistency of the various thermodynamic properties computed in alternative ways[22] or more directly by comparison with Monte Carlo calculations.[32]

The first complete solutions to the PY and HNC approximations for the primitive-model electrolyte (charged hard spheres in a dielectric continuum) were obtained by Carley.[21] Shortly thereafter, another version of these approximations, derived by Allnatt[33][5] for electrolytes with their peculiarities in mind,[34] was studied, and the superiority of the HNC equation was established.[22] When considered as problems in numerical analysis, these approximations appear, at first, not to be much less simpler than the parent virial or Mayer expansion from which they are derived. The added gain in the automatic summation of dominant terms is offset by the appearance of a nonlinear integral equation in matrix form, but fortunately this can be solved with the aid of packaged programs[6] and electronic computers. Apart from its accuracy, especially for lower-valence electrolytes, the HNC approximation and the numerical methods used to solve it are not specialized to a given short-range potential, so that refinements to a model are readily incorporated into the general computer programs.[27-30] In a sense, therefore, this approximation is a powerful tool in the study of various models for electrolytes.

Three other theories which appeared recently are characterized by their remarkable simplicity and comparative accuracy, although the results available are specialized to the primitive-model electrolyte. The first of these is the mean spherical approximation[7] (MS) which has been solved by Waisman and Lebowitz[24] for charged hard spheres of the same size (restricted primitive model). Their results for the excess energy and the osmotic coefficients are available in simple analytic form. Although the self-consistency of the various osmotic coefficients is not very good, the excess energy $E^{ex'}$, the osmotic

---

[5] Allnatt's HNC and PY approximations[33] and Mayer's ionic cluster expansion[17] for the free energy are related in the same manner that the original HNC and PY equations are related to the virial expansion for a gas. Both sets of expansions are usually derived from the corresponding density expansions for the radial distribution functions. Meeron[18] solved the problem of reclassification and resummation of the individual cluster integrals in the expansion of $g_{ij}(r)$ for ionic solutions in order to avoid the divergences due to coulomb forces. The result

$$g_{ij}(r) = \exp(-\beta u_{ij}^* + q_{ij} + \alpha_{ij})$$

contains the Debye-shielded potential $q_{ij} = -e_i e_j e^{-\kappa r}/\varepsilon k T$ as a leading term. The quantity α_{ij} is a complicated function for which approximations were derived by Allnatt. These are the analogues of the HNC and PY equations solved numerically in ref. 22. The analogue of the PY equation is referred to as the PYA equation to distinguish it from the usual PY equation.[10a] Such a distinction between the two versions of the HNC approximation is unnecessary since they are equivalent and should give the same numerical results. When u_{ij}^* and α_{ij} are neglected and $\exp(q_{ij})$ is linearized, the Debye–Hückel limiting law results (see discussion).

[6] The most useful package (IBM Share No. 3465) is the fast Fourier transform program FORT developed by Cooley and Tukey [Math. Computation 19, 297 (1965)]. With 1024 points in the discrete representation of the Fourier transform, the time for computation is reduced by a factor of 60 over the corresponding time for trapezoid rule computation.[22] Since these integral equations are usually solved iteratively by Fourier transformation, the use of FORT gives an enormous saving in computer time.

[7] The MS approximation for the primitive-model electrolyte is defined by the relations

$$g_{ij}(r) = 0 \qquad\qquad r < a_{ij}$$
$$C_{ij}(r) = -e_i e_j/\varepsilon k T r \qquad r > a_{ij}$$

coefficients ϕ_E obtained via the energy equation, and the corresponding mean activity coefficient[32f] $\gamma_{\pm,E}$ are especially accurate for 1-1 electrolytes:

$$E^{ex}/ckT = -x[1 + x - (1 + 2x)^{1/2}]/4\pi a^3 c \qquad (14)$$

$$\phi_E = \phi^{HS} + (4\pi a^3 c)^{-1}[x + x(1 + 2x)^{1/2} - \tfrac{2}{3}(1 + 2x)^{3/2} + \tfrac{2}{3}] \qquad (15a)$$

$$\ln \gamma_{\pm,E} = \ln \gamma^{HS} + E^{ex'}/ckT \qquad (15b)$$

where $x = \kappa a$ and the superscript HS denotes the properties of an uncharged system of hard spheres of radius $a/2$ according to the Percus–Yevick theory.

The second development is the mode expansion theory of Andersen and Chandler.[25] The twin problems of divergent integrals due to the coulomb potential and the slow convergence of the cluster expansion are nicely avoided here. They observed that since the coulomb potential has a Fourier transform, divergence difficulties arising from the long-range part of this potential can be circumvented by working in transform space. The divergences at small r are avoided by treating the coulomb interactions as perturbations to a reference system which are dominated by short-range repulsions. The free energy $F^{ex,*}$ of the reference system is assumed to be known, and F^{ex} for the electrolyte is expressed as an infinite series

$$F^{ex}/kT = F^{ex,*}/kT + \sum_{n=1}^{\infty} a_n \qquad (16)$$

in which the a_n are determined by the perturbing potentials and the distribution functions (two-, three-, and higher-particle) for the reference system, which are also assumed to be known.[8]

When the reference system consists of uncharged hard spheres, the per-

---

where $C_{ij}(r)$ is the direct correlation function which is related to the radial distribution function $g_{ij}(r)$ by

$$g_{ij} - 1 = C_{ij} + \sum_{k=1}^{\sigma} c_k \int (g_{ik} - 1)C_{kj}\,d\{k\}$$

Here $\{k\}$ denotes the coordinates of particle k, whose concentration is c_k, and $\int \ldots d\{k\}$ signifies integration over these coordinates. The solution to the MS approximation lies in determining $g_{ij}(r)$ for $r > a_{ij}$ or, equivalently, $C_{ij}(r)$ for $r < a_{ij}$. When all the a_{ij}s are zero, the MS approximation gives the Debye–Hückel radial distribution function. When all the charges (e_i, e_j, etc.) are zero, i.e., when $C_{ij}(r) = 0$ for $r > a_{ij}$, the MS approximation reduces to the PY approximation for uncharged hard spheres. The MS and PY approximations are not the same for the primitive-model electrolyte, a point which we emphasize here only because the opposite is implied in statements that have appeared recently in the literature.[49] The MS $g_{ij}(r)$ functions are also different from the corresponding Debye–Hückel functions. For instance, because of the inclusion of hard-sphere interactions in the former, it gives rise to oscillations in the charge density $\rho_i(r)$ at high concentrations, while the Debye–Hückel theory does not. Attempts to bring the thermodynamic properties of the two theories into agreement by adjusting the ion-size parameter would therefore seem to be difficult to justify.[49]

[8] Equivalent results can be obtained by applying the γ-ordering and Γ-ordering techniques of LSB [J. L. Lebowitz, G. Stell, and S. Baer, *J. Chem. Phys.* 6, 1282 (1965)] to ionic solutions. In the LSB formalism, the a_n are given by a graphical γ-ordering scheme while in the Andersen–Chandler theory they are defined in terms of certain collective modes.[42]

turbation within the hard core can be chosen arbitrarily. Chandler and Andersen exploited this flexibility and chose the perturbation for $r < a$ in such a way as to gain satisfactory convergence of the series within two modes (i.e., a_n up to $n = 2$). For symmetrically charged primitive-model electrolytes, the calculation of the first two modes in Andersen and Chandler's theory is a simple matter in comparison to corresponding efforts needed to solve the HNC equation. The integrals involved in the mode expansion up to $n = 2$ are all one-dimensional, and only the free energy and pair distribution function for the reference system are needed. For unsymmetrical electrolytes, however, three- and four-particle correlation functions are also required in the evaluation of the second mode (a_2), which increases the difficulty of computation considerably and explains why results for 2–1 and 3–1 electrolytes are not readily available. The results for 1–1 electrolytes are comparable in accuracy to the HNC approximation.

In a later paper,[25c] Andersen and Chandler also describe a criterion for optimizing the convergence of the mode expansion when only the first few terms are taken into account in numerical calculations. This criterion ensures that the distribution functions for physically inaccessible regions ($r < a$) are zero. The perturbation series truncated at $n = 1$ is called the random-phase approximation (RPA), and when Andersen and Chandler's optimization criterion is applied to this, it turns out that the results of the optimized random phase approximation (ORPA) are equivalent to the mean spherical approximation provided the reference system obeys the Percus–Yevick theory exactly. The addition of the next mode (a_2) hence represents an improvement over the mean spherical approximation, and we refer to the theory which includes this as the MEX theory. We can summarize the relationships between the various optimized theories as follows[9]:

$$\begin{array}{l} \text{ORPA} + a_2 \text{ with the Percus–Yevick} \\ \text{theory for the reference system} \end{array} \equiv \text{MS} + a_2 \simeq \text{MEX} \equiv \text{ORPA} + a_2$$

The assumption about the reference system obeying the PY approximation is very good for most electrolytes (e.g., models describing the alkali halides in aqueous solution, but not the larger tetraalkylammonium halides) since a salt concentration of 1 M corresponds to a very low reduced ionic concentration, which implies that almost any reasonable approximation for the reference part of the interaction would be adequate. Since the thermodynamic properties in the MS approximation are available in analytic form [Eqs. (14) and (15)] only the second mode (a_2) needs to be calculated numerically, which is easy for symmetrical electrolytes. The problems associated with the difficulty in computing a_2 for unsymmetrical electrolytes remain. The accuracy of the mode expansion is improved by optimization, and although Andersen and Chandler's numerical results are for the restricted primitive model, it is

[9] If the same optimization is used in the LSB theory (see footnote 8), and the first- and second-order Γ-ordered approximations are abbreviated FOGA and SOGA, respectively, then FOGA $\equiv$ ORPA, and SOGA $\equiv$ ORPA $+ a_2 \equiv$ MEX.[42]

probable that comparable accuracy can be obtained for other models as well by exploiting some of the techniques developed recently in the theory of fluids by which the thermodynamics of a real fluid system is expressed in terms of the properties of hard spheres.[12, 20, 35]

A useful feature of the mode expansion theory is that convergence of the series can usually be assessed by comparing the magnitudes of successive modes. Reliance on expensive Monte Carlo calculations or tedious self-consistency tests for confirmation of accuracy become less essential. When this test is applied to 2–2 electrolytes, it is found that the convergence is less rapid; the osmotic coefficients, for instance, are not very accurate. A similar reduction in the accuracy of ϕ, when the ions become more highly charged, also occurs in the HNC approximation, but at least this theory is qualitatively correct in showing the same anomalous effects that 2–2 valent electrolytes exhibit at high dilution in aqueous solution. They are also seen in the DHLL + B_2 approximation but do not occur in the MS (or ORPA) or MEX (i.e., ORPA + a_2) theories. Hence the second virial coefficient B_2 contains an ingredient, related to the observed properties of 2–2 valent aqueous electrolytes, which the second mode (a_2) lacks. Recognizing this, Andersen, Chandler, and Weekes[36] appear to have combined the best of both worlds in their new ORPA + B_2 approximation.[10] This is the third of the new theories mentioned earlier.

Preliminary Monte Carlo calculations of the excess energies of 2–2 electrolytes by Card and Valleau indicate that the ORPA + B_2 and HNC approximations are of comparable accuracy (see Table I). Since Monte Carlo results for 2–2 electrolytes below 0.0625 M are not yet available (and may be difficult to obtain[11]), it remains to be seen whether the energy and other thermodynamic properties from ORPA + B_2 are numerically accurate in the

[10] We do not mean to imply that the ORPA + B_2 approximation was obtained by a simple fusion of ORPA and B_2. This approximation is embedded in a more elaborate perturbation theory in which some of the techniques introduced by Mayer (summation of rings and chains) are utilized in writing down equations for F^{ex} and $g(r)$. These are formally related to the work of Stell and Lebowitz,[19] but the technique of improving convergence by an optimal choice of the perturbation within the hard core is again exploited to obtain ORPA + B_2 as the leading terms in the cluster expansion for F^{ex}. The B_2 term is derived using this particular choice of the perturbation in physically inaccessible regions, and hence it is not the same as B_2 referred to in DHLL + B_2. When the perturbations within and outside the hard cores are deliberately chosen to be coulombic, then there is no difference between the B_2's. In terms of the LSB (see footnote 8) and SL[19] graph-theoretic descriptions, the ORPA + B_2 approximation could be characterized as an optimized second-order nodal-ordered approximation (SONA) where the nodal order of a graph is the number of its vertices or hypervertices. The optimized first-order nodal-ordered approximation is identical to FOGA (see footnote 9).

[11] Usually the accuracy of Monte Carlo calculations for charged systems diminishes when the concentration decreases because of an accompanying decrease in the shielding (κ becomes smaller) which controls the range of the forces between charged particles. This problem is particularly acute for 2–2 valent and other highly charged electrolytes. None of the currently available Monte Carlo calculations are accurate enough at low concentrations ($c < 0.025$ M) to show the anomalous behavior of 2–2 electrolytes.

Table I. Excess Energies[a] for a 2–2 Restricted Primitive-Model Electrolyte
$(a = 4.2$ Å, $\varepsilon = 78.358$, $t = 25°C)$

$$-E^{ex\prime}/ckT$$

		Error (MC theory)			
Molarity[b]	MC[c]	MS[d]	MEX[e]	ORPA + B$_2$[f]	HNC[g]
0.0625	1.893 ± .017	0.43	0.43	−0.13	0.15
0.250	2.473 ± .019	0.29	0.29	−0.08	0.10
0.5625	2.822 ± .008	0.18	0.16	−0.08	0.06
1.000	3.091 ± .011	0.11	0.08	−0.10	0.05
2.000	3.509 ± .016	0.13	0.04	−0.06	0.02

[a] Note that (1) $E^{ex\prime}$ is the excess energy for the primitive model assuming $\partial\varepsilon/\partial T = 0$. (2) MS $\simeq$ ORPA; MEX $\equiv$ ORPA + a_2 $\simeq$ MS + a_2.
[b] Conventional salt concentration (C_2) or half the total ionic concentration (c) for symmetrically charged electrolytes. [See Eq. (9) for definition of c)].
[c] Monte Carlo calculations of Card and Valleau (quoted in ref. 25c). The errors are one standard deviation.
[d] Mean spherical approximation (ref. 24).
[e] Optimized mode expansion (ref. 25c).
[f] Optimized random phase + B$_2$ approximation (ref. 36c).
[g] Hypernetted-chain approximation (ref. 23b).

region of high dilution where it could be useful, for instance, in the extrapolation of the ZnSO$_4$ cell to infinite dilution. It is already known[23b] from the self-consistency requirements that the osmotic coefficients of 2–2 electrolytes derived from the HNC approximation are not very accurate in this region (see Fig. 9). The accuracy of the energy in the same region from this approximation is unknown.

Before concluding this section on current electrolyte theory, we would like to mention two other developments. Stillinger and Lovett[37] have developed a theory of electrolytes in which the ions are formally paired, according to a prescribed scheme, into uncharged dipolar molecules. They studied the dielectric properties of this system by considering its response to an applied spatially periodic electrostatic field, which led them to a new condition that must be satisfied by the exact pair distribution functions, called the second-moment condition.

$$-6 \sum_{i=1}^{\sigma} c_i e_i^2 = \kappa^2 \int \sum_{i=1}^{\sigma} \sum_{j=1}^{\sigma} c_i c_j e_i e_j g_{ij}(r) r^2 \, d\mathbf{r} \qquad (17)$$

This, and the condition for local electroneutrality (zeroth-moment condition)

$$-e_i = \int \sum_{j=1}^{\sigma} c_j e_j g_{ij}(r) \, d\mathbf{r} \qquad (18)$$

are useful in testing the consistency of various theories of electrolytes. For instance, the Debye–Hückel limiting law satisfies, as expected, both conditions, and Groeneveld[39] has shown that the HNC and MS approximations belong to a general class of theories for which the second-moment condition holds exactly. Numerical results for the ion-pair theory are not available, but another interesting result, derived in an extension to this work by Stillinger and White,[38] is that the local charge density ρ_l

$$\rho_l(r) = \sum_{j=1}^{\sigma} c_j e_j g_{lj}(r) \qquad (19)$$

has an asymptotic form which does not decay exponentially but varies as r^{-8}. From this, they conclude that the distribution functions $g_{lj}(r)$ could also be expected to have similar tails. None of the theories which involve chain summations (e.g., the HNC approximation) can conceivably give such an asymptotic form for $\rho_l(r)$, but the numerical error in the thermodynamic properties caused by not having the right shape of tail of $g_{lj}(r)$ is probably negligible.

The other theory, which we would like to mention briefly, deals with Outhwaite's[26, 40] improvements to the Poisson–Boltzmann equation in which errors due to the omission of a fluctuation term are corrected. The errors due to the omission of an excluded volume term (both of these were first identified by Kirkwood[41]) are ignored. The modified Poisson–Boltzmann (MPB) equation, as it is called, gives energies as good as can be obtained from the HNC equation for 1–1 electrolytes when they are treated as charged hard spheres. The osmotic coefficients are less accurate, and Burley, Hutson, and Outhwaite[26, 40] suggest that the difficulty is due to the omission of short-range effects in their treatment. This difficulty has recently been corrected.[26b]

It may appear from the foregoing discussion of newer theories of electrolytes that there are now so many that are successful, that there is an embarrassment of riches in this field. While this is perhaps true, it is unlikely that any one theory will be more useful than the others in all ionic solutions (e.g., various charge types, different solvent media and temperatures, etc.). The success of a theory is more likely to depend on the particular problem at hand. The Mayer expansion, for instance, and the Stell–Lebowitz expansion are expansions in two different ordering parameters, and the theory of choice in a given situation should be the one for which the corresponding parameter is small.

The convergence of these expansions can be improved by using the optimization trick, and it is likely that this innovation,[12] introduced by Andersen and Chandler to ionic solutions, will be seen more frequently in the future in a variety of different contexts. It is an interesting fact that when the Stell–Lebowitz expansion is optimized in the same way as the mode expansion, or if the latter expansion is left unoptimized, that is to say, if the perturbation within the hard cores is also chosen to be the coulomb potential in both theories, then the two theories look more nearly alike in the sense that the first few terms are identical for a symmetrical electrolyte.[42] Apparently, optimization, or some other judicious readjustment of the perturbation within the hard

---

[12] It appears to have been first used in magnetic models where optimization turns out to be equivalent to the imposition of the mean spherical constraint. The continuum analogue of this constraint is called the mean spherical approximation and its close relationship to the optimized approximations ORPA and FOGA is therefore understandable.

cores, makes all the difference in numerical estimates of the thermodynamic properties.

3. NUMERICAL RESULTS

In this section we review the numerical results for some simple model electrolytes. The emphasis, for the primitive-model electrolyte, will mainly be on the mean spherical (MS) approximation, the optimized mode expansion (MEX), and the hypernetted chain (HNC) theory. A few scattered results in the ORPA + B_2 approximation are also discussed. The HNC properties for this model will be compared with experiment, and a minor modification to the short-range potential $u_{ij}^*(r)$ to bring the properties of the model in line with experiment for the alkali halides will be considered. What emerges from one molecular interpretation of this change is a picture of the environment around an ion which is similar to that proposed by Frank and Evans in 1945.[43]

The primitive-model electrolyte[22, 30] (charged hard spheres in a dielectric continuum) is characterized by

$$u_{ij}^*(r) = \infty \quad \text{if } r < a_{ij}$$
$$= 0 \quad \text{if } r > a_{ij} \tag{20}$$

where a_{ij} is the distance of closest approach of the (i,j)-th pair. The restricted primitive model specializes this to hard spheres of the same diameter a. If we take the radii of the spheres to be typically about 2.1 Å, the ordering parameters ca^3 and $e_- e_+/\varepsilon kTa$ are 0.09 and 1.7, respectively, for a 1–1 salt in aqueous solution at a concentration of 1 m with the temperature at 25°C. The reduced ionic concentration ca^3 is so small that the equilibrium properties of the corresponding uncharged system are well described by the PY or HNC approximations and, for that matter, even by the virial expansion for the excess free energy $F^{ex, \, HS}$ carried no further than the third virial coefficient. The potential-energy functions $u_{ij}(r)$ for this model depend on the temperature T and the pressure P_0 even when the diameters of the spheres are fixed for all temperatures and pressures because the solvent dielectric constant ε is a function of these two variables.

In Fig. 4 we have the osmotic coefficients ϕ_v and excess energies $E^{ex'}$ calculated from the HNC approximation. Comparisons with other theories, including various internal tests of the self-consistency in the same theory, are made in Figs. 5–10. Wherever possible, we also compare these with the Monte Carlo calculations of Card and Valleau.[32] The main conclusions are the following.

The HNC and MEX osmotic coefficients are both excellent for 1–1 electrolytes (Fig. 5); the energies in both theories are also accurate (Fig. 6), although the HNC predictions are slightly superior in the concentration range where the MEX theory converges to the MS approximation. This discrepancy, which occurs at low concentrations, is accentuated for higher-valence electrolytes (Table I). For 1–1 electrolytes, the ORPA + B_2 approximations for ϕ and $E^{ex'}$ agree almost exactly with the Monte Carlo calculations up to salt concentrations of 2.0 m.[36a]

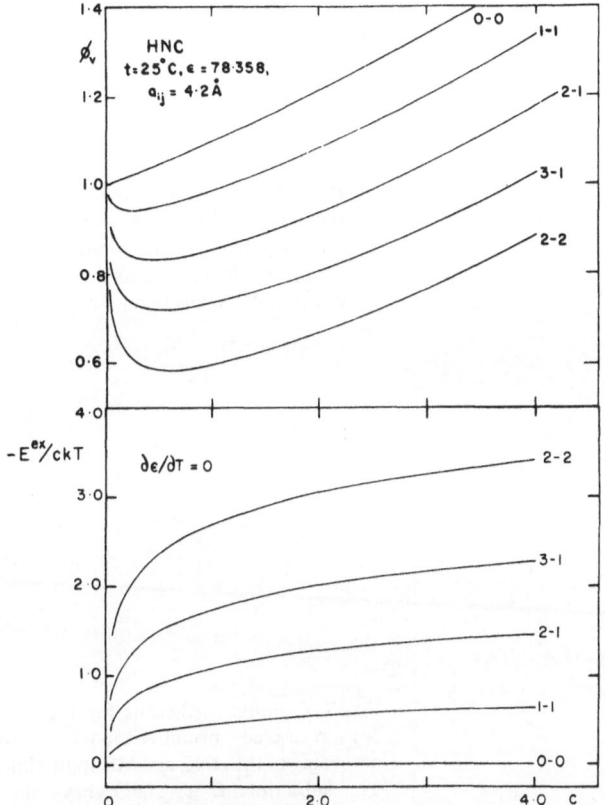

Fig. 4. Thermodynamic properties of charged hard spheres of radii 2.1 Å according to the HNC approximation. The curves are labeled according to the charges on the ions. The osmotic coefficients were all obtained via the virial equation. The concentration c is the total ionic concentration. From ref. 23a.

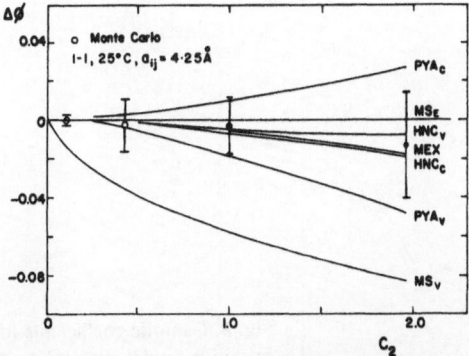

Fig. 5. Osmotic-coefficient results for a 1–1 electrolyte compared on an enlarged scale as the difference from the mean spherical prediction calculated from the energy equation [Eq. (15)]. The subscripts E, v, and c refer to the energy, virial, and compressibility equations which provide alternate routes to the osmotic coefficients. The Monte Carlo error bars show three standard deviations. The quantity c_2 is the conventional electrolyte concentration; for a 1–1 electrolyte the total ionic concentration $c = 2c_2$. From ref. 32f.

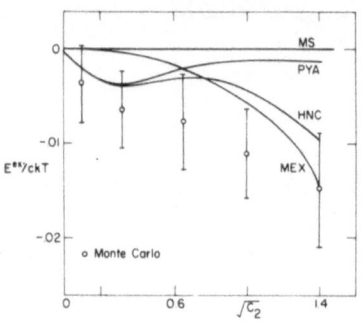

Fig. 6. Excess-energy results for a 1–1 electrolyte compared on an enlarged scale as the difference from the MS approximation. The Monte Carlo error bars are ±3 standard errors. Adapted from ref. 32f.

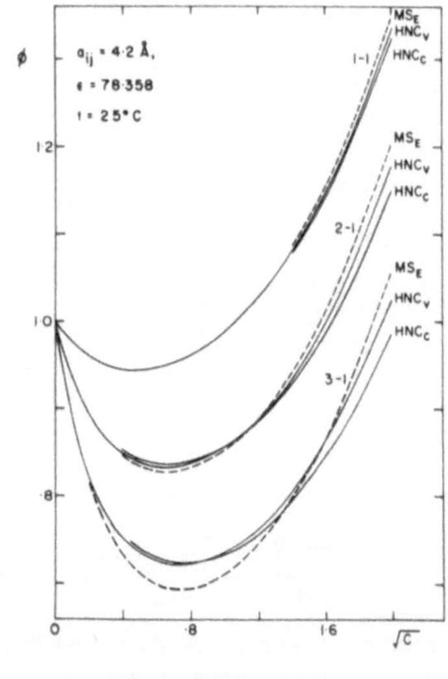

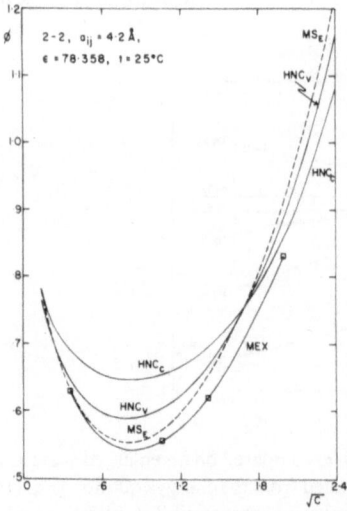

Fig. 7. Osmotic coefficients for 1–1, 2–1, and 3–1 restricted primitive-model electrolytes according to the HNC and MS approximations. The subscripts E, v, and c have the same meaning as in Fig. 5. The concentration c is the total ionic concentration. From ref. 23b.

Fig. 8. Osmotic coefficients for a 2–2 restricted primitive-model electrolyte according to the HNC, MS, and MEX theories. The subscripts have the same meaning as in Fig. 5. From ref. 23b.

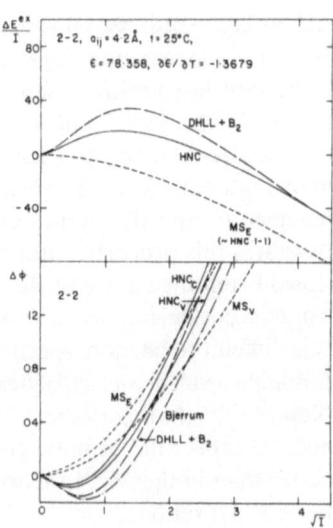

Fig. 9. Deviations from the Debye–Hückel limiting law for the energy and the osmotic coefficients of a 2–2 electrolyte according to several theories. $\Delta\phi = \phi(\text{theory}) - \phi(\text{DHLL})$ and $\Delta E^{ex} = E^{ex}(\text{theory}) - E^{ex}(\text{DHLL})$. The results for the MEX theory are indistinguishable, in this concentration range, from the curves labeled MS_E. The MS approximation for E^{ex}/I does not distinguish between 1–1 and 2–2 electrolytes. The quantity I is the ionic strength. $E^{ex} = E^{ex\prime}(1 + \partial \ln \varepsilon / \partial \ln T)$. From ref. 23b.

The accuracy of ϕ in all of these theories diminishes as the charges on the ions are increased. For example, the discrepancy between ϕ_v and ϕ_c at a total ionic concentration c of 2.0 m increases from 0.002 to 0.007 to 0.01 in the HNC approximation when the charges on the ions are changed progressively to correspond to 1–1, 2–1, and 3–1 valence types, respectively (Fig. 7). These numbers pertain to the same restricted primitive-model electrolyte in which $a = 4.2$ Å, $t = 25°C$, and $\varepsilon = 78.358$. The difference between ϕ_v and ϕ_c for the corresponding 2–2 electrolyte (Fig. 8) is, of course, still greater, but the HNC equation produces the correct sign for the deviations of ϕ from the Debye–Hückel limiting law (Fig. 9). Neither the MS approximation nor the

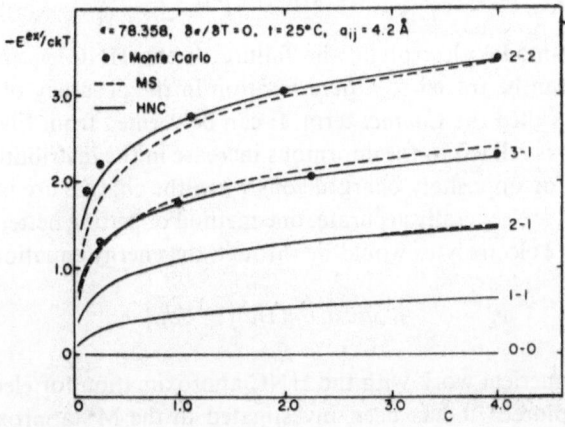

Fig. 10. The excess energy calculated from the HNC and MS approximations. Monte Carlo calculations of Card and Valleau for 2–2 and 3–1 electrolytes are also shown. From ref. 23b.

MEX theory possesses this property which is usually regarded as a manifesta-
tion of ion association. Unlike Bjerrum's treatment of this problem in which
a more-or-less arbitrary separation of ions into associated and free ions is
assumed, the HNC approximation, and certain others discussed below,
produce qualitatively the same results as the classical association theories by
allowing the ions "to do their own thing." That is to say, no association is
assumed or introduced into the model. The DHLL + B_2 approximation also
possesses this property, although its accuracy is limited at higher concentrations. From the nature of the ORPA + B_2 approximation, it is clear that the
osmotic coefficients for 2–2 electrolytes must also be qualitatively correct.
It is difficult to be more specific about the accuracy of the osmotic coefficients
for higher-valence electrolytes, since no detailed estimates are available.[13]
Numerically the theory should be an improvement over the MS approximation,
and we expect the osmotic coefficients for highly charged electrolytes to be
better than in the MEX theory because B_2 unlike a_2 does not break down at
low concentrations.

Returning now to the energies, the HNC and ORPA + B_2 predictions
appear to be generally very good (Table I and Fig. 10). For 2–2 electrolytes
this conclusion is at present restricted to concentrations above 0.0625 M, the
lowest concentration at which Monte Carlo results are currently available
for direct comparison (see footnote 11). While it is safe to assume that the
accuracy in the energy persists all the way to infinite dilution for less highly
charged ions, it is by no means safe to draw the same conclusion for 2–2
electrolytes because of the irregularities at lower concentrations. What is at
issue here is the magnitude of these irregularities; none of the qualitatively
correct theories depicted in Fig. 9 seem to agree on this point. Going back to
the energies above 0.06 M, since

$$\phi_v = (2\pi/3c) \sum_{i=1}^{\sigma} \sum_{j=1}^{\sigma} c_i c_j g_{ij}(a_{ij}+) a_{ij}^3 + E^{ex}/ckT \qquad (21)$$

for the primitive-model electrolyte, the failure of the HNC ϕ_v for highly
charged species can be traced to a deterioration in the accuracy of the first
term of Eq. (21), called the *Contact* term. It can be verified from Fig. 11 that
this deterioration is related to the enormous increase in the distribution functions at contact for oppositely charged ions when the charges are increased.
Since the energies are generally accurate, one method of getting better osmotic
coefficients for 2–2 electrolytes would be through the energy equation:

$$\phi_E = \phi^{HS} + (\partial/\partial \ln c)[(1/c)\int_0^{\beta} E^{ex'} d\beta] \qquad (22)$$

This aspect of numerical work with the HNC approximation for electrolytes
has not been explored; it has been investigated in the MS approximation

---

[13] The ORPA + B_2 osmotic coefficients and excess energies of higher-valence electrolytes,
including 2–1 and 3–1 charge types, are being analyzed by S. Hudson and H. Andersen
(private communication from H. Andersen).

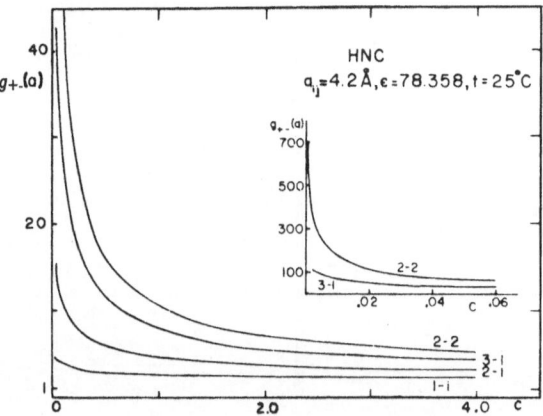

Fig. 11. Distribution functions at contact for oppositely charged ions calculated in the HNC theory as a function of the total ionic concentration c. From ref. 23b.

where it is known that the energy equation [Eq. (15)] gives the best results. A similar conclusion has been drawn in applications of the PY theory to Lennard–Jones fluids.[44]

It is important to realize that our criticism of the relative merits of various theories is actually derived from the study of models that are useful in describing the properties of aqueous electrolytes consisting of fairly small ions at room temperature. Our analysis does not necessarily apply to other systems also; for instance, the osmotic coefficients from the HNC theory deteriorate more rapidly at higher concentrations if the ions are very large. Nevertheless, the theory has been found useful and accurate enough in the investigation of models which predict the thermodynamic properties of tetraalkylammonium halides at concentrations below 0.4 M.[45]

One other aspect of the equilibrium properties of electrolytes which has been investigated is the oscillations in the charge density $\rho_i(r)$ which have been predicted to occur at some critical concentration.[46, 26] The second moment condition also implies this when there are hard-sphere interactions,[37] although the prediction there is more an upper limit to the critical concentration at which oscillations must appear. Hard-sphere interactions are absent in the limiting Debye–Hückel theory, hence no oscillations appear even though both moment conditions are satisfied. Oscillations in $\rho_i(r)$ have been found in the HNC approximation[23] and also in the modified Poisson–Boltzmann equation of Outhwaite.[26] In Fig. 12 we demonstrate these oscillations in the HNC theory for various ionic charges in the same primitive-model electrolyte at the same total ionic concentration. Since the radii of positive and negative ions have been chosen to be the same, the oscillations in the reduced charge density s_i^* defined by

$$s_i^* = -a\rho_i(r)4\pi r^2/e_i \qquad (23)$$

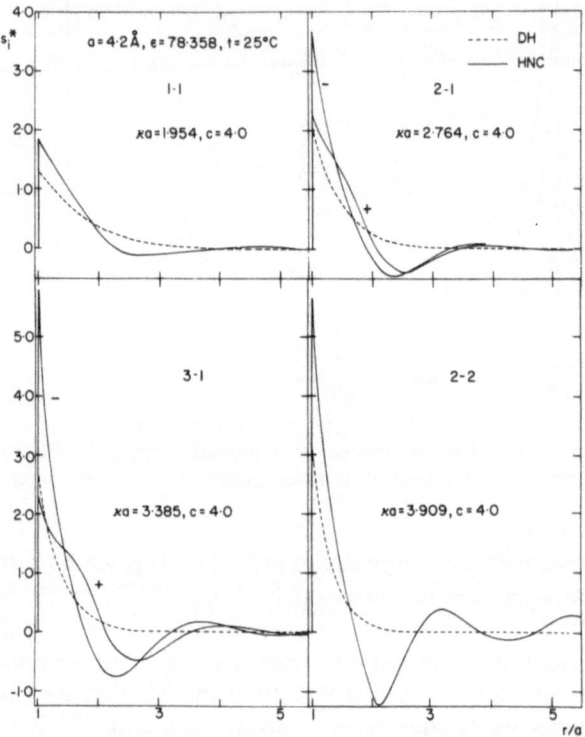

Fig. 12. Oscillations in the charge density for four restricted primitive-model electrolytes at the same total ionic concentration of 4 M. The electrolytes differ only in the charges on the ions. From ref. 23b.

cannot be due to sorting or filling in of interstitial holes, the behavior that one might expect in a mixture of spheres with grossly different radii. The origin of these oscillations lies in the attractive and repulsive coulomb interactions superimposed on the hard-sphere repulsions that are assumed to be the same for all pairs of ions.

We turn now to a consideration of the relevance of various models to the properties of real electrolytes. In doing so, we recognize the ready adaptability of the HNC equation to modifications of the short-range potential. A cursory examination of the osmotic coefficients for the primitive model, with various ion sizes (Fig. 13), shows that it is a plausible description of some simple real electrolytes, namely, the alkali halides in aqueous solution. The primary effect in this model, as expected for a given set of ionic charges, is determined by the distance of closest approach a_{+-} of oppositely charged ions. An infinite number of combinations of additive ionic radii which correspond to the same a_{+-} is possible, but they are only relevant to the magnitudes of the osmotic coefficients at higher concentrations, where a disparity in the sizes of the ions induces a proportionately larger excluded-volume contribution to ϕ. On choosing

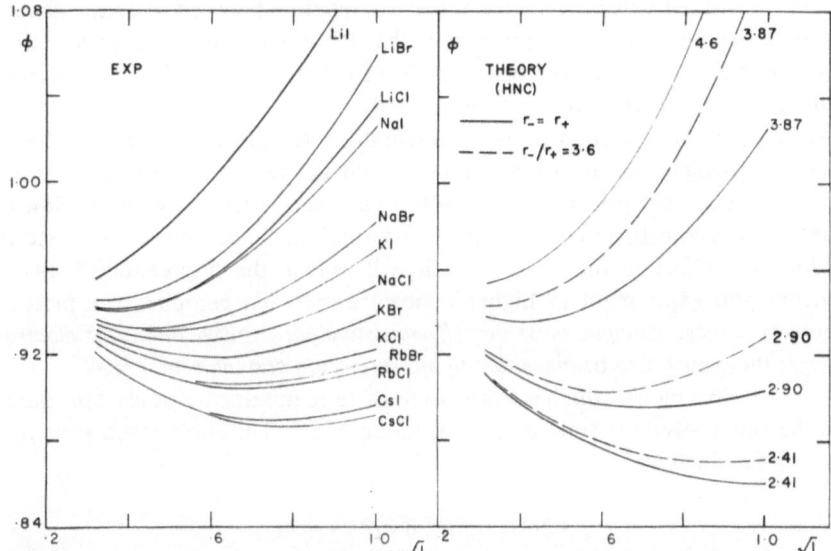

Fig. 13. The theoretical (HNC) osmotic coefficients ϕ for a range of ion-size parameters in the primitive model compared with experimental data for aqueous solutions of alkali halides at 25°C. The curves are labeled according to the value assumed for $r_+ + r_- = a_{+-}$. The experimental results have been corrected to the McMillan–Mayer standard states. From ref. 30.

a_{+-} to fit the experimental data at low concentrations, it is found that accurate calculations give osmotic coefficients that are too large at higher concentrations. This is in contrast to the DHLL + B$_2$ and other less accurate approximations which predict osmotic coefficients that are too low (Fig. 14). Since the introduction of unequal radii (having the correct $r_+ + r_-$ to fit the data at low

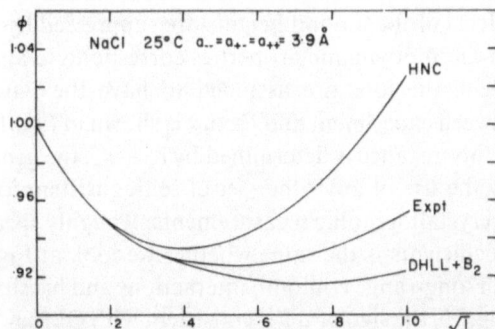

Fig. 14. Osmotic coefficients for the primitive-model electrolyte compared with the experimental data for NaCl in aqueous solutions at 25°C. The a_{+-} parameters in both the HNC and DHLL + B$_2$ approximations have been chosen to fit the data below 0.05 M. The ϕ's from the DHLL + B$_2$ approximation are similar to those obtained in the extended Debye–Hückel theory. From ref. 30.

concentrations) will only increase the discrepancy between accurate theory and experiment, the indication here is that the model must be changed. This is so not only for NaCl but also for all the other alkali halides[22a, 30] except the fluorides, which have not been studied.

Now there is a great range of possibilities for acceptable models, but it seems expedient, as in any preliminary and costly exploration, to proceed systematically according to some plan. The plan is to retain the notion of harsh repulsions between the ions at very small distances but to introduce an additional effect further away which will reduce the discrepancy between theory and experiment at higher concentrations. To complete the picture, however, these *changes must be self-consistent for a given family of electrolytes*; they must also be *amenable to interpretation at a molecular level*.

A simple model which is found to fit these requirements, at least partially, is the square-well (or square-mound) model.[29, 30] The short-range potential $u_{ij}^*(r)$ is defined by

$$
\begin{aligned}
u_{ij}^*(r) &= d_{ij} \quad \text{if } a_{ij} < r < b_{ij} \\
&= \infty \quad \text{if } r < a_{ij} \\
&= 0 \quad \text{if } b_{ij} < r
\end{aligned}
\tag{24}
$$

Also, d_{ij} is positive for a mound and negative for a well. If we assume that the distance a_{ij} is the sum of Pauling radii for the ions i and j and that $b_{ij} - a_{ij} = 2.76$ Å, which is approximately the width of a water molecule, then the only adjustable parameters are the mound heights d_{ij}. These may be correlated with the free-energy changes accompanying the displacement of water between two ions when they make contact in a sea of solvent molecules containing no other ions. The empirical finding is that the osmotic coefficients are determined primarily by d_{+-}. Hence, in effect, there is only one adjustable parameter in the calculation of the osmotic coefficients for this model.

When the effects of the mound heights are suppressed by setting them all equal to zero, the thermodynamic properties correspond to a primitive-model electrolyte in which the ions are assumed to have the Pauling radii. The difference $\Delta\phi$ between experiment and theory is shown in Fig. 15, and since the primary effect in this instance is determined by $r_+ + r_-$, the general picture will not be altered by the use of any other set of self-consistent ionic radii which are derived from crystallographic measurements. Roughly speaking, the order of the osmotic coefficients is the same whether we look at Fig. 13 or Fig. 15. Since the effects of long-range coulomb interactions and harsh repulsions have been deleted in Fig. 15, the sign of $\Delta\phi$ is consistent with Gurney's hypothesis[47] that the order in which the osmotic coefficients at a given concentration decrease within a family of electrolytes is also the order in which the cospheres around oppositely charged ions change from dissimilar to similar character. In any event, it is obvious that what is required for closer agreement with experiment in the square-well model is a positive d_{+-} when $\Delta\phi > 0$ and a

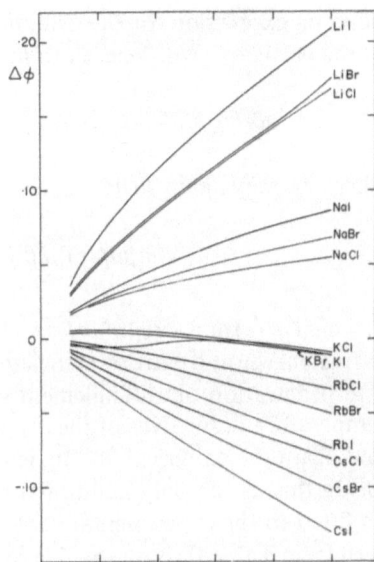

Fig. 15. The quantity $\Delta\phi$ as a function of ionic strength I at 25°C, where $\Delta\phi = \phi(\text{exp}) - \phi(\text{Pauling})$, where $\phi(\text{Pauling})$ is the theoretical osmotic coefficient in the primitive model with the ion radii equal to the Pauling radii. From ref. 30.

negative one when $\Delta\phi < 0$. Figure 16 gives us some idea of the success that can be achieved in fitting the osmotic coefficients of the alkali halides using d_{+-} as the only adjustable parameter.[30] The magnitudes of d_{+-} are all plausible, and it is natural to interpret a large positive d_{+-} as signifying strong hydration for at least one of the ions. More generally, of course, these parameters may be taken to reflect changes that occur during the overlap of cospheres around the ions. The temperature coefficients of the d_{ij} parameters will be expected to contribute to the excess energy E^{ex} which is related to the heat of dilution. This arises from the fact that E^{ex} is determined by the integral of $\partial[\beta u_{ij}(r)]/\partial\beta$ times the distribution function $g_{ij}(r)$ multiplied by r^2, whereas the corres-

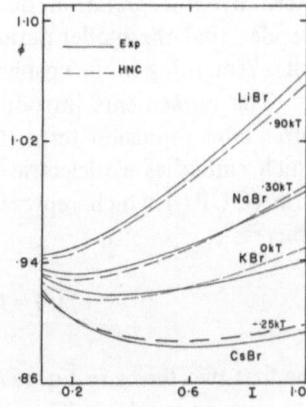

Fig. 16. Theoretical (HNC) osmotic coefficients for the square-well model compared with experimental data for aqueous solutions of the alkali bromides at 25°C. The parameters assumed for this model are $a_{ij} = r_i(\text{Pauling}) + r_j(\text{Pauling})$, $d_{++} = d_{--} = 0$, and d_{+-} as indicated in the figure. From ref. 30.

ponding expression for the osmotic coefficient contains $\partial[u_{ij}(r)]/\partial\ln r$ instead of the derivative with respect to $\beta = 1/kT$. Now

$$\partial(\beta u_{ij})/\partial\beta = (e_i e_j/\varepsilon r)\,[1 + (\partial\ln\varepsilon/\partial\ln T)] + \partial[\beta u_{ij}^*(r)]/\partial\beta \qquad (25)$$

where we may again write

$$\partial[\beta u_{ij}^*(r)]/\partial\beta = u_{ij}^*(r) - T\partial[u_{ij}^*(r)]/\partial T \qquad (26)$$

in which $u_{ij}^*(r)$ and $-\partial u_{ij}^*(r)/\partial T$ are the short-range contributions to the changes in free-energy and entropy associated with the overlap of the cospheres and the interpenetration or entanglement of two ions. In the square-well model the temperature derivatives of the d_{ij} parameters determine the entropy changes, and again the empirical finding is that the primary effect is due to $\partial(d_{+-})/\partial T$. Using this as the only additional parameter, it is found that the theory can be fitted to the experimental heats of dilution.[30] The agreement, as may be seen from Table II, is reasonable for all except the lithium salts. The difficulty, however, is with the sign of $\partial d_{+-}/\partial T$. They are all positive, which signifies a negative contribution to the entropy of overlap—the opposite of what one might expect from a process which involves the expulsion of the water of hydration into the bulk solvent. This paradoxical sign for the entropy change is also found in more refined models for the same alkali halides.[27] One explanation of this is that since a possible second layer of disrupted water molecules around the inner hydration sheath is not represented explicitly in the square-well model, it does so implicitly in the sign of $\partial d_{+-}/\partial T$ when the model is forced to fit the experimental data.[30] This explanation supports the picture due to Frank and Evans[43] that there are generally two concentric regions of frozen and melted water around an ion. The outer regions then contribute overwhelmingly to E^{ex} and the inner regions contribute predominantly to ϕ for all the alkali halides except cesium salts ($d_{+-} < 0$). An alternative interpretation, due to Ramanathan and Friedman,[27] is based on the idea that the model parameters also reflect changes in properties of the water remaining in the cospheres outside the overlap region.

The refinements introduced by Ramanathan and Friedman include a softer core repulsion term (COR_{ij}) varying as r^{-9}, a cavity term (CAV_{ij}) which embodies a dielectric repulsion effect varying as r^{-4}, and a Gurney term (GUR_{ij}) which represents the contributions from the overlap of cospheres:

$$u_{ij}^*(r) = COR_{ij} + CAV_{ij} + GUR_{ij} \qquad (27)$$

The first two terms in Eq. (27) are repulsive, while the Gurney term may be attractive or repulsive. The effect of the cavity term, as formulated by Rama-

Table II. A comparison of E^{ex}/I for Aqueous Solutions of the Alkali Halides at 25°C with the HNC Results for the Square-Well Model[a,c]

Molarity		LiCl	LiBr	NaCl	NaBr	NaI	KCl	KBr	KI	CsCl
$k^{-1}T^{-1}d_{+-}$		0.90	0.90	0.25	0.30	0.35	0	0	0	-0.20
$k^{-1}(\partial d_{+-}/\partial T)$		1.39	1.57	1.03	1.22	1.28	0.65	0.81	0.88	0.60
0.1	Exp.[b]	108	100	84	75	72	80	67	57	55
	HNC	108	100	84	75	72	80	67	57	55
	Diff.	0	0	0	0	0	0	0	0	0
0.4	Exp.	193	183	69	49	32	67	31	-11	-8
	HNC	152	130	78	51	43	66	22	-6	-6
	Diff.	41	53	-9	2	-11	1	9	-5	-2
0.7	Exp.	250	239	26	-5	-33	27	-33	-98	-87
	HNC	163	148	46	4	-10	26	-40	-88	-94
	Diff.	87	91	-20	-9	-23	1	7	-10	-7
1.0	Exp.	300	284	-28	-64	-105	-20	-110	-184	-172
	HNC	165	118	8	-50	-68	-19	-119	-171	-180
	Diff.	135	166	-36	-14	-37	-1	-1	-12	8

[a] The quantity E^{ex}/I is the heat of dilution in the McMillan–Mayer system, cal-mole^{-1}.

[b] Interpolated at each molarity from the heats of dilution compiled by V. B. Parker (U.S. National Bureau of Standards, Washington, D.C. 1965), NJRDS-NBS 2.

[c] From ref. 30. The results for LiCl and LiBr were transposed in Table II of ref. 30. They are correctly given here.

nathan and Friedman, is small. The only adjustable parameters in the calculation of the osmotic coefficients from this model are the coefficients A_{ij} of the Gurney term

$$GUR_{ij} = A_{ij} V_w^{-1} V_{mu}(r) \tag{28}$$

where $V_{mu}(r)$ is the mutual volume of overlap of the cospheres and V_w is the molar volume of the pure solvent. The A_{ij} coefficients correspond roughly to the d_{ij} parameters in the square-well model. The temperature derivatives $-\partial A_{ij}/\partial T$ are also required in the calculation of the excess energy, while the set $\{\partial A_{ij}/\partial P_0\}$ is the additional requirement in estimating the excess volumes. The thermodynamic properties of aqueous alkali halide solutions can be fitted very nicely to this model. The model has also been applied to the study of mixed electrolytes (LiCl)[28a] and (NaCl–MgCl$_2$)[28b] and exhibits certain empirically established regularities (Harned's rule, Young's rule, and the observation that the sum of the Harned α coefficients is a linear function of the ionic strength). The properties of this model, calculated from the HNC approximation, have been extensively reviewed elsewhere,[31] and will not be discussed further except to mention an interesting new application to tetraalkylammonium halides.[45]

The Gurney coefficients A_{ij} required to fit the osmotic data for two tetraalkylammonium bromides are reproduced in Table III together with the parameters $S_{ij} = -\partial A_{ij}/\partial T$ and $V_{ij} = \partial A_{ij}/\partial P_0$ required to fit the excess energies and volumes. A_{--} and its derivatives with respect to temperature and pressure are assumed to be zero. The model fits the experimental data obtained by J. C. Ku[48] in Prof. Frank's laboratory, shown in Fig. 17. The interpretation of the Gurney parameters is complicated since the extent to which chain deformation and entanglement and the interpenetration of ions contribute implicitly to the Gurney term is unknown. What is known is that the Gurney term overshadows the coulomb term at distances close to contact.

Table III. Gurney Parameters for Two Tetraalkylammonium Bromides (Fig. 17)[a]

Electrolyte	Et$_4$NBr	Bu$_4$NBr
A_{++}	-120	-93
A_{+-}	-173	-210
A_{--}	0	0
TS_{++}	160	280
TS_{+-}	-78	-6
V_{++}	-9.6	-14
V_{+-}	2.1	0.2

[a] Energies in calories per mole and volumes in microliters per mole. A_{--} and its derivatives with respect to T and P_0 are assumed to be zero. (From ref. 45.)

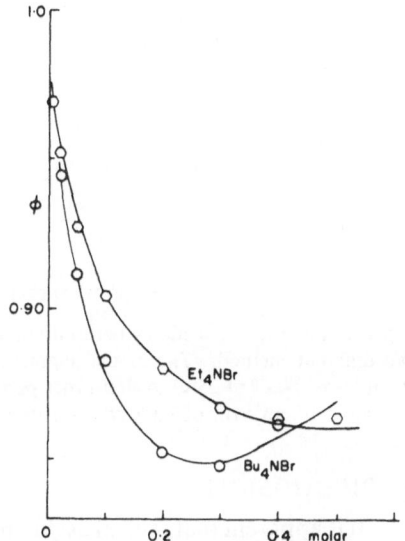

Fig. 17. The theoretical (HNC) osmotic coefficients in a refined model compared with the experimental data of Ku[48] *for two tetraalkylammonium bromides.* The data are represented as full lines and the HNC values for the model by discrete points (from P. S. Ramanathan, C. V. Krishnan, and H. L. Friedman, ref. 45).

It is useful to inquire to what extent some of the refinements to the primitive model can be elucidated by methods that are simpler than those described here. The HNC approximation, though generally accurate, especially for lower-valence electrolytes, is expensive and requires the services of a computer with a fairly extensive memory. When the results for the primitive model are at hand, however, the significance of various modifications to the short-range potential can be determined by a simple first-order perturbation theory.[20] We illustrate this by returning to the square-well model, where we consider the well depth to be the perturbation to a reference system consisting of charged hard spheres. The perturbing potential is equal to d_{ij} when $a_{ij} < r < b_{ij}$ and is zero otherwise. Then an upper bound for the excess free energy per unit volume F^{ex} in the square-well model is given by[20]

$$F^{ex}/ckT \leqslant F^{ex,0}/ckT + (2\pi/ckT) \sum_{i=1}^{\sigma} \sum_{j=1}^{\sigma} c_i c_j d_{ij} \int_{a_{ij}}^{b_{ij}} g_{ij}^0(r,c)r^2\,dr \qquad (29)$$

where the superscript zero refers to the properties of the charged hard-sphere system. Since accurate HNC computations for both these models are available, the upper bound given to the right of the inequality sign above can be compared with the essentially "exact" HNC results for the square-well model. This is done in Fig. 18; the results are from ref. 20. At $2\,M$ the two calculations differ by 3.3%; the error could be reduced further by calculating the next term in the perturbation theory, but this would be more difficult to compute. Instead of using the HNC approximation for the reference system, it may be more convenient in future applications to use some other variant such as the MS approximation or the ORPA + B_2 (for $F^{ex,\,0}$) in conjunction with the exponential approximation [for $g_{ij}^0(r)$] developed by Andersen, Chandler, and Weeks.[36]

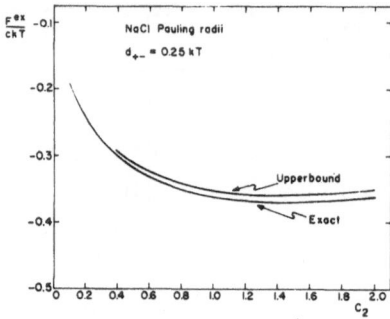

Fig. 18. The excess free energy per unit volume F^{ex} in the square-well model calculated by two different methods. The model parameters fit the experimental F^{ex} data for aqueous solutions of NaCl at 25°C. A first-order perturbation theory gives an upper bound for F^{ex} with a maximum error of 3.3 % at a stoichiometric concentration c_2 of 2 M. From ref. 20.

4. DISCUSSION

It is apparent that several useful methods are now available for elucidating the equilibrium properties of model electrolytes. Because of the accuracy of these methods, it is possible to conclude that discrepancies between theory and experiment are due to deficiencies in the model rather than deficiencies in their statistical treatment. The generality of the methods described also makes it possible to consider changes in the model. So far, these changes have been influenced by what we have learned about the weaknesses of simple models, especially when they are applied to a wide range of thermodynamic properties that are influenced in different ways by different parts of $u_{ij}(r)$ and its derivatives. Changes in the model have also been made on the basis of our (as yet incomplete) knowledge of the exact $u_{ij}(r)$ and also on the basis of intuition. The latter has undoubtedly been strongly influenced by speculations about the behavior of ions in solution, which have helped to systematize the thermodynamic properties of electrolytes. Nothing like an *a priori* calculation of the potential of average force $u_{ij}(r)$ has yet been possible, but if it were possible, then the methods described earlier can be relied upon to reproduce the thermodynamic properties of the model with a fair degree of certainty.

One of the shortcomings of the primitive model (and all of its refinements discussed here) is that the granularity of the solvent has been ignored. The solvent is represented as a continuous dielectric medium characterized by a macroscopic dielectric constant ε, and the effect of the detailed molecular properties of the solvent and ions have been included in refinements only in an approximate way. Friedman[16][14] has recently discussed the existence of a contribution to the solute potential of average force from the packing requirements of solvent molecules. We will consider very briefly this and other aspects of the problem in electrolyte theory.

[14] For an earlier discussion, see E. A. Guggenheim, *Disc. Faraday Soc.* **15**, 66 (1953).

To simplify the numerical part of the problem, we treat the ions and solvent molecules as hard spheres of equal radii. Typically, we can take the diameters of these molecules to be 2.76 Å, which is equal to the width of a water molecule and is close to the sum of the Pauling radii for sodium and chloride ions. If the reduced solvent density is ρ_s^*, then solvent granularity requires Eq. (20) to be replaced by

$$u_{ij}^*(r) = v_{ij}^{HS}(r) + w_{ss}^{HS}(\rho_s^*, r) \qquad (30)$$

$$= -kT \ln g^{HS}(\rho_s^*, r) \qquad (31)$$

where $v_{ij}^{HS}(r)$ is the solute–solute hard-sphere potential, $g^{HS}(\rho_s^*, r)$ is the hard-sphere distribution function at a reduced solvent density ρ_s^*, and the perturbation $w_{ss}^{HS}(r)$ to the primitive model is defined by Eqs. (30) and (31). [$g^{HS}(\rho_s^*, r)$ at $\rho_s^* = 0.7$ is drawn in Fig. 19.] The generalization of Eqs. (30) and (31) to other systems (not necessarily hard spheres) in which all solute–solvent interactions are the same as solvent–solvent interactions is straightforward but evidently for all of these systems, including the specialized one described here, $\varepsilon = 1$, since there is no charge-dipole interaction. Stell (see footnote 4) has recently included this contribution also in his discussion of a model which he describes as the exact ion–solvent interaction model (EISIM). The solvent–solvent interactions are idealized as in the primitive model, but what is of special interest is that there is a cavity term (varying asymptotically as r^{-4}) and a Gurney term associated with ion–dipole–ion interactions which contribute to $u_{ij}(r)$ (recall the indirect interactions between two isolated ions discussed earlier in Sec. 2).

One way to determine the effect, on the thermodynamic properties, of additional terms in the potential of average force $u_{ij}(r)$ is by simple perturbation theory in which the primitive model, for instance, is the reference system. This exactly parallels the treatment of the charged square-well model by

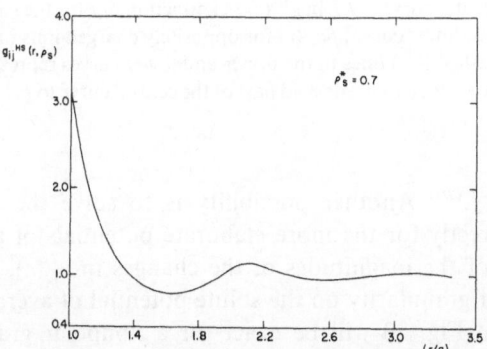

Fig. 19. Monte Carlo radial distribution functions for hard spheres at a reduced density $\rho_s^* = 0.7$ which corresponds crudely to liquid water regarded as hard spheres of diameter 2.76 Å. Monte Carlo data from Barker and Henderson. (See ref. 50.)

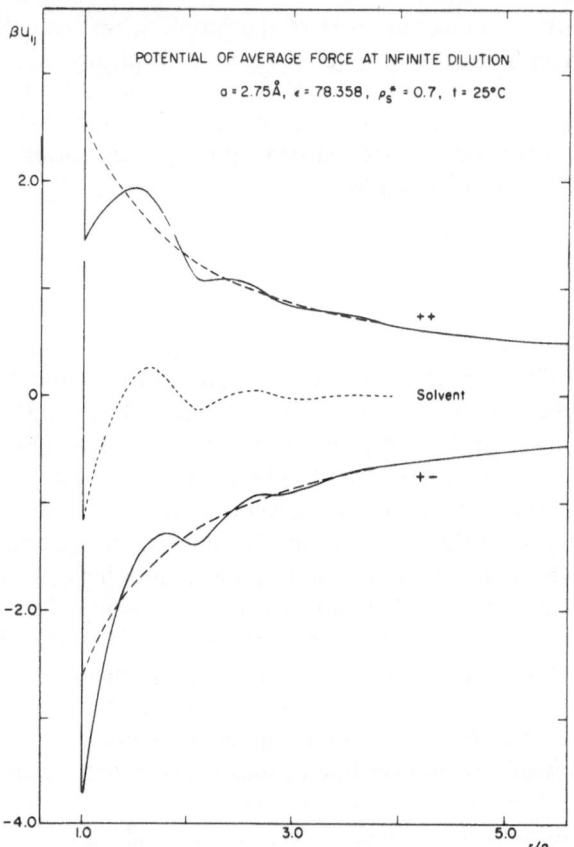

Fig. 20. Modification of the potential of average force for two ions due to solvent granularity. Both the ions and the solvent molecules (when granularity is invoked) are assumed to be spheres of diameter 2.76 Å. Upper curve: $\beta u_{ij}(r)$ for like ions in continuous (---) and granular (——) solvent media; center curve: $-kT \ln g^{HS}(r)$ as a function of r for the solvent at $\rho_s^* = 0.7$ (see caption for Fig. 19); lower curve: $\beta u_{ij}(r)$ for oppositely charged ions in continuous and granular media. The dashed (---) lines in the upper and lower curves represent the coulomb potential $(e_i e_j / \varepsilon r)$. This is added to the dashed part of the central curve to get the modification due to granularity.

perturbation theory.[20] Another possibility is to solve the HNC or MS approximations directly for the more elaborate potentials of average force. To get some idea of the magnitudes of the changes in $u_{ij}(r)$, we present in Fig. 20 the effect of granularity on the solute potential of average force. The solute potentials in Fig. 20 will be exact for a nonpolar granular solvent $(\varepsilon = 1)$ containing two ions of the same size as the solvent molecules but with charges equal to $e_i/78.358^{1/2}$. For a polar solvent $(\varepsilon = 78.358)$ containing ions of charge e_i, we may consider Fig. 20 to depict only that part of the deviations from the primitive model caused by solvent granularity.[15] Since there are attractive and repulsive contributions from this source, the net effect on the osmotic coefficients has yet to be learned from detailed calculations.[49]

ACKNOWLEDGMENTS

I wish to thank Professors Henry S. Frank, Harold L. Friedman, and George Stell for their contributions to my understanding of electrolytes, not to mention Dr. M. A. V. Devanathan who started me in this field. I am also indebted to Hans C. Andersen, David Chandler, John Weeks, John P. Valleau, D. N. Card, and C. W. Outhwaite for an exchange of information and lively correspondence.

REFERENCES

1. R. A. Robinson and R. H. Stokes, *Electrolyte Solutions* (Butterworths, London, 1965), 2nd ed.
2. J. C. Rasaiah. Ph.D. Thesis, University of Pittsburgh, 1965.
3. I. A. Cowperthwaite and V. K. LaMer, *J. Am. Chem. Soc.* **53**, 4333 (1931).
4. U. B. Bray, *J. Am. Chem. Soc.* **49**, 2372 (1927).
5a. H. S. Frank and P. T. Thompson, "A Point of View on Ion Clouds," in *The Structure of Electrolyte Solutions*, W. T. Hamer, ed. (Wiley, New York, 1959), Chap. 8.
5b. H. S. Frank and P. T. Thompson, *J. Chem. Phys.* **31**, 1086 (1954).
6. A. R. Allnatt, *Mol. Phys.* **8**, 409 (1970).
7. P. Debye and E. Hückel, *Z. Physik.* **24**, 185 (1923).
8. E. A. Guggenheim and J. E. Prue, *Physicochemical Calculations* (Interscience, New York, 1955), p. 215.
9. R. A. Robinson and R. S. Jones, *J. Am. Chem. Soc.* **58**, 959 (1936).
10. a. H. L. Frisch and J. L. Lebowitz, eds., *The Equilibrium Theory of Classical Fluids* (Benjamin, New York, 1964); b. J. E. Mayer, *Equilibrium Statistical Mechanics* (Pergamon Press, Oxford, 1968).
11. H. L. Friedman, *Ionic Solution Theory* (Interscience, New York, 1962).
12. J. A. Barker and D. Henderson, *Accts. Chem. Res.* **4**, 303 (1971).
13. a. B. J. Alder and W. G. Hoover, in *Physics of Simple Liquids*, H. N. V. Temperley, J. S. Rowlinson, and G. S. Rushbrooke, eds. (North Holland, Amsterdam, 1968); b. W. W. Wood, *ibid.*, c. I. R. McDonald and K. Singer, *Quart. Rev. Chem. Soc.* **24**, 238 (1970).
14. W. G. McMillan and J. E. Mayer, *J. Chem. Phys.* **13**, 276 (1945).
15. J. C. Poirier, *J. Chem. Phys.* **21**, 965 (1953).
16. a. H. L. Friedman, *J. Chem. Phys.* **32**, 1351 (1951); b. H. L. Friedman, *J. Solution Chem.* **1**, 387 (1972).
17. J. E. Mayer, *J. Chem. Phys.* **18**, 1426 (1950).
18. E. Meeron, *J. Chem. Phys.* **26**, 804 (1957).
19. G. Stell and J. L. Lebowitz **49**, 3706 (1968).
20. J. C. Rasaiah and G. Stell, *Mol. Phys.* **18**, 249 (1970).
21. D. D. Carley, *J. Chem. Phys.* **46**, 3783 (1967).
22. a. J. C. Rasaiah and H. L. Friedman, *J. Chem. Phys.* **48**, 2742 (1968); b. J. C. Rasaiah and H. L. Friedman, *J. Chem. Phys.* **50**, 3965 (1969).
23. a. J. C. Rasaiah, *Chem. Phys. Letters* **7**, 260 (1970); b. J. C. Rasaiah, *J. Chem. Phys.* **56**, 3071 (1972).

---

[15] The other terms, mentioned earlier in connection with Stell's EISIM, must, of course, be added to $u_{ij}^*(r)$ in any self-consistent evaluation of the solvent effect.

24. a. E. Waisman and J. L. Lebowitz, *J. Chem. Phys.* **52**, 4307 (1970); b. E. Waisman, Ph.D. Thesis, Yeshiva University, 1970. c. E. Waisman and J. L. Lebowitz, *J. Chem. Phys.* **56**, 3086, 3093 (1972).
25. a. H. C. Andersen and D. Chandler, *J. Chem. Phys.* **53**, 547 (1970); b. D. Chandler and H. C. Andersen, *J. Chem. Phys.* **54**, 26 (1971); c. H. C. Andersen and D. Chandler, *J. Chem. Phys.* **54**, 1497 (1971).
26a. D. M. Burley, V. C. L. Hutson, and C. W. Outhwaite, *Mol. Phys.* **23**, 867 (1972).
26b. D. M. Burley, V. C. L. Hutson, and C. W. Outhwaite, *Mol. Phys.* (to appear). Private communication from C. W. Outhwaite.
27. P. S. Ramanathan and H. L. Friedman, *J. Chem. Phys.* **54**, 1086 (1971).
28a. H. L. Friedman and P. S. Ramanathan, *J. Phys. Chem.* **74**, 3756 (1970).
28b. H. L. Friedman, A. Smitherman, and R. De Santis, *J. Solution Chem.* **2**, 59 (1973).
29. J. C. Rasaiah and H. L. Friedman, *J. Phys. Chem.* **72**, 3352 (1968).
30. J. C. Rasaiah, *J. Chem. Phys.* **52**, 704 (1970).
31. H. L. Friedman, Computed Thermodynamic Properties and Distribution Functions for Simple Models of Ionic Solutions, in *Modern Aspects of Electrochemistry*, J. O'M. Bockris and B. E. Conway, eds. (Plenum Press, New York, 1971), Vol. 6, Chap. 1.
32. a. P. N. Voronstov-Vel'yaminov, A. M. El'yashevich, and A. K. Kron, *Electrokhimiya* **2**, 708 (1966); b. P. N. Voronstov-Vel'yaminov and A. M. El'yashevich, *Electrokhimiya* **4**, 1430 (1968); c. D. N. Card and J. P. Valleau, *J. Chem. Phys.* **52**, 6232 (1970); d. J. P. Valleau and D. N. Card, *J. Chem. Phys.* **57**, 5457 (1972); e. P. N. Voronstov-Vel'yaminov, A. M. El'yashevich, J. C. Rasaiah, and H. L. Friedman, *J. Chem. Phys.* **52**, 1013 (1970). f. J. C. Rasaiah, D. N. Card, and J. P. Valleau, *J. Chem. Phys.* **56**, 248 (1972).
33. A. R. Allnatt, *Mol. Phys.* **8**, 533 (1964).
34. a. E. Meeron, *J. Chem. Phys.* **27**, 1238 (1957); b. E. Meeron, *J. Chem. Phys.* **28**, 630 (1958).
35. H. C. Andersen, D. Chandler, and J. D. Weeks, *J. Chem. Phys.* **56**, 3812 (1972).
36. a. H. C. Andersen and D. Chandler, *J. Chem. Phys.* **57**, 1918 (1972); b. D. Chandler and H. C. Andersen, *J. Chem. Phys.* **57**, 1930 (1972); c. H. C. Andersen, D. Chandler, and J. Weeks, *J. Chem Phys.* **57**, 2626 (1972).
37. a. F. H. Stillinger and R. Lovett, *J. Chem. Phys.* **48**, 3858 (1968); b. R. Lovett and F. H. Stillinger. *J. Chem. Phys.* **48**, 3869 (1968); c. F. H. Stillinger and R. Lovett, *J. Chem. Phys.* **49**, 1991 (1968); d. F. H. Stillinger, *Proc. Nat. Acad. Sci. U.S.* **60**, 1138 (1968).
38. F. H. Stillinger and R. J. White, *J. Chem. Phys.* **54**, 3395 (1971); b. F. H. Stillinger and R. J. White, *J. Chem. Phys.* **54**, 3405 (1971).
39. a. J. Groeneveld, private communication; b. J. Groeneveld, quoted in ref. 24.
40. C. W. Outhwaite, *Mol. Phys.* **20**, 705 (1971), and references therein.
41. J. G. Kirkwood, *J. Chem. Phys.* **2**, 767 (1934).
42. G. Stell, *J. Chem. Phys.* **55**, 1485 (1971).
43. H. S. Frank and M. W. Evans, *J. Chem. Phys.* **13**, 507 (1945).
44. D. Henderson, J. A. Barker, and R. O. Watts, *IBM J. Res. Develop.* **14**, 668 (1970); see also ref. 12.
45. P. S. Ramanathan, C. V. Krishnan, and H. L. Friedman, *J. Solution Chem.* **1**, 237 (1972).
46. J. G. Kirkwood and J. C. Poirier, *J. Chem. Phys.* **58**, 591 (1954).
47. R. W. Gurney, *Ionic Processes in Solution* (Dover Publications, Inc. New York, 1963).
48. J. C. Ku, Ph.D. Thesis, University of Pittsburgh, 1971.
49. R. H. Stokes, *J. Chem. Phys.* **56**, 3382 (1972).
50. J. A. Barker and D. Henderson, *Mol. Phys.* **21**, 187 (1971).

DISCUSSION

Professor K. S. Pitzer (*University of California, Berkeley*). In looking over some of this very impressive work reported by Prof. Rasaiah, I have noted the transfer of the new statistical-mechanical methods from gases to solutions. It occurred to me that it might be interesting—and I hadn't seen it done except numerically by Card and Valleau—to put the Debye–Hückel distribution in the pressure (or osmotic) equation in statistical mechanics. This is a way of including the hard-sphere effects. The old charging-process method could not deal with the hard-sphere kinetic effects, and therefore it was completely defective in that term. Also, one uses the usual power series expansion in which, as the exponential is expanded, the first term gives a hard-core effect but no electrostatic effect. The second term gives the typical Debye–Hückel effect, but this effect is not numerically or quantitatively the same if one uses the pressure equation instead of the charging process, although the limiting law is the same. For symmetrical electrolytes without any inconsistency, the third term in the exponential expansion which gives a hard-core contribution can be included. It is a quite simple analytical formula which, in fact, agrees well within the computational error of the Monte Carlo results up to about 0.5 m and probably up to about 1 m. Also, for the case of the Monte Carlo calculation with a 4.25 Å distance of closest approach, I notice that this fits the empirical data for HBr essentially accurately.

The formula for the osmotic coefficient is

$$\phi - 1 = -z^2 \, l\kappa/6(1 + \kappa a) + c[2\pi a^3/3 + \pi a z^4 \, l^2/3(1 + \kappa a)^2]$$

where $\kappa = (4\pi l)^{1/2} z c^{1/2}$ and $l = e^2/DkT$. Thus, a three-term formula is obtained, and the third term involves an ionic-strength dependence of the second virial coefficient. It is a very simple result which is implicit numerically in the presentation of Card and Valleau of their Monte Carlo results but has this simple analytical form which I think may be of some use.

The other topic which I hope we could discuss is the question of how close the primitive model with adjustable radius will come if you take a little bit of liberty in the most dilute region. I have checked out HBr with the 4.25 Å and it fits really quite well. I was wondering, for other solutes and for slightly smaller radii, if you do not commit yourself to perfect fit of the apparent data around 0.01 m, how close a fit can be obtained at a little higher concentration?

Professor Rasaiah (*University of Maine, Orono*). In their paper on Monte Carlo calculations for 1–1 electrolytes, Card and Valleau [*J. Chem. Phys.* **52**, 6232 (1970)] compared their results with several approximate theories. The HNC approximation was found to be very satisfactory, but they also discovered that the nonlinear Debye–Hückel approximation (DHX) was quite good (for 1–1 electrolytes) when the osmotic coefficients were calculated using the virial equation. The DHX approximation assumes that the radial distribution function $g_{ij}(r)$, at distances larger than the contact distance, is given by

$$g_{ij}(r) = \exp[-z_j e\phi_i(r)/kT] \tag{D1}$$

where $\phi_i(r)$ is the average electrostatic potential calculated in the Debye–Hückel theory. The term $\phi_i(r)$ is obtained as the solution to the linearized Poisson–Boltzmann equation which is derived by combining the Poisson equation with an approximation for $g_{ij}(r)$ given by the first two terms in the series expansion of Eq. (1). The Debye–Hückel limiting law results when the analysis is completed in a self-consistent way, but if the Debye–Hückel $\phi_i(r)$ is used in Eq. (1), the results for the thermodynamic properties calculated in various ways are not concordant. For instance, the osmotic coefficients obtained with the compressibility equation would be quite different from those obtained using the virial theorem or one of the charging processes. I think the extension to the Debye–Hückel theory proposed by Prof. Pitzer will run into the same difficulties since he proposes using the Debye–Hückel $\phi_i(r)$ with the first three terms in the expansion of Eq. (1).

In reply to the second point raised by Prof. Pitzer, it might be possible to fit the primitive-model osmotic coefficients to experiment quite nicely at higher concentrations if one does not pay too much attention to the closeness of fit at low concentrations; but over the whole preparative concentration range, the primitive model seems to be a poor representation of most electrolytes.

Professor H. S. Frank (*University of Pittsburgh*). There could be different purposes to be entertained here. One has to do with the kind of formalism that fits the data and the other is to describe what really takes place. I expect that what really is taking place is closer to what Professor Friedman was talking about this morning regardless of how well the formalisms fit the data.

Professor Pitzer. The point, if I may explain it in more detail, is that more terms can be taken in the exponential expansion for symmetrical electrolytes without inconsistency. Once the ions have been summed over to get electrical neutrality, the second term is the only term that effects the Poisson equation since the third term has no effect on the Poisson equation for symmetrical electrolytes. Therefore, one can carry this third term along without any logical inconsistency with the linearized Poisson equation for a symmetrical electrolyte but not for an unsymmetrical one. That is the point, and this third term leads to some interesting results which, as far as I know, no one has ever paid any attention to.

Professor Rasaiah. You are talking about doing essentially the same thing that Card and Valleau did, except that they used the complete exponential form.

Professor Pitzer. Indeed, they took the complete exponential form, but much of the interesting part of it can be obtained from this third term, which is not inconsistent for a symmetrical electrolyte.

Professor R. H. Stokes (*University of New England, Armidale, Australia*). In a calculation of transport properties about two years ago, I included that second term. For a symmetrical electrolyte the second-order term vanishes from the charge density, so that it does not affect the activity coefficient, but it still appears in the diffusion coefficient.

Professor Frank. But it still would not give the same free energy when the Güntelberg or the Debye–Hückel charging processes are used, which means that there is a fundamental self-inconsistency there, even with that term.

Professor Stokes. It does give the same free energy, I think.

Professor Pitzer. That third term does not give anything in the charging process because it only gives a hard-core effect; it gives no electrostatistics. We are now saying the pressure equation gives a different result from the charging process. It is just an approximation, but a rather interesting point.

Professor Frank. The term $z_j \varepsilon \phi_i$ is not a proper potential of average force.

Professor H. L. Friedman (*State University of New York, Stony Brook*). I wonder whether the difficulty of the application of the HNC equation to 2–2 electrolytes might be relieved if one would use a soft-core potential because then the cusp in g_{+-} for the same radii is rounded off.

Professor Rasaiah. Yes, that is possible, but I cannot be certain because I have not done it.

Professor Frank. From the standpoint of the extrapolation, I am going to take advantage of my position and point out that if you are willing to mix a little art with your science (as a matter of fact, everybody does; there are always personal choices one makes in the form of an equation to fit this, that, and the other thing), you can take Professor Rasaiah's E' and subtract an E_0, put in the proportionality factor, divide by $\sqrt{m}$, and plot his experimental points against $m^{1/2}$ for some assumed E_0. If we have the right E_0, we know where this has to come in (we are treating this the way we would a ϕ, you see; ϕ is customarily extrapolated this way). Figure 21 (this was done on the blackboard) shows such a plot coming in to the correct theoretical intercept. What creates the problem in the first place is that the experimental points for $ZnSO_4$ go down to low-enough concentrations to show that the quantity $(E' - E_0)/km^{1/2}$ rises above the theoretical limiting value, so that, in going to $m^{1/2} = 0$, the curve is going to have to turn back down again. But how a trial extrapolation will actually go will depend on what trial value of E_0 has been

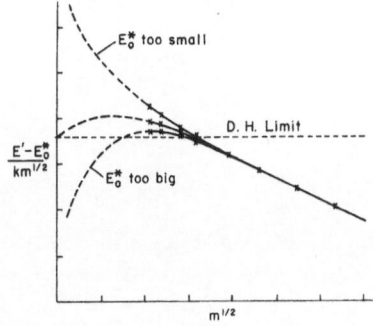

Fig. 21

chosen. If it has been chosen too big, the quantity $(E' - E_0)/km^{1/2}$ will be everywhere too negative, but the more so the smaller $m^{1/2}$, till it goes to negative infinity in the limit, as the figure shows. (The symbol E_0^* is used to show that this quantity is a trial value.) If E_0 has been chosen too small, the opposite catastrophe occurs, the curve lying everywhere too high and going positively infinite at $m^{1/2} = 0$, as also shown. It is only if E_0 is "just right," like the third little bear's porridge, that the curve can be imagined to turn over and come in to the correct theoretical intercept (as also shown). That is why I say that if you are willing to put a little art into your science—try different E_0's and see which gives a curve which your artistic eye tells you is going to come in at the right place—you can set limits on what E_0 must be without any theory except the limiting law, which we know is right. With Rasaiah's $ZnSO_4$ data, it seems to be not too hard to guess to within 30 or 40 μV where E_0 has to come, and we may have to make do with this until someone comes up with a better method.

Professor Stokes. Dr. Rasaiah's final figure (Fig. 20) seems to me of great interest because here the radial distribution function is considered for two uncharged solute particles which are hard spheres of the same diameter as the solvent. If you follow this up, you will be exactly where Debye was. In effect, Dr. Frank, what you are saying is that the best way to describe the solution is to consider it as an ideal solution, which is exactly what the hard-sphere system plus the coulomb term would be. This is what Debye used. It was very good indeed.

Concentrated Electrolyte Solutions at High Temperatures and Pressures[1]

E. U. Franck[2]

Received February 7, 1973

A survey is given of recent experimental results obtained from high-temperature, high-pressure investigations with water, aqueous solutions, and ionic fluids. Data on the static dielectric constant of water to 550°C and 5 kbar are given and discussed with respect to their relation to water structure. Infrared and Raman spectra of HDO in pure water have been obtained to 400°C and 4 kbar, which give information on hydrogen bonding. Xe–H$_2$O and CO$_2$–H$_2$O mixtures were investigated in the infrared. Ni(II) and Cu(II) complexes were investigated by absorption spectroscopy in aqueous solutions of high chloride content to 350°C and 2–6 kbar. The gas–liquid critical point of ammonium chloride was found at 880°C and 1635 bars. This fluid appears to be predominantly ionic even in the critical region. The possibility of converting pure polar fluids such as ammonia and water into concentrated ionic solutions by self-ionization at very high pressures is mentioned.

KEY WORDS: High temperature; high pressure; static dielectric constant; infrared and Raman spectra; critical point; water; aqueous solutions; ammonia; HDO; Xe; CO$_2$; NH$_4$Cl; Ni(II) complexes; Cu(II) complexes.

1. INTRODUCTION

The term "concentrated electrolyte solutions" will here be used for fluid mixtures with ion concentrations comparable to the concentrations of the nonionic components. This is an intermediate range of ion concentrations between those of the "normal," relatively dilute electrolyte solutions and the highly or completely ionized fused salts. These concentrated electrolyte solutions are not yet very thoroughly investigated, although their peculiar

---

[1] This paper was presented at the symposium, "The Physical Chemistry of Aqueous Systems," held at the University of Pittsburgh, Pittsburgh, Pennsylvania, June 12–14, 1972, in honor of the 70th birthday of Professor H. S. Frank.

[2] Institute of Physical Chemistry and Electrochemistry, University of Karlsruhe, Germany.

properties are of considerable fundamental and practical interest. Investigations in this range often require, however, application of elevated temperatures and pressures in order to obtain the necessary solubilities or sufficient ranges of stability of certain complex ionic species. Thus knowledge of relevant properties of dense polar solvents at high temperatures is desirable for the discussion of the concentrated electrolytic solutions.

Recent investigations have provided new information of this kind which will be presented below. The dielectric constant has been measured and examined in dense supercritical water and hydrogen chloride. The OH stretching vibration was investigated as an indicator for water association in pressurized pure water and concentrated solutions by infrared and Raman techniques. The stability of heavy metal complexes, particularly of copper complexes, was investigated spectroscopically in solutions to 350°C. The ionization of pure liquid ammonium chloride to the critical point and of very highly pressurized fluid water and ammonia was studied by conductivity measurements.

Figure 1 gives a temperature–density diagram of water which extends to 1000°C and 1.6 g-cm^{-3}. The critical point CP (374°C, 221 bars) and the triple point TP are indicated on the gas–liquid coexistence curve in the lower left part. The points of the broken line extending to the right from TP denote the transitions between the different high-pressure modifications of ice. To about 10 kbar the isobars are based on static experiments.[1, 2] At pressures above 25 kbar, water densities have been derived from shock-wave experiments.[3]

In order to evaluate the properties of water at high pressures and temperatures as an electrolytic solvent, knowledge of the dielectric constant, infrared and Raman spectra, and the viscosity are particularly useful. The cross-

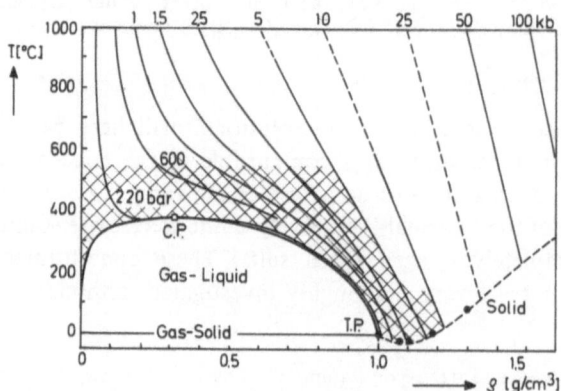

Fig. 1. Temperature–density diagram of water. —— Isobars measured; ------ isobars interpolated.

hatched area in Fig. 1 approximately indicates the region where these phenomena have been investigated experimentally in recent years. Dielectric constant and spectra will be discussed below. The viscosity has been measured to
500°C and 3.5 kbar.[4] At 500°C and 0.2 and 0.8 g-cm^{-3}, respectively, the
viscosity has been found to be 3.8×10^{-4} and 10.5×10^{-4} P. This means that
the viscosity at these conditions is lower than that of liquid water at room
temperature by a factor of 10 or 20. Diffusion coefficients and ion mobilities
are correspondingly high. Viscosity and other physical properties of water
at high pressures and high temperatures are reviewed elsewhere.[5]

2. DIELECTRIC CONSTANT

The static dielectric constant of water is to a large extent determined by
the peculiar structural properties caused by hydrogen bonds. Thus this quantity is not only interesting as such but also as a means of obtaining structural
information. Earlier measurements of the dielectric constant to 400°C and
2 kbar[5, 6, 7] were made and discussed. Recent experiments were performed
to 550°C and 5 kbar.[8] The capacity of a condenser of gold–palladium
mounted inside an autoclave was determined at frequencies between 0.1 and
1 MHz. The geometry of the condenser could be changed at high temperatures
and pressures. Figure 2 gives a compilation of results as curves of dielectric
constants superimposed on isobars of a temperature–density diagram. Within
a wide range of supercritical temperatures and densities, the constant has
values between 5 and 25. This corresponds to dielectric properties of certain
polar organic liquids at normal conditions. To calculate the dielectric constant

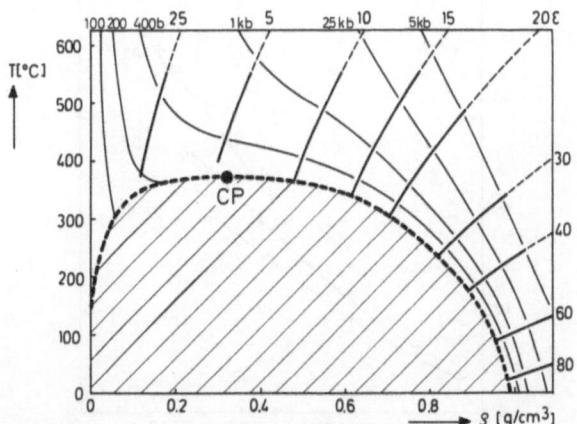

Fig. 2. Static dielectric constant ε of water as a function of temperature and density ρ. - - - - -
Measured values; - - - - - calculated values; —— isobars.

ε of liquid water at moderate temperatures, Kirkwood[9] used the equation

$$(\varepsilon - 1)(2\varepsilon + 1)/9\varepsilon = (4\pi\rho/3)\alpha + (4\pi\rho\mu^2/9kT)g_K \tag{1}$$

Here ρ is the number of particles per unit volume, α the isotropic polarizability, and μ the strength of the molecular dipoles within the dielectric environment. The quantity g_K is the *Kirkwood correlation factor*, introduced to take into account short-range ordering between dipoles. Extensive discussions of Eq. (1) and of methods of estimating the dipole strength within the dielectric and of calculating g_K in liquid water are given elsewhere.[9, 10]

If neighboring dipoles are uncorrelated, one should find $g_K = 1$. At 0°C in liquid water g_K is about 2.9. The evaluation of g_K from the experimental ε data of Fig. 2 shows that g_K decreases with increasing temperature and decreasing water density, as would be expected.[8, 11] It appears, however, that even at the critical density and at supercritical temperatures of 400° and 500°C, g_K is still about 1.6, indicating a considerable degree of dipole correlation.

Very recent experimental results obtained for the dielectric constant of supercritical dense hydrogen chloride with a similar technique[12] showed that, at corresponding supercritical conditions, g_K is only about 1.05 for this

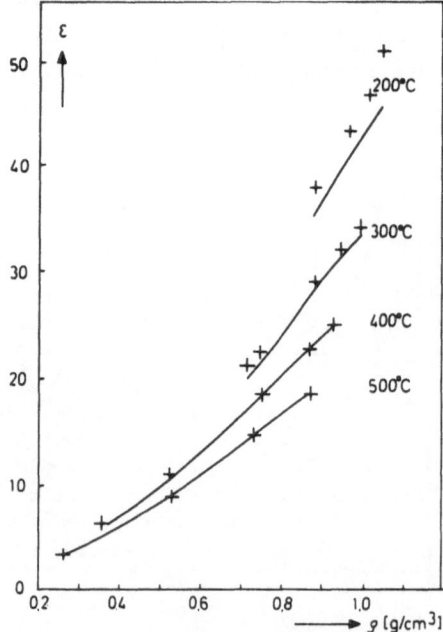

Fig. 3. Static dielectric constant of water ε as a function of density and temperature. + Experimental values; —— calculated.

fluid. One might conclude that the relatively high Kirkwood factors indicate a considerable degree of association by hydrogen bonding in dense super-critical water. Caution is necessary, however, since it has been demonstrated that the experimental ε data of water between 200° and 500°C and 0.1 and 1.0 g-cm^{-3} can be calculated rather well without using the concept of hydrogen bonding.[13, 14] No adjustable parameters had to be used besides the dipole moment, polarizability, and diameter of water molecules. For such calcula-tions, the general expressions for the dielectric constant of a fluid of hard cores with embedded point dipoles presented by Wertheim[15] have been used. This is shown in Fig. 3.

3. INFRARED SPECTRA

More detailed information about the association of water by hydrogen bonding can be obtained from infrared and Raman spectra. Particularly well suited for this purpose is the study of the absorption of the OD stretching vibration around 2500 cm^{-1} of HDO diluted in H$_2$O because of the absence of interference of other vibrations in this frequency range. Figure 4 gives several absorption curves for this vibration of HDO in ordinary water at a constant density of about 1 g-cm^{-3} from 30° to 400°C and at 400°C down to 0.01 g-cm^{-3}.[16] The frequency of the maximum shifts from 2507 to 2720 cm^{-1}, the value for the Q Branch of the OD vibration in dilute HDO gas. This shift of the maximum frequency, as well as the decrease of intensity, has been considered as an indication of decreasing association by hydrogen bonds.

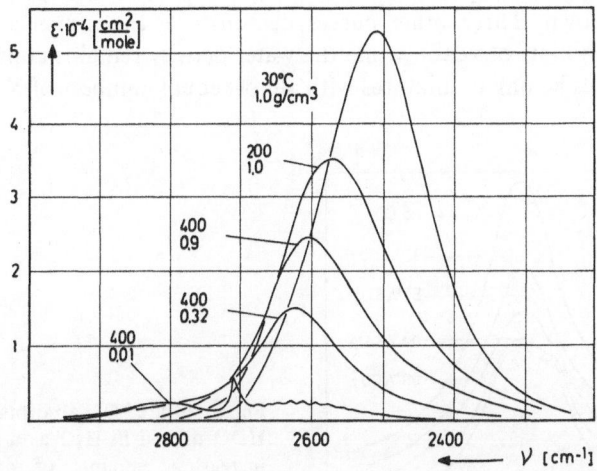

Fig. 4. Infrared OD vibration bands of 9.5 mole % HDO in H$_2$O at different temperatures and densities.

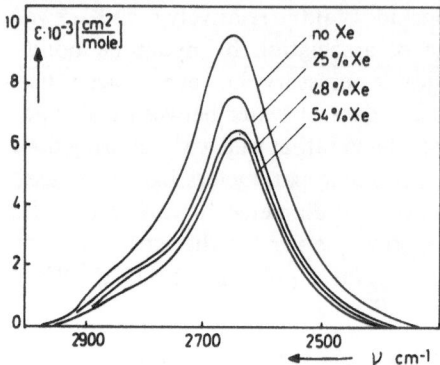

Fig. 5. Infrared OD vibration bands of HDO diluted in H_2O as a function of increasing amounts of added xenon. The temperature is 400°C, and the total water density for all curves is 0.17 g-cm^{-3}.

It is interesting to investigate the question as to what extent this OD bond at dense supercritical conditions will be affected by the addition of very high concentrations on nonionic but highly polarizable second components to the fluid phase. Experiments of this kind have been made recently with Xe–HDO–H_2O and CO_2–HDO–H_2O mixtures.[17] Xe and CO_2 have been chosen because both particles have similar size, although for CO_2, in contrast to Xe, a specific interaction with water molecules even at 400°C could be expected. Both Xe and CO_2 are completely miscible with water at 400°C. Figures 5 and 6 give selected absorption curves at 400°C, which should be compared with the relevant curves of Fig. 4.

In Fig. 5 the absorption of HDO in H_2O at a total water density of 0.17 g-cm^{-3} is shown. Three other curves demonstrate the influence of added increasing amounts of xenon while the water density remains constant. Two of these curves belong to mixtures with almost equal numbers of Xe and H_2O

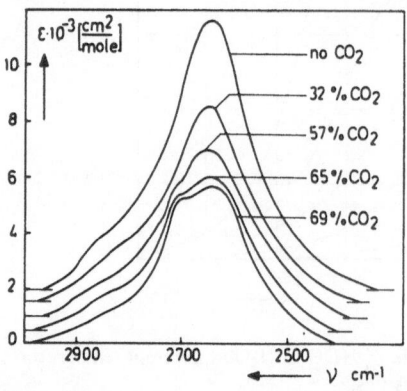

Fig. 6. Infrared OD absorption bands of HDO diluted in H_2O as a function of increasing amounts of added CO_2. The temperature is 400°C, and the total water density for all curves is 0.17 g-cm^{-3}.

particles. Apart from a certain reduction of intensity, the effect of the xenon on the character of the band is insignificant. The bands of Fig. 6, obtained in analogous experiments with CO_2, are different. A shoulder at 2700 cm^{-1} appears at 57 mole % CO_2 and develops into a separate band at 69 mole %. This band may be produced by deuterated carbonic acid molecules. PVT data of CO_2-H_2O mixtures at high pressures and supercritical temperature do not exclude formation of such carbonic acid molecules.[18] This assignment is not yet conclusive, however. Investigation of the absorption in the region of C–O vibrations (1400 cm^{-1}) would be desirable, but the sapphire windows of the high-pressure infrared cell are opaque at such wavelengths. Perhaps Raman measurements would be possible.

It would be interesting to observe the hydrogen–oxygen stretching vibration in a binary system of water and a simple salt from pure water through the region of concentrated solutions to the pure fused salt. Most simple salts have melting points that are too high. Sodium hydroxide, however, melts at 318°C. Measurements of the absorption of the stretching vibration of normal water in liquid NaOH–H_2O mixtures have been made at present at Karlsruhe to 2 kbar and 350°C, where the two compounds are completely miscible. It appears as if even at this temperature the whole range of compositions from pure H_2O to pure NaOH can be investigated with sapphire windows, although with some difficulty.

4. RAMAN SPECTROSCOPY

The Raman spectra of pure water at high pressure,[19] and the conclusions to be drawn from these, have been thoroughly discussed already by G. E.

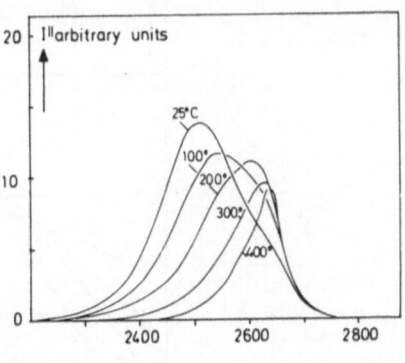

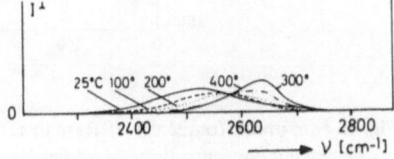

Fig. 7. Raman band of the OD vibration of HDO diluted in H_2O at a constant water density of 1.0 g-cm^{-3} from 25° to 400°C. Upper part: $I^{\parallel}$, parallel polarization relative to laser beam. Lower part: $I^{\perp}$, vertical polarization.

Walrafen in his paper. If the Raman spectra of the OD stretching vibration of HDO in H_2O are shown again in Fig. 7, it is mainly to draw attention to the $I^\perp$ curves in the lower part, which appear to exhibit an isosbestic point at $2575\ cm^{-1}$. This may support the assumption of two distinguishable hydrogen-bonded and non-hydrogen-bonded states of the OD groups.

Figure 8 demonstrates the influence of the addition of increasing amounts of an electrolyte on the maximum frequency of the OD vibration. KI was used mainly because of the large ion sizes and good solubility. Curve a, derived from Fig. 7, gives the values for pure HDO–H_2O mixtures. At low temperatures, a pronounced increase of the maximum frequency with KI concentration is observed, while at 400°C the salt acts in the opposite direction and reduces the maximum frequency, although to a lesser extent. Around 180°C there is a crossover region. The effect at lower temperatures has been observed and discussed earlier. The present results at 25°C are in agreement with those of previous authors.[20] A quantitative discussion of the results of Fig. 8 may not be justified without a decomposition of the asymmetric bands into components and a proper estimate of the specific interaction between iodide ions and water molecules. The "structure-breaking" effect of the electrolyte up to about 200°C is very obvious, however. At 400°C, where pure water does not have much structure any more, a certain "structure-making" influence of the ions seems to predominate, perhaps by ion hydration.

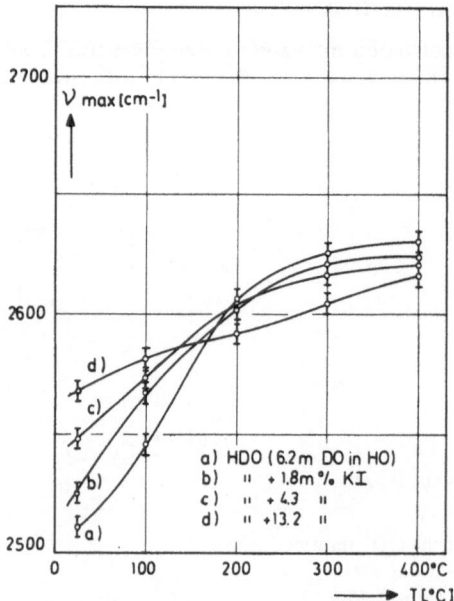

Fig. 8. Maximum frequency of Raman OD vibration band of HDO diluted in H_2O with increasing concentrations of added KI. Total solution density constant at 1.0 g-cm³.

Raman spectroscopy can also be used to examine existence ranges of dissolved metal complexes, which cannot be very well investigated in the visible or ultraviolet regions.[21] This applies particularly to corrosive high-temperature, high-pressure solutions. Many of these are of interest for power-plant corrosion problems and as hydrothermal fluids in geochemistry. Lead chloride and zinc chloride solutions, for example, have considerable geochemical interest, and $ZnCl_2$ has been extensively investigated as fused salt and in low-temperature, concentrated aqueous solutions. It is possible to extend laser Raman investigations of concentrated aqueous $ZnCl_2$ solutions to 400°C.

Figure 9 gives several first results obtained at a constant pressure of 2.5 kbar with a high-pressure Raman cell equipped with sapphire windows and irradiated with an argon-ion laser beam. Observation was vertical to the beam direction. The two vertical bars indicate major bands found with molten $ZnCl_2$.[22] For the melt, the band at 305 cm^{-1} is ascribed to the monomer $ZnCl_2$, and the band at 266 cm^{-1} to a complex $ZnCl_4{}^{2-}$. From Fig. 9 it appears as if in the concentrated aqueous solutions with increasing temperature a shift occurs from mainly fourfold-coordinated ionic complexes towards a predominance of lower-coordinated molecular monomers.

5. SOLUTION SPECTRA IN THE VISIBLE AND NEAR ULTRA-VIOLET

In the zinc chloride solutions, an increase of temperature seems to favor the less highly coordinated complexes. Similar behavior has been observed with several other complex-forming metals at high temperatures by means of absorption spectra in the visible and ultraviolet regions. Complexes of

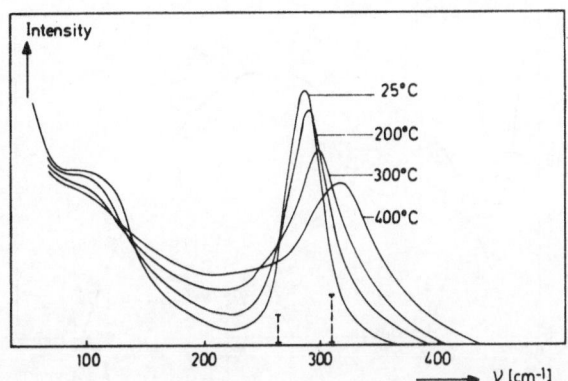

Fig. 9. Raman bands of aqueous $ZnCl_2$ solutions (7 *m*) at 2.5 kbar between 25° and 400°C. ------ Fused ZnCl².

bivalent cobalt, nickel, and copper are examples which to some extent can be used as a kind of probe to obtain additional information on water properties in solutions at unusual conditions.

One example, the absorption of $NiCl_2$ in a concentrated aqueous NaCl solution, is shown in Fig. 10.[23] The curve for 25°C corresponds to the normal solutions of light green color with two absorption bands caused by two electron transitions of octahedral hexaquo complexes. The pressure of 500 bars at room temperature has little influence on the spectrum. Temperature increase to 300°C at the same pressure produces blue solutions with a strong, broad band at 680 nm. It is caused by a combination of bands from octahedral and tetrahedral complexes. The tetrahedral complexes appear to prodominate. Additional spectra indicate that this kind of complex becomes the only stable form in 10 m lithium chloride solutions at this temperature and at pressures between 150 and 300 bars. A detailed analysis of the magnitude of the extinction coefficients suggests that trichloromonoaquo and dichlorodiaquo tetrahedral complexes are the most abundant types. It can be shown from Fig. 10 that pressures of several kbar increase the range of stability of higher-coordinated aquo complexes to temperatures which may even be above the critical temperature of pure water.

Recently, absorption spectra of copper solutions were also obtained to 400°C and 2 kbar.[24] A special high-pressure optical cell had to be designed in which the solution samples were only in contact with sapphire and Teflon at high temperatures. Increasing amounts of LiCl (up to 14 m) have been added.

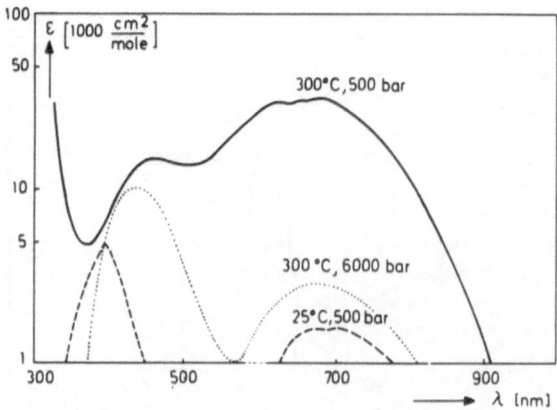

Fig. 10. Absorption spectrum of $NiCl_2$ (0.025 m) in aqueous NaCl solution (4.0 m) at high temperature and pressure.

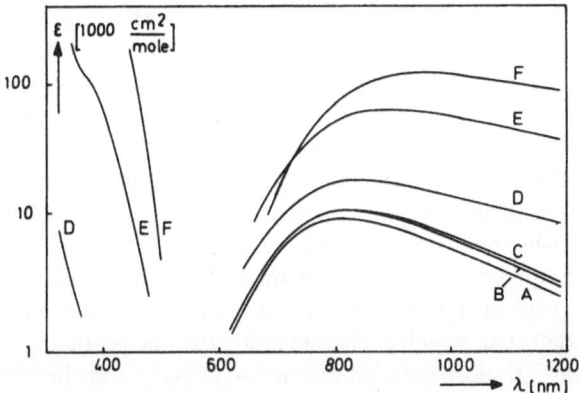

Fig. 11. Absorption spectra of $CuCl_2$ and $Cu(ClO_4)_2$ in water and aqueous LiCl solutions of various concentrations at 25°C and 2 kbar. (A) $Cu(ClO_4)_2$, 0.050 m; (B) $CuCl_2$, 0.0499 m; (C) $CuCl_2$, 0.0124 m + LiCl, 0.10 m; (D) $CuCl_2$, 0.0100 m + LiCl, 1.0 m; (E) $CuCl_2$, 0.0095 m + LiCl, 5 m; (F) $CuCl_2$, 0.0097 m + LiCl, 10 m.

In Figs. 11 and 12 a series of absorption curves at 2 kbar is given for 25° and 350°C. Absorption changes caused by pressure variations between 500 and 2000 bars did not exceed the range of experimental uncertainty of the data. Higher pressures could not be applied for technical reasons. Curves A give the absorption of dilute copper perchlorate solutions for comparison.

Curves B to F belong to copper(II) chloride solutions with increasing amounts of lithium chloride. Below 600 nm, portions of the "charge-transfer bands" are visible. At higher wavelengths, one has the "d–d bands," caused by transitions between d-levels, which are mainly considered here.

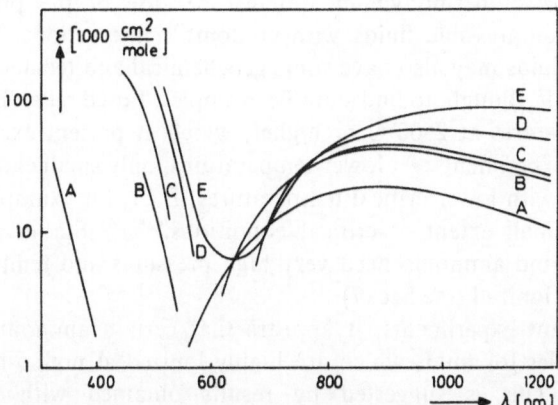

Fig. 12. Absorption spectra of $CuCl_2$ and $Cu(ClO_4)_2$ in water and aqueous LiCl solutions of various concentrations at 350°C and 2 kbar. (A) $Cu(ClO_4)_2$, 0.050 m; (B) $CuCl_2$, 0.0499 m; (C) $CuCl_2$, 0.0124 m + LiCl, 0.10 m; (D) $CuCl_2$, 0.0100 m + LiCl, 1.0 m; (E) $CuCl_2$, 0.0095 m + LiCl, 5.0 m.

Comparison of the 25°C curves of Fig. 11 with the results and discussions in the literature shows that curves A, B, and C for dilute solutions indicate a predominance of hexaquo complexes. In the yellow solutions F which contain 10 m LiCl the tetrahedral complex $(CuCl_4)^{2+}$ clearly prevails even at 25°C. Temperature increase to 350°C at 2 kbar (Fig. 12) causes enhancement of intensity and red shift, particularly for the solutions with lower chloride content. This is attributable to stronger distortion of the octahedra and increased copper-ligand distances as well as to the exchange of water ligands by chloride. The appearance of the spectra suggests a successive exchange of H_2O by Cl^- with rising temperature and Cl^- concentration, although no band can be detected, which is characteristic for one particular type of aquochloro complex. A more detailed discussion is given elsewhere.[24] Formally, the sequence of transitions from hexaaquo complexes to tetrachloro complexes with increasing temperature and chloride content can be described by

$$[Cu(H_2O)_6]^{2+} \rightleftharpoons Cu[H_2O)_5Cl]^+ \rightleftharpoons Cu[(H_2O)_4Cl_2] \rightleftharpoons$$
$$Cu[(H_2O)_3Cl_3]^- \rightleftharpoons Cu[(H_2O)_2Cl_4]^{2-} \rightleftharpoons Cu[Cl_4]^{2-}$$

6. AMMONIUM CHLORIDE

Although some of the solutions discussed so far, with electrolyte molalities of 10 or more, have very high ion concentrations, they are still quite different from fused salts where there may be only ions in the fluid. One might ask whether it is possible to study pure ionic fluids at temperatures where the intermolecular distance and possibly the degree of ionization can be changed continuously over wide ranges by variation of pressure. This would be of considerable basic thermodynamic interest because of the possibility of investigating compressible fluids with coulombic interactions. Such dense gaseous ionic fluids may also have some geochemical and technical interest.

It is difficult, though, to find suitable examples. Fused alkali halides have critical temperatures at 2500°C or higher, which at present excludes static high-pressure experiments. At lower temperatures, only small expansions are possible. Salts with lower critical temperatures, $BiCl_3$ for example, are only ionized to a small extent at critical conditions.[25, 26] Stable polar fluids such as water and ammonia need very high pressures and temperatures to become higher ionized (see Sec. 7).

From recent experiments, it appears that certain ammonium halides may be examples for fluids which are highly ionized at not-so-high critical temperatures. This is suggested by results obtained with ammonium chloride.[27] NH_4Cl has a solid–liquid–gas triple point at 520°C and 48 bar. Using several types of internally heated corrosion-resistant cells, made mainly of sapphire and gold–platinum, it can be shown that NH_4Cl has a single vapor-pressure curve which extends from the triple point to a critical point at

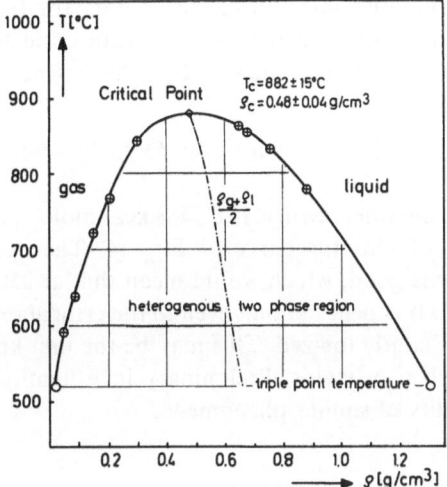

Fig. 13. Gas–liquid coexistence curve of NH_4Cl in a temperature–density diagram.

$882 + 15°C$ and 1635 ± 20 bars. The coexistent gaseous and liquid densities can be measured and are shown in Fig. 13.

At this triple point, the specific conductance of liquid NH_4Cl is 2.07 ohm^{-1}-cm^{-1}, only 5% lower than that of liquid KCl at the melting point. Liquid NH_4Cl near the triple point is therefore to be considered as an ionic melt. The electrolytic conductance can be measured in the liquid at saturation condition to 850°C—about 32° below the critical temperature.[28] Since the liquid at 850°C has expanded already to about twice its molar volume at the triple point, a decrease of ionization and conductance can be expected. Actually, however, the conductance increases, and the ratio of specific conductance over molar volume, the molar conductance Λ at 850°C, is three times higher than Λ_T at the triple point (see Fig. 14). Reasonable

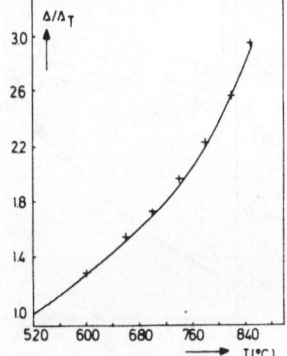

Fig. 14. Relative molar conductivity Λ/Λ_T of liquid NH_4Cl at saturation conditions. Λ_T is the molar conductivity of liquid NH_4Cl at the triple point.

estimates of ion mobilities and utilization of data for fused alkali halides suggest the approximate relation[2] for the temperature and density dependence of Λ for a fully ionized fluid. The quantity E_a is the activation energy for

$$\Lambda(T, \rho) = \text{const.} \times e^{-E_a/R_T} \rho^{-5/6} \tag{2}$$

ionic conductance at constant density. If $E_a = 3$ kcal-mole^{-1}, as for comparable alkali halides, one obtains the curve of Fig. 14. The agreement with the experimental points is good, which would mean that at 850°C liquid NH_4Cl is still an ionic fluid. It is believed that even in the critical region at 882°C the fluid is still predominantly ionized. This may be the first known nonmetallic fluid exhibiting such a behavior. Preliminary investigations with NH_4HF_2 indicate the possibility of similar phenomena.

7. CONCLUSION

It cannot be excluded that pure fluids of small, polar, and stable molecules may become very concentrated electrolyte solutions if self-ionization can be increased by many orders of magnitude. Self-ionization is favored by increasing temperatures and high pressures. From recent static conductance measurements with pure ammonia,[29] which were extended to 500°C and 40 kbar, it follows that the ion product at these conditions should be between 10^{-4} and 10^{-3} mole^2-liter^{-2}. This is still far from a concentrated ionic solution. Similar conductance measurements with water, however, have been made with shock waves and with a static method to 1000°C and more than 100 kbar.[30, 31] An ion product in the vicinity of 10^{-2} mole^2-liter^{-2} for water has been derived from these experiments.

Figure 15, taken from Hamann and Linton,[31] gives a compilation of ion-product data for pure water as a function of water density obtained from

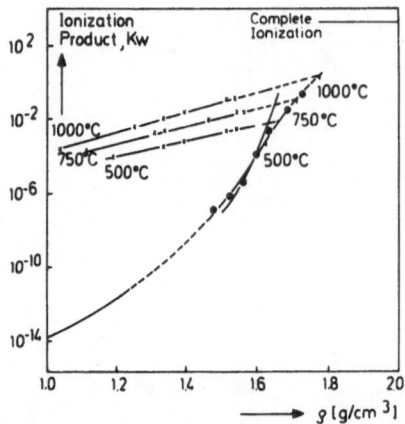

Fig. 15. Ionization product of compressed water as a function of density (after Hamann and Linton[31]. —●—●—●— From shock-wave experiments; ×—×—×—× from static experiments.

the shock-wave and static measurements. The agreement is good, considering the difficulty of the experiments. The authors have made the well-justified proposition that complete ionization of water at high supercritical temperatures would be achieved if it could be compressed to a density of 1.8 to 2.0 g-cm^{-3}, that is, twice the normal density. This range is indicated in the upper right part of Fig. 15. It is interesting to note that under these conditions the molar volume of $(H_2O)_2$ would be similar to the molar volume of fused NaOH at normal pressure.

REFERENCES

1. H. Köster and E. U. Franck, *Ber. Bunsenges. Physik. Chem.* **73**, 716 (1969).
2. C. W. Burnham, J. R. Holloway and N. F. Davis, *Am. J. Sci.* **267A**, 70 (1969).
3. M. H. Rice and J. M. Walsh, *J. Chem. Phys.* **26**, 824 (1957).
4. K. H. Dudziak and E. U. Franck, *Ber. Bunsenges. Physik. Chem.* **70**, 1120 (1966).
5. K. Tödheide, Water at High Temperatures and Pressures, in *Water, a Comprehensive Treatise*, F. Franks, ed. (Plenum Press, New York, 1972).
6. E. U. Franck, *Z. Physik. Chem. (Frankfurt)* **8**, 107 (1956).
7. A. S. Quist and W. L. Marshall, *J. Phys. Chem.* **69**, 3165 (1965).
8. K. Heger, thesis, Institute of Physical Chemistry, University of Karlsruhe, 1969.
9. J. G. Kirkwood, *J. Chem. Phys.* **7**, 911 (1939).
10. D. Eisenberg and W. Kauzmann, *The Structure and Properties of Water* (Clarenton Press, Oxford, 1969).
11. E. U. Franck, *Pure Appl. Chem.* **24**, 13 (1970).
12. W. D. Harder, thesis, Institute of Physical Chemistry, University of Karlsruhe, 1972.
13. V. M. Jansoone and E. U. Franck, *Ber. Bunsenges. Physik. Chem.* **76**, 945 (1972).
14. V. M. Jansoone, *Oesterr. Physiker-Z.*, in press.
15. M. S. Wertheim, *J. Chem. Phys.* **55**, 4291 (1971).
16. E. U. Franck and K. Roth, *Disc. Faraday Soc.* **49**, 108 (1967).
17. W. U. v. Osten, thesis, Institute of Physical Chemistry, University of Karlsruhe, 1971.
18. K. Tödheide and E. U. Franck, *Z. Physik. Chem. (Frankfurt)* **37**, 26 (1963).
19. H. A. Lindner, thesis, Institute of Physical Chemistry, University of Karlsruhe, 1970.
20. T. T. Wall and D. F. Hornig, *J. Chem. Phys.* **47**, 784 (1967).
21. G. J. Janz, *J. Electroanal. Chem.* **29**, 107 (1971).
22. D. F. Irish and T. F. Young, *J. Chem. Phys.* **43**, 1765 (1965).
23. H. D. Lüdemann and E. U. Franck, *Ber. Bunsenges. Physik. Chem.* **72**, 514 (1968).
24. B. Scholz, H. D. Lüdemann, and E. U. Franck, *Ber. Bunsenges. Physik. Chem.* **76**, 406 (1972).
25. E. U. Franck, *Ber. Bunsenges. Physik. Chem.* **76**, 341 (1972).
26. G. Treiber and K. Tödheide, *Ber. Bunsenges. Physik. Chem.*, to be published.
27. M. Buback and E. U. Franck, *Ber. Bunsenges. Physik. Chem.* **76**, 350 (1972).
28. M. Buback and E. U. Franck, *Ber. Bunsenges. Physik. Chem.*, to be published.
29. D. Severin, thesis, Institute of Physical Chemistry, University of Karlsruhe, 1971.
30. W. Holzapfel and E. U. Franck, *Ber. Bunsenges. Physik. Chem.* **70**, 1105 (1966).
31. S. D. Hamann and W. Linton, *Trans. Faraday Soc.* **65**, 2186 (1969).

DISCUSSION

Dr. F. H. Stillinger (*Bell Laboratories, New Jersey*). I am afraid that seeing this $\beta = \frac{1}{2}$ is very disturbing—I am referring to the coexistence curve for ammonium chloride. I guess I did not see it on the slide, but it turns out that even a substance as unusual as water has $\beta = 0.34$, and this has been established very accurately. How close to the critical point were these ammonium chloride density measurements made.? Recalling your slide, it seemed that one was only within roughly 10 out of 1100°K, in other words, a reduced temperature difference $\Delta T/T_c$ of roughly 10^{-2}. Normally, when one makes precise critical-exponent measurements on something like the coexistence-curve exponent β, one is obliged to push $\Delta T/T_c$ to something like 10^{-4}, two more decades than you have achieved.

Professor Franck (*University of Karlsruhe, Germany*). This is true, but we have investigated six or seven substances of different kinds and derived the β values from the known data over a wider range of the coexistence curve. It turns out that one gets the right values of β, usually 0.34 or so, not only if the densities are taken in the close vicinity of the critical point but also for the parts of the coexistence curve which are at a greater distance from the critical point. The exponent 0.5 from the van der Waals' equation, however, is also open to discussion since there are assumptions, for example, concerning the Taylor expansion, on which this value depends. It is not necessarily correct to infer from the fact that since one finds 0.5, this is support for the basic concept of van der Waals.

Dr. Stillinger. It is of course conceivable that an imprecision in determination of T_c (on the high side) could make an apparent parabolic shape appear for this coexistence curve, whereas more precise measurements would in fact indicate a greater flatness. Has T_c, the critical temperature, been well determined for ammonium chloride?

Professor Franck. We believe that the maximum uncertainty of our experimental T_c value is $\pm15°$K. We feel, however, that the probable uncertainty is below 10°K. It appears, however, that for a number of comparable substances one gets reasonable values for β even if one discards the density values which are close to T_c.

Professor H. L. Friedman (*State University of New York, Stony Brook*). In 1963, Widom of the University of Rochester told me that he thought that for an ionic system like this, β might be $\frac{1}{2}$. At the time I had just repeated some experiments on systems which might be good for studying this, but the suggestion has never been followed up. The point is, there are a number of electrolyte solutions which, under terrestrial conditions, form two liquid phases [*J. Phys. Chem.* **66**, 1595 (1962)]. For example, a solution of Na_4GaCl in ether has two liquid phases at 25°C at about 1 m. The proposition was to follow it up in temperature to find the shape of the coexistence curve. I did not realize that maybe it really is a crucial point. It seems to me that for a system of that kind one could determine the coexistence curve with much more accuracy.

Professor C. A. Angell (*Purdue University*). Many alkali metal ammonium solutions in which one might expect long-range forces to be important have two-phase regions for which the coexistence curves are also supposed to be accurately parabolic.

Dr. Stillinger. They are also very quantum-mechanical.

Professor Franck. Both those are two-component systems. I was talking here of only single-component systems. We saw that for pure sodium one also finds 0.50.

Professor R. H. Stokes (*University of New England, Armidale, Australia*). I believe you said Jansoone and you had made calculations of the dielectric constant on the basis of Wertheim's theory and found good agreement with your experimental data. You said that the hard-core diameter used was 2.1 Å. If this figure is correct, I am delighted, since it seemed too small when I obtained it from some work on the interaction of the dipolar groups of alcohol molecules in nonpolar solvents. A value of 2.13–2.18 Å is essential to fit the curvature and the limiting values of the heats of dilution. I thought it was too small because we know water is about 2.8 Å.

Professor Franck. If the viscosity of dilute steam is considered in terms of a Stockmayer potential and a diameter σ is derived formally, I think 2.5 Å is obtained. Compared to that, 2.10 seems a little small, and I would rather have it somewhat higher. However, I don't think this is so essential at this point of the discussion. The main point here is that besides starting with a reasonable σ value, one polarizability value α (which is given by the square of the refractive index), and a gaseous dipole moment of 1.84, there is one additional adjustment that can be made—the dipole moment can be increased because of interaction in the fluid. This increase cannot go on indefinitely with density as there is a threshold. The position of this threshold was fixed to about 2.33 Debye units, and certainly this is equivalent to an assumption of a saturation effect. The point is, however, that it is possible, with such a simple model, to represent the data over a range of about 300°C (which is almost half the absolute temperature and over practically the whole range of density) and to do this without introducing specifically the concept of hydrogen bonding.

Professor Stokes. This was the point that interested me, and it was just the assumption I was making in treating the alcohol-molecule interactions.

Professor Franck. I do not mean that there are no hydrogen bonds, but it is possible to represent the data over a wide range without using them.

Dr. G. E. Walrafen (*Bell Laboratories, New Jersey*). What sort of apparatus was employed in the high-pressure, high-temperature ion-product work?

Professor Franck. The static measurements were carried out in a very tiny platinum–iridium cell, conically shaped, with two electrodes. A certain amount of water was frozen down to −30°C in order to be below the melting-point minimum of water. The desired pressure (say, 90 kbar) was then applied and the water afterwards heated up *in situ* by little graphite beads which served as heating elements. One of the electrodes actually was a thermocouple which was introduced sidewise into the cell. By simultaneously recording the temperature and resistance, one could observe the behavior over the whole range. At the highest temperatures, we believe to have obtained the actual specific conductance of the pressurized fluid. Certainly, this has to be accompanied by a number of calibration measurements by other substances. For instance, the melting pressure curve of a few salts have been used for calibration. Such conductance measurements are very crude compared to conductance measurements that we normally make. We are happy if a conductance is correct within a factor of 2 or 3. This is tolerable because the total range covers over 10 orders of magnitude.

Dr. Walrafen. Is there any hope of conducting spectroscopic investigations at such high pressures?

Professor Franck. Not yet.

Dr. Walrafen. Did you use a tetrahedral press?

Professor Franck. It was done in a uniaxial press—the technique of supporting rings and anvils somewhat similar to the technique of Drickamer. That has been done with ammonia too, and we could reach quite a high conductance of ammonia. Dr. Severin did this, and there is a good reason to assume for pure ammonia that the ion product goes up by about 25 orders of magnitude.

Professor Angell. The visible spectra of nickel in concentrated sodium chloride solutions appeared rather broad compared to other spectra of nickel–chlorine coordinated complexes I have seen under similar conditions. I was wondering if this is an experimental problem or something to do with the composition?

Professor Franck. This is just one slide of a set of six or ten, and they all show this broadness. This may be due to several effects, such as Jahn–Teller effects, certainly. The geometry of the complex is less well defined for these than those at low temperature. Whether it is the effect of the high temperature or the effect of the large amount of salt that is in the solution, I cannot tell. We have to add that much salt in order to get into the required range.

Professor Friedman. I felt the same thing about the $CoCl_4$ complex. In its usual spectra, one can see the vibrational fine structure in that same region. In your case, it showed up in the nickel–chlorine complex but not in the cobalt–chloride complex.

Professor Franck. I cannot answer that. We did not have the best spectrophotometer. These measurements were done with a Zeiss PNQ2, but I do not believe this is the reason. I believe the character of the spectra is determined by the high temperature and the strong interaction between solution partners.